IMAGE AND TEXT COMPRESSION

THE KLUWER INTERNATIONAL SERIES IN ENGINEERING AND COMPUTER SCIENCE

COMMUNICATIONS AND INFORMATION THEORY

Consulting Editor:
Robert Gallager

Other books in the series:

Digital Communication. Edward A. Lee, David G. Messerschmitt
ISBN: 0-89838-274-2

An Introduction to Cryptology. Henk C.A. van Tilborg
ISBN: 0-89838-271-8

Finite Fields for Computer Scientists and Engineers. Robert J. McEliece
ISBN: 0-89838-191-6

An Introduction to Error Correcting Codes With Applications.
Scott A. Vanstone and Paul C. van Oorschot
ISBN: 0-7923-9017-2

Source Coding Theory. Robert M. Gray
ISBN: 0-7923-9048-2

Switching and Traffic Theory for Integrated Broadband Networks.
Joseph Y. Hui
ISBN: 0-7923-9061-X

Advances in Speech Coding, Bishnu Atal, Vladimir Cuperman and Allen Gersho
ISBN: 0-7923-9091-1

Coding: An Algorithmic Approach, John B. Anderson and Seshadri Mohan
ISBN: 0-7923-9210-8

Third Generation Wireless Information Networks, edited by Sanjiv Nanda and David J. Goodman
ISBN: 0-7923-9128-3

Vector Quantization and Signal Compression, by Allen Gersho and Robert M. Gray
ISBN: 0-7923-9181-0

IMAGE AND TEXT COMPRESSION

edited by

James A. Storer
Brandeis University

KLUWER ACADEMIC PUBLISHERS
Boston/Dordrecht/London

Distributors for North America:
Kluwer Academic Publishers
101 Philip Drive
Assinippi Park
Norwell, Massachusetts 02061 USA

Distributors for all other countries:
Kluwer Academic Publishers Group
Distribution Centre
Post Office Box 322
3300 AH Dordrecht, THE NETHERLANDS

Library of Congress Cataloging-in-Publication Data

Image and text compression / edited by James A. Storer.
p. cm. -- (Kluwer international series in engineering and computer science. Communications and information theory)
Includes bibliographical references and index.
ISBN 0-7923-9243-4
1. Data compression (Telecommunication) 2. Text processing (Computer science) 3. Image processing--Digital techniques.
I. Storer, James A. (James Andrew). 1953- . II. Series.
TK5102 . 5 . I48 1992
005 . 74 ' 6--dc20 92-17339
CIP

Printed on acid-free paper.

Printed in the United States of America

CONTENTS

Introduction

James A. Storer
Computer Science Dept.
Brandeis University
Waltham, MA 02254

Data compression is the process of encoding a body of data to reduce storage requirements. With *lossless* compression, data can be decompressed to be identical to the original, whereas with *lossy* compression, decompressed data may be an acceptable approximation (according to some fidelity criterion) to the original. For example, with digitized video, it may only be necessary that the decompressed video look as good as the original to the human eye. The two primary functions of data compression are:

Storage: The capacity of a storage device can be effectively increased with data compression software or hardware that compresses a body of data on its way to the storage device and decompress it when it is retrieved.

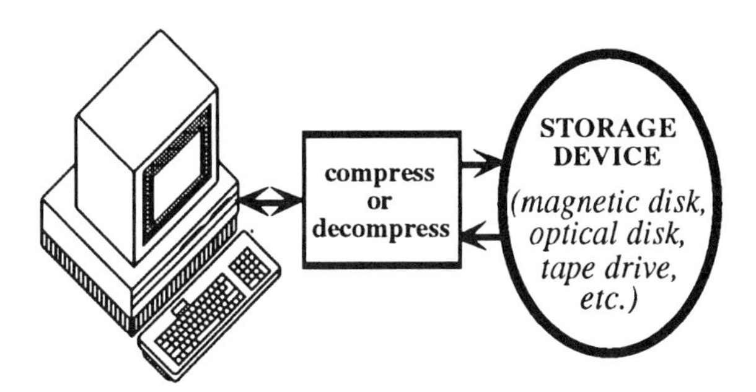

Communications: The bandwidth of a digital communication link can be effectively increased by compressing data at the sending end and decompressing data at the receiving end. Here it can be crucial that compression and decompression can be performed in real time.

Key types of data to which compression technology is currently being applied include natural language text, computer source and object code, bit-maps, numerical data, graphics, CAD data, map and terrain data, speech, music, scientific and instrument data, fax / half-tone data, gray-scale or color images, medical data and imagery, video, animation, and space data.

An important issue with all types of data is how much speed is *required* to process the data in real time; the following list gives an idea of the diverse range of speeds for different practical applications:

Text sent over a modem ~ *2,400 bits per second*
(Depending on the cost of the modem, commonly used speeds range from 1,200 bits per second to 9,600 bits per second.)

Speech ~ *100,000 bits per second*
(One government standard uses 8,000 samples per second, 12 bits per sample.)

Stereo Music ~ *1.5 million bits per second*
(A standard compact disc uses 44,100 samples per second, 16 bits per sample, 2 channels.)

Picture Phone ~ *12 million bits per second*
(A low resolution black and white product might require 8 bits per pixel, 256x256 pixels per frame, 24 frames per second.)

Black&White Video ~ *60 million bits per second*
(A medium resolution product might use 8 bits per pixel, 512 by 512 pixels per frame, 30 frames per second.)

HDTV ~ *1 billion bits per second*
(A proposed standards has 24 bits per pixel, 1024 by 768 pixels per frame, 60 frames per second.)

Although complete compression systems often employ both lossless and lossy methods, the techniques used are typically quite different. The first part of this book addresses lossy image compression and the second part lossless text compression. The third part addresses techniques from coding theory, which are applicable to both lossless and lossy compression.

The chapters of this book were contributed by members of the program committee of the *First Annual IEEE Data Compression Conference*, which was held in Snowbird Utah, April, 1991.

Part 1: IMAGE COMPRESSION

1

Image Compression and Tree-Structured Vector Quantization

Robert M. Gray, Pamela C. Cosman, Eve A. Riskin *

Abstract

Vector quantization is one approach to image compression, the coding of an image so as to preserve the maximum possible quality subject to the available storage or communication capacity. In its most general form, vector quantization includes most algorithms for data compression as structured special cases. This paper is intended as a survey of image compression techniques from the viewpoint of vector quantization. A variety of approaches are described and their relative advantages and disadvantages considered. Our primary focus is on the family of tree-structured vector quantizers, which we believe provide a good balance of performance and simplicity. In addition, we show that with simple modifications of the design technique, this form of compression can incorporate image enhancement and local classification in a natural manner. This can simplify subsequent digital signal processing and can, at sufficient bit rates, result in images that are actually preferred to the originals.

1 Introduction

Image compression maps an original image raster into a bit stream suitable for communication over or storage in a digital medium so that the number of bits required to represent the coded image is smaller than that required for the original image. If the original image is analog, it would require infinite precision to represent it digitally. Ideally one would like the coded image to require as few bits as possible so as to minimize the storage space or communication time. One may also require in some applications that the original image be perfectly recoverable from the coded form. Unfortunately, this latter goal is often not possible. For example, if the original image is an analog photograph, then it is impossible to recreate it exactly from a digital representation regardless of how many bits are used. As one might expect, however, using enough bits in a digital representation should result in a coded image that is perceptually indistinguishable from the original. "Enough bits," however, can be too many to fit into available storage or to communicate over an available link in reasonable

*Robert M. Gray and Pamela C. Cosman are with the Information Systems Laboratory, Electrical Engineering Department, Stanford University, Stanford, CA 94305. Eve A. Riskin is with the Electrical Engineering Department, FT-10, University of Washington, Seattle, WA 98195. The research reported here was partially supported by the National Institutes of Health, the National Science Foundation, and ESL, Inc.

time. Furthermore, information might still be missing that could conceivably be important, even though the untrained eye might not detect its absence. The definition of quality is therefore strongly dependent on the application. A mathematically tractable measure of quality may be used to help a computer optimize a code, and a more complicated perceptually derived subjective test might be used to validate the code designed for a particular application.

The development of compression systems optimized for a particular application has proceeded along a variety of parallel paths. Early systems were based on engineering intuition suggesting that predicting or transforming a signal would "remove redundancy" and lead to more efficient simple quantization on the resulting linearly transformed signal components. High rate or asymptotic quantization theory provided an approximate theory for analyzing such systems when the bit rate was high. Shannon showed how digital information such as quantized images (or inherently digital images) could be compressed in an invertible fashion provided enough bits were available. The "lossy" compression techniques typified by quantization were early combined with the "lossless" techniques introduced by Shannon by simply cascading them to provide an overall compression system. Such a system typically consisted of linear operations, followed by quantization of the transformed or predicted data using several quantizers with differing bit rates, followed by an invertible code. This basic approach has worked well and still dominates in practice.

Along with invertible coding, Shannon also introduced what he called "source coding with a fidelity criterion." This referred to the mapping of blocks or vectors produced by a signal into binary codewords so as to form a coded representation that was optimum in the sense of minimizing the average of some mathematically well defined distortion measure between the original signal and the digital reproduction. Such coding schemes for vectors have become known as block quantizers or vector quantizers. Shannon provided a theory of how good such codes could be and quantified their optimal achievable performance for certain random processes, but he did not describe how to design such codes.

We begin the next section with a description of the general image compression problem from a Shannon point of view, pointing out how most popular compression schemes can be considered as vector quantizers with specific code design algorithms and structural constraints. The distinction between lossless and lossy codes is emphasized and a variety of specific coding techniques are briefly described and compared in terms of their differing structural constraints and design algorithms.

Section 3 reviews the fundamental ideas of vector quantization and Section 4 outlines a variety of design approaches, including several traditional techniques as special cases. Our emphasis is on clustering algorithms in general and the Lloyd algorithm in particular. The complexity of "vanilla" full search vector quantization leads to a consideration of a variety of structured codes aimed at lowering the complexity required for a given fidelity. We briefly describe a menagerie of codes, but quickly focus on two examples: predictive vector quantization and tree-structured vector quantization.

Section 5 describes a simple variation of predictive vector quantization suitable for image compression. Section 6 sketches the ideas underlying tree-structured vector quantization and the principal design algorithms for such codes, and Section 7 provides some examples of applications of tree-structured vector quantization.

The remainder of the paper is devoted to recent results combining image compression with other signal processing, especially image enhancement and image classification. Section 8 provides an example of signal processing accomplished off-line on stored code vectors instead of real time computation on the reconstructed image. Section 9 illustrates more complicated distortion measures that allow combined compression and enhancement or classification. The basic idea is to allow an automatic algorithm or a human expert to designate certain input vectors as being more important than others for a particular application. The code is then designed to weight those vectors more when measuring overall distortion. Section 10 considers the effects of nonrectangular vector shapes with the intent of lessening some of the visual artifacts due to blocking an image.

As our goal is a general overview of fundamentals and variations, not much time is spent on detail. The interested reader is referred to [10] for a thorough study of the fundamentals of vector quantization, an extensive bibliography, and to the cited papers for further details of specific examples.

2 Image Compression

The basic goal of image compression is the conversion of an original sampled continuous or high bit rate image into a "compressed" image with a binary coding having R bits per pixel (bpp) so that the reproduction has the best possible fidelity. Fidelity can be judged by quantitative criteria such as average squared error, by subjective criteria such as viewer quality ratings, or by statistically based acceptability tests for specific applications, such as diagnostic accuracy in a specific medical imaging application. Image compression aims at attaining the best possible quality for the available storage or communication capacity. There are a variety of ways that compression can be useful:

- Compression conserves storage space which allows larger inventories. In some applications such as medical imaging it means that studies can be retained in local, on-line storage longer before being archived in a cheap but remote off-line medium.

- Compression reduces the data rate. For example, compressed images can be squeezed into smaller bandwidth and one can have more video channels on fiber networks.

- Good compression techniques permit progressive transmission, producing an increasingly good reconstruction image as bits arrive. This is especially useful for telebrowsing, selection of important parts of image, and "quick-look" decisions from remote sensors.

- Compression systems can mean reduced complexity signal processing. For example, enhancement algorithms can be built into codes, potentially resulting in compressed images that are *better* subjectively than original with greatly reduced processing. This attribute will be explored at some length here. Compression can also perform various image processing operations such as edge detection or halftoning.

- Compression systems can incorporate encryption for data security.

Although bandwidth is becoming larger in many applications thanks to fiber optics and storage is becoming cheaper thanks to optical and magnetic media, compression remains of interest because

- there will always be relatively low capacity links that people wish to use, e.g., satellite communication links that cannot handle scientific data from remote sensors, hand-held communication devices that must use limited spectrum, low cost low rate modems, and marine radio on small boats with small antennas,

- even fiber optic links can fill up fast with digital multichannel HDTV and cine mode medical images,

- enormous image data bases will need to be browsed in reasonable time and good compression algorithms can provide efficient data structures, speed communication, and lower required on-line memory.

In addition to the basic application of reduced bit rate for communication and storage, compression can assist in subsequent signal processing. One reason for digitizing images is to permit digital signal processing (DSP). Common examples include enhancement, classification, and scene analysis. An image compression scheme often serves as a front end of such DSP. Fewer bits can mean simpler processing, provided the compression system does not lose essential information. As we shall demonstrate, however, sometimes such DSP operations can be incorporated into the compression system itself, yielding a "smart" front end and reducing the need for postprocessing.

Lossless and Lossy Compression

There are two basic types of compression:
- *Lossless compression* or noiseless coding, data compaction, entropy coding, or invertible coding, where the original image can be perfectly recovered from the digital representation. Lossless compression is only applicable for already digital images and it requires variable length coding techniques. The basic idea of such codes is to use long codewords for unlikely inputs, short for likely. The codes are explicitly designed so that the average number of bits per input pixel is as small as possible. The most popular lossless coding techniques are Huffman, adaptive Huffman, Ziv-Lempel, and arithmetic codes [14, 9, 34, 35, 30, 21, 28, 29, 10]. Typical compression ratios for lossless codes on still frame images run from expansion (poor) to 4:1 compression (unusually good). 1.7:1 to 2:1 compression is common in medical imaging.

It is widely believed that lossless coding is necessary in many applications and hence a great deal of effort has been spent fine tuning these algorithms and developing hardware for their implementation. The necessity for no loss is obvious in some applications such as the compression of computer programs and arbitrary binary files, but, as we shall argue, it is less clear that it is necessary in medical imaging applications (where legal issues arise) and in scientific applications (where pristine sensor data may seem desirable). Lossless coding is not possible if the original data is analog and common scalar quantization is inherently lossy. If one must have some loss, one might do better than to simply cascade a scalar quantizer with a noiseless coder.

The ultimate limits to lossless compression are determined by the Shannon source entropy (kth order entropy if k-dimensional vectors are coded). It should be emphasized that loss is *inevitable* if the bit rate R in bits per pixel is smaller than the entropy of the source. One cannot use lossless codes if there is insufficient capacity. Lossy codes can always be used so as to do as well as possible over available communication or storage links.

• *Lossy Compression* where the original pixel intensities cannot be perfectly recovered is the case in simple quantization or A/D conversion. The goal is to minimize an average distortion such as average squared error for a given bit rate. Typical compression ratios range from 4:1 to 32:1 for still frame images using current codes using common techniques, and much better compression is reported for some more recent techniques.

Even in cases where lossless codes are required for legal reasons or simply to be safe, lossy coding may be useful for progressive transmission in an eventually lossless system (such as in telebrowsing or quick-look systems).

Lossy Compression

Lossy compression systems can be clustered into several basic overlapping types.

- Scalar Quantization (PCM) with entropy coding [16]. This system is simple, very good when used with lots of bits and memoryless sources, and its design is well understood.

- Predictive Coding with scalar quantizers [16]. This is a basic differential PCM (DPCM) system with perhaps a complicated predictor. The quantizer acts on the closed loop residual error formed by taking the difference between the new scalar input and a prediction of the input based on the digitally coded past. The system is efficient when the bit rate is high (or the signal is oversampled) and there is strong correlation between samples.

- Transform and Subband Coding [31, 5, 22, 32]. Here input vectors are transformed in a linear fashion and the transformed scalars are scalar quantized using smart bit allocation. The transformation can be accomplished by matrix operations or by linear filtering. The technique is popular, easily made adaptive, and well understood after many years of research and practice. Many variations of the approach exist and some are being standardized (e.g., CCITT, JPEG, MPEG video, p*64). Fine tuned variations of these techniques typically provide high quality at around 1 bpp on still frame monochrome images.

  The most popular transform is the discrete cosine transform (DCT), but many others have been proposed and used, including the Fast Fourier,

Hartley, Walsh, Wavelet, and Karhunen-Loeve transforms. Subband filters, including quadrature mirror filters (QMF) have also proved useful for producing the transformed signal. The wavelet transforms are still relatively young and have shown promise, as have the closely related QMF subband coders. In most such codes, the decoder or decompressor must undo the transform by either inverse transforming or recombining the separate components of a decomposed signal.

- "Second Generation" Codes [17]. This class of compression systems operates by extracting important features such as edge and texture information from the input signal and then coding them. Some techniques use wavelet representations for signals. Most reported so far are quite complicated, but can have very good compression ratios. These algorithms often involve implicit or explicit classification of portions of the image.

- Multiresolution Codes. These code an image at several spatial and intensity resolutions, usually in a nested or embedded fashion. Perhaps the best known example is Burt and Adelson's pyramid codes [3], which form a Laplacian pyramid representation of the image by downsampling, interpolating, and forming a residual in repeated layers. The various layers of the pyramid are then separately quantized. The reconstruction inverts the process of constructing the pyramid, but uses the coded values instead of the residuals. This technique has a useful data structure for other signal processing. The recent wavelet techniques provide a similar multiresolution structure.

- Vector Quantization (VQ) [1, 22, 10]. In the abstract this is simply Shannon's model for source coding. The idea is to compress a group of pixels together instead of one at a time. In this generality it includes transform codes that block transform and then scalar quantize the coefficients. By operating directly on vectors, however, quantizers can be designed specifically to minimize average distortion given an assumed code structure.

  Vector quantization has several other potential advantages:

  - By coding vectors instead of scalars, one can have fractional bit allocations for the vector coordinates.
  - Clustering techniques can be incorporated into code design to provide good quality.
  - Tree-structured codes can be used to give efficient implementations.
  - Tree-structured codes have a natural progressive transmission or successive approximation structure and the resulting codes are embedded.
  - Both clustering and classification tree design methods have been used for enhancement and classification. Since extensions of these techniques are used to design codes, the codes can incorporate simple enhancement and classification.
  - Vector quantization permits optimal (without asymptotic approximations) bit allocation and interpolation/estimation.

- Vector quantizers have fast software decompression based on table lookup and little, if any, computation.
- Vector quantization can be incorporated into other techniques, such as transform coding (the coefficients or the possible bit allocation vectors can be vector quantized).

3 Vector Quantization

A VQ for image compression is depicted in Figure 1. The image is parsed into a

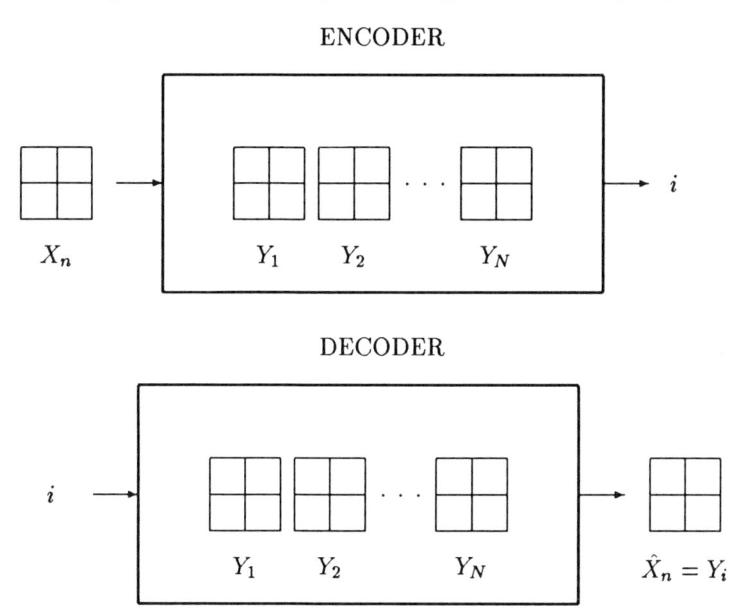

Figure 1: Vector Quantizer

sequence of groups of pixels, often 2×2 squares as shown in the figure, but larger squares and rectangles are also used. Later we shall consider nonrectangular vectors which potentially can lessen the visual artifacts caused by rectangular blocks. The encoder views an input vector X_n at time n and produces a channel codeword i, which is a binary R-tuple if the code has R bits per vector and the system is fixed rate. We will later consider the situation where the indices i can have variable dimension. The decoder is a table lookup: on receiving a channel codeword i, the decoder puts out a stored codeword or template $\hat{Y}_i$, that is, a word in memory indexed by the channel codeword. Given a codebook containing all 2^R of the possible codewords the decoder is completely described, but we have yet to describe how the encoder maps input vectors into binary words. The basic Shannon source code model provides a means of operation that is optimal for a given codebook if the goal is to minimize an average distortion. If we assume $d(X, \hat{X}) \geq 0$ measures the distortion or cost of reproducing an input vector X as a reproduction $\hat{X}$ and if we further assume that the overall distortion (or lack of fidelity) of the system is measured by an average distortion, then the optimal encoder for a given codebook selects the vector Y_i if

$$d(X_n, Y_i) \leq d(X_n, Y_j), \forall j,$$

that is, the encoder operates in a nearest neighbor or minimum distortion fashion. For the moment the code is not assumed to have any structure, that is, it can be an arbitrary collection of code words and the encoder is assumed to be capable of finding the best one by exhaustive search.

If the codebook of reproductions is fixed for all input vectors, then the VQ operates in a memoryless fashion on each input vector; that is, each vector is encoded independently of previous inputs and outputs of the encoder. In general an encoder can have memory by varying the codebook according to past actions. Predictive and finite-state vector quantizers are examples of VQ with memory. Predictive VQ in fact uses a fixed codebook, but the codebook is for residual error vectors, not for the incoming raw input vectors. The system is equivalent to a time varying codebook for the input vectors formed by combining a prediction with a fixed residual codebook.

If we do not insist on the optimal encoder, all compression schemes based on block transforms or mappings followed by quantization fit this general model.

The choice of distortion measure permits us to quantify the performance of a VQ in a manner that can be computed, used in analysis, and used in design optimization. The theory focuses on average distortion in the sense of a probabilistic average or expectation. Practice emphasizes the time average or sample average distortion

$$D = \frac{1}{L} \sum_{n=1}^{L} d(X_n, \hat{X}_n)$$

for large L. This is frequently normalized and a logarithm taken in order to form a signal-to-noise ratio (SNR) or peak signal-to-noise ratio (PSNR). With well defined stationary random process models instead of real signals, the sample average and expectation are effectively equal.

We confine interest to the class of input-weighted squared error distortion measures of the form

$$d(X, \hat{X}) = (X - \hat{X})^* B_X (X - \hat{X}), \qquad (1)$$

where B_X is a positive definite symmetric matrix [11]. The most common B_X is the identity matrix, yielding a simple squared error. In several of our examples we consider more general weightings. Such weighting has the potential advantage of counting the distortion to be more or less according to how important the input vector is judged to be. For example, B_X can be used to adjust the distortion according to the input X being classified in some fashion (e.g., as being active, textured, bright, dim, important, unimportant). All of the basic VQ design methods work for such input-weighted quadratic distortion measures.

Although a tractable distortion measure is needed for optimizing the design, other explicit or implicit measures of distortion may be required to validate the worth of a compression system. For example, in medical imaging it may be far more important to verify that diagnostic accuracy remains as good or better with compressed images than it is to compare SNRs.

A final important attribute of a VQ is its implementation complexity, but this is not easy to quantify or incorporate into algorithm design. Instead we will content ourselves to make comparisons among the various code structures.

4 Code Design

There are many approaches to vector quantizer design. Some commonly seen are:

- The Traditional Approach: Perform a linear transformation of the input vector and then use scalar quantization. This approach includes transform codes and some predictive codes.
- Randomly populate a codebook from a training set or probability distribution, possibly pruning useless codewords. This technique is simple and has proved useful for large vector sizes and noise-like signals (e.g., residual coding in CELP speech compression).
- Lattice codes and lattice-based codes. Here the codebook is a lattice or a subset of a lattice concentrated on regions of high probability. These codes have also proved quite good in the medium to high bit rates. They are multidimensional generalizations of uniform quantization, perhaps confined to a subset of vector space.
- Clustering techniques such as the Lloyd (Forgey, Isodata, k-means) algorithm or the pairwise nearest neighbor algorithm [8] can be used to design codebooks. This is the approach emphasized here.
- Simulated annealing, deterministic annealing, and stochastic relaxation techniques can be used in some cases to find provably optimum codes, but these techniques can be very slow.

One of the simplest clustering techniques is the Lloyd algorithm. It iteratively improves a codebook by alternately optimizing the encoder for the decoder (using a minimum distortion or nearest neighbor mapping) and the decoder for the encoder (replacing the old codebook or collection of templates by generalized "centroids"). For squared error, centroids are the Euclidean mean of the input vectors yielding a given channel binary codeword. They are well defined, but slightly more complicated, for the general input weighted squared error. Code design is based on a training set of typical data, not on mathematical models of the data. For example, in medical imaging applications one might train on several images of the modality and organ scanned for a particular application. This avoids the necessity of modeling the data and bases the design on an empirical distribution as does the popular Bootstrap algorithm of statistics. Unlike the Bootstrap, however, there is no resampling in the Lloyd algorithm. The Lloyd algorithm has been described in detail in a variety of places (see, e.g., [1] [10]), so we here content ourselves to this brief description.

Structured VQ

Shannon theory states that VQ can perform arbitrarily close to the theoretical optimal performance for a given rate if the vectors have a sufficiently large dimension, but unfortunately the complexity of the codes grows exponentially with vector dimension. The practical solution to this "curse of dimensionality" is to constrain the structure of codes. This may result in codes that are not mathematically optimal, but may provide much better performance with implementable codes for all bit rates. The following lists provide some of the most common constrained code structures, including the two that will occupy the remainder of this paper: tree-structured VQ and predictive VQ.

- Constrained Memoryless VQ
  - Lattice-based codes. Codewords are constrained to lie on the points of a subset of a multi-dimensional regular lattice. These codes have fast codeword selection algorithms, but the selection of binary indexes for the codewords can require effort.
  - Classified VQ. A classifier distinguishes different modes of behavior of the input vectors and a separate codebook is designed for each. For image compression, interesting classes might be edges (or edges and orientation), texture, and background.
  - Tree-Structured VQ (TSVQ). Here the codebook is designed to have a tree structure to permit fast searches. The encoder acts like a classification tree, viewing an input vector and "classifying" it according to the best available reproduction.
  - Multistep VQ. Also called cascade VQ and residual VQ, this special case of TSVQ forms a residual based on the reproduction thus far constructed at each stage of the tree, and quantizes the residual. The principal difference between general TSVQ and multistep VQ is that the latter has only a single codebook for each layer of the tree rather than a separate codebook for every node.
  - Product Codes (Gain/Shape and Mean-Removed). Product codes typically extract features from a vector and code them in a separate, but interdependent, fashion. The main examples are (1) the gain/shape code which first codes the shape of a vector by a maximum correlation match (no normalization is done), and then given the shape codeword selects a scalar gain to minimize the overall distortion; and (2) Mean-removed VQ, which forms a sample mean of a pixel block (effectively low pass filtering it), subtracts it from the block, and codes the resulting residual.
  - Transform VQ. In any transform code the transformed coefficients can be vector quantized instead of scalar quantized. This in particular permits fractional bit allocations to low energy coefficients without throwing them away entirely.
  - Hierarchical VQ. This includes a variety of techniques that operate in stages on an image, usually in a multiresolution fashion.

- Recursive VQ
  - Predictive VQ (vector DPCM). One can predict the current block based on previously encoded blocks, form the residual, and then VQ.
  - Finite State VQ. Like classified VQ one can have multiple codebooks for separate modes of behavior, but now we choose the codebook to be used for the current vector based on previously encoded vectors. This can be viewed as a finite state machine approximation to a predictive VQ.

5 Predictive VQ

Suppose that the encoder has already encoded blocks above, to the left, and to the upper left of the current block (i.e., blocks are arriving in a raster scan format). A prediction of the current block can be formed by taking a linear combination of the three previously encoded contiguous vectors. This prediction can be subtracted from the actual input vector and the resulting residual vector quantized. The decoder will then reverse the operation, forming the current reproduction by adding the quantized residual to the linear prediction of the current block (which is the same as at the encoder). An example is shown in Figure 2 where pixels $[X_1, X_2, X_3]$ are used to predict pixel Y. The predictor

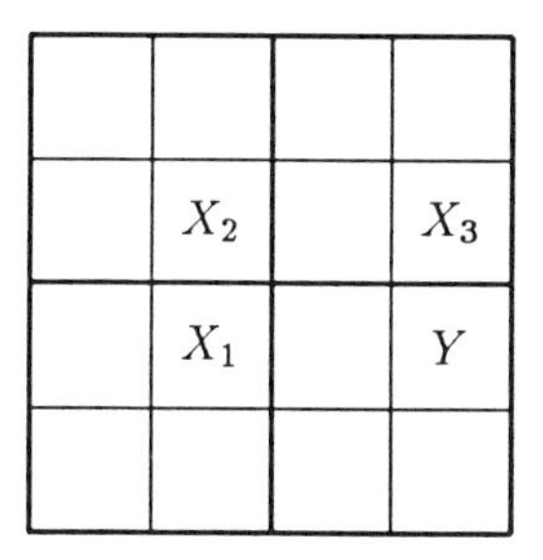

Figure 2: Image Prediction

design is analogous to the usual one-dimensional technique for predictive quantization. One solves a Wiener-Hopf equation for the optimal predictor based on the first and second order statistics of the training set. This open-loop predictor is then applied to the reproductions to close the loop.

6 Tree-Structured Vector Quantization

The key idea in tree-structured VQ (TSVQ) is to perform a tree-search on a codebook designed for such a search rather than perform a full search of an unstructured codebook. Figure 3 depicts two simple trees.

In both cases, the code word is selected by a sequence of binary decisions. The search begins at the root node where the encoder compares the input vector to two possible candidate reproductions and picks the one with minimum

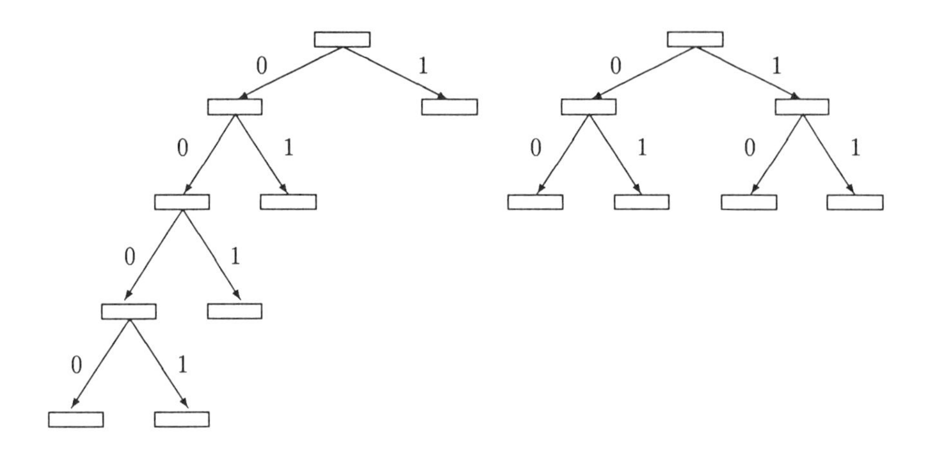

Figure 3: Unbalanced and Balanced Trees

distortion. This is a full search of a simple binary codebook. If the squared error is used, the selection is equivalent to a maximum correlation test or a hyperplane test. The encoder advances to the selected node. If the node is not a terminal node or leaf of the tree, the encoder continues and chooses the best available node of the new pair presented. The encoder produces binary symbols to represent its sequence of decisions. For example, a 0 indicates the left child is selected and a 1 indicates the right child. The binary codeword is then a path map through the tree to the terminal node, which is labeled by the final codeword. The two trees differ in that one is *balanced* and all binary channel codewords have the same length, say R. This tree yields a fixed length or fixed rate code. The other is *unbalanced* and has binary channel codewords of differing length. Here the instantaneous bit rate, the number of bits per input vector or pixel, changes, but the average is constrained. Like a lossless code, this gives the code the freedom to allocate more bits to active areas where they are needed, and fewer bits to uninteresting areas such as background. The goal in lossy compression, however, is to choose long or short codewords so as to minimize average distortion for a given bit rate, not to match probable or improbable vectors.

The search complexity of a balanced tree is linear in bit rate instead of exponential, at the cost of a roughly doubled memory size. For unbalanced trees the search complexity remains linear in the average bit rate, but the memory can be considerably larger unless constrained.

Since the search is constrained, it usually does not find the nearest neighbor in the full codebook. Experience has shown, however, that the loss of performance is small in many applications, while the gain in speed and the reduction of computation are enormous.

Potentially a great advantage of the general code structure is that if properly designed, each successive decision should further refine the reproduction. Such a code will have a built-in successive approximation or progressive character.

As more bits describing a given vector arrive, the quality of the reproduction improves. This progressive structure also means that the rate can be adjusted according to available communication capacity. If the available rate is cut, one codes less deeply into the tree, and conversely. The same tree is used regardless of the allowed rate, and the quality is as good as possible for the allowed rate. How is a tree-structured code designed so as to have these properties? This is accomplished by combining clustering with ideas from classification and regression tree design [2]. Classification trees are effectively a sequence of questions or tests on an input in order to classify it. The general philosophy of classification tree design is a gardening metaphor: first *grow* a tree and then *prune* it.

Growing Trees

Growing a tree in our context means beginning with the root node, which can be considered to be labeled by the optimum rate 0 codeword which is the centroid of the learning set or distribution. One splits the node into two new child nodes. The split can involve simply perturbing the root node label slightly and using the root node label and its perturbation as the two new child labels, or it can be more clever such as forming two new labels along the axis produced by a principal components analysis [33]. One then runs a clustering algorithm on the pair to produce a good one bit codebook as level 1 of the tree. There are now two quite different options.

Split all terminal nodes:

Given a tree, one way to extend it is to split simultaneously all current terminal nodes, which at the moment amount to two nodes, and then cluster. This results in a fixed rate code. Thus for each current terminal node, all of the learning set (or the conditional probability) surviving to that node will be used in the clustering algorithm to design a good one bit code which forms the given node's children. When the clustering algorithm converges for all the split nodes, one will have a new tree with twice as many nodes. As we began this step with a balanced tree with two nodes in the first level, we now have a balanced tree with four nodes in the second level. One can continue in this fashion. All terminal nodes are split, the clustering algorithm is run simultaneously for all of the new nodes, and one has a new balanced tree. This technique is depicted in Figure 4.

Balanced trees have two clear drawbacks:

- As the tree grows, some nodes may become quite sparse in training vectors (or low in conditional probability) and hence the resulting clusters cannot be trusted. Some nodes may even be devoid of training vectors and will be a waste of bits.
- Since all binary codewords have the same length, one will likely have too many bits to represent inactive vectors and too few to represent active vectors.

Split one node at a time:

An alternative design paradigm is to split nodes one at a time rather than an entire level at a time. After the level one codebook has converged, we choose

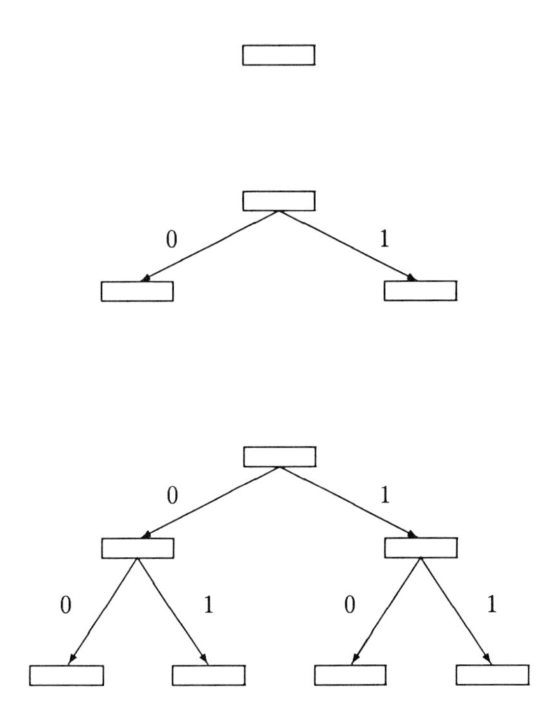

Figure 4: Growing a balanced tree

one of the two nodes to split and run a clustering algorithm on that node alone to obtain a new, unbalanced tree. We then repeat, splitting one node at a time until the tree reaches the desired average rate. How do we choose which node to split? The usual answer is not optimal, but it is optimal in an incremental or greedy fashion and it is simple and effective. Splitting a node will cause an increase in average rate, ΔR, and a decrease in average distortion, ΔD. We choose to split the node that maximizes the magnitude slope $|\Delta D/\Delta R|$, thereby getting the largest possible decrease in average distortion per increase in average bit rate [26]. This turns out to be a natural extension of a fundamental design technique for classification and regression tree design as exemplified in the CART™ algorithm of [2].

Pruning

Whether balanced or unbalanced, the growing algorithm greedily optimizes for the current split only, not looking forward to the impact the current split can have on future splits. Furthermore, even the unbalanced tree can result in sparsely populated or improbable nodes which cannot be fully trusted to typify long run behavior. A solution to both these problems is to take a grown tree and prune it back using a similar strategy. Pruning by removing a node and all its descendents will reduce the average bit rate, but it will increase the average distortion. So now the idea is to minimize the increase of average distortion per decrease in bit rate, that is, to minimize the magnitude slope $|\Delta D/\Delta R|$. Now, however, we can consider the effect of removing entire branches rather than individual nodes. This permits one to find the optimal subtrees of an initial tree in the sense of providing the best rate-distortion tradeoff. The key property that makes such pruning work is that the optimal subtrees of decreasing rate are nested, that is, the optimal TSVQs formed by pruning an initial tree form embedded codes. This means in particular that these codes indeed have the successive approximation character: the distortion decreases on average as the bit rate increases. Figure 5 is a simple depiction of a sequence of pruned subtrees.

A TSVQ designed by growing and pruning is called a pruned TSVQ or PTSVQ [4]. To summarize its good points: PTSVQ

- yields lower distortion than fixed rate full search VQ for a given *average* rate,
- has a simple encoder: it is a sequence of binary decisions,
- has a simple design algorithm: grow and prune,
- has a natural successive approximation (progressive) property,
- is well matched to variable-rate environments such as storage or packet communications,
- provides as a byproduct an automatic segmentation into active and inactive regions by bit rate, and

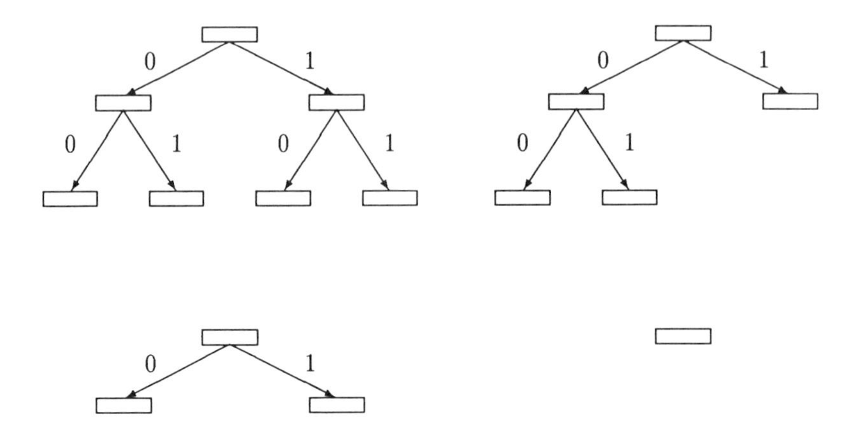

Figure 5: Pruned Subtrees

- can have its tree tailored by an input-dependent importance weighting by using weighted distortion measures. This permits the incorporation of enhancement and classification into the tree.

The final attribute will be explored shortly. Having listed the good points of PTSVQ, we should admit the principal bad points. In some applications one needs to maintain a fixed rate communications link and the buffering required to send a variable rate bit stream over a fixed rate link is not justified. One might still desire the PTSVQ advantages of fast encoding/fast table-lookup decoding and good performance. A simple solution due to [18] for an unbalanced tree structured VQ is to simply relabel the nodes using fixed size binary words [19]. For the unbalanced tree in Figure 3 one can perform the mapping:

$$\begin{array}{rcl} 1 & \rightarrow & 000 \\ 01 & \rightarrow & 001 \\ 001 & \rightarrow & 010 \\ 0001 & \rightarrow & 011 \\ 0000 & \rightarrow & 100 \end{array}$$

This relabeling preserves the fast search and table lookup decoder, but it loses some in rate-distortion performance and, more importantly in some applications, it loses the successive approximation property.

The second drawback of PTSVQ is that the total codebook size can become quite large, even though the average rate is constrained to be small. This problem can be lessened by trading average distortion off against a linear combination of rate and the number of nodes in the tree so that the latter is included in the cost of the tree.

7 Compression Examples

One of the most common databases for testing image compression techniques is the USC data base. A PTSVQ was designed using 10 images from this set not including Lenna. The vector used was a 16-dimensional 4×4 square and the bit rate was .5005 bpp. The resulting PSNR ($10 \log_{10} 255^2/D$) was 30.63 dB. The image is shown in Figure 6. The bit rate is low enough to show some of the

Figure 6: Memoryless PTSVQ: Lenna at .5005 bpp

common problems with VQ. The blocking effect resulting from square vector shapes is clearly visible, especially in the sawtooth artifacts in the edges of the arm and nose and in the gentle shading transitions in the shoulder and band of the hat that appear edge-like. Although this effect can be diminished with the use of predictive PTSVQ, we postpone such an example to the medical images. This example is primarily for comparison with the later alternative vector shapes.

One of the earliest applications of VQ to medical images was the compression of magnetic resonance (MR) images using PTSVQ [13] [27][25]. A training sequence of 5 MR mid-sagittal brain scans of 5 different subjects was blocked into 2×2 vectors, and an unbalanced tree was grown to an average depth of 2 bpp. Pruned back to .75 bpp, it was used to encode a test image not in the training set. Figure 7 shows the original test image at 8 bpp together with a compressed image at .75 bpp using a memoryless PTSVQ. Figure 8 provides a similar comparison between the original and a predictive PTSVQ at the same rate.

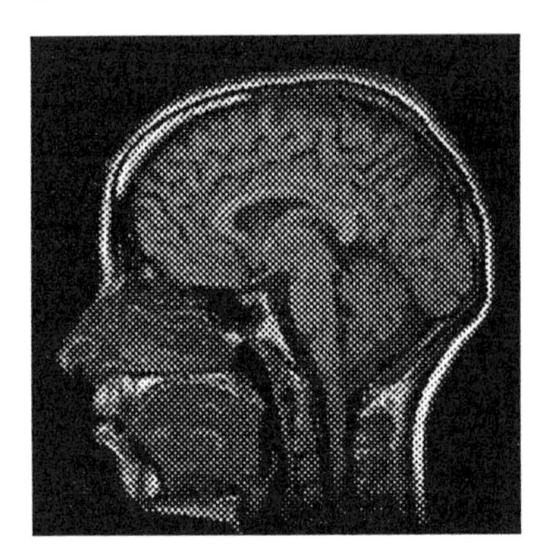 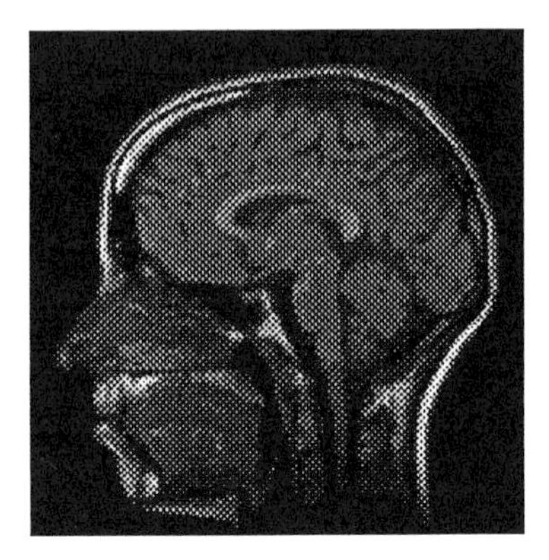

Figure 7: Original Image and Memoryless PTSVQ

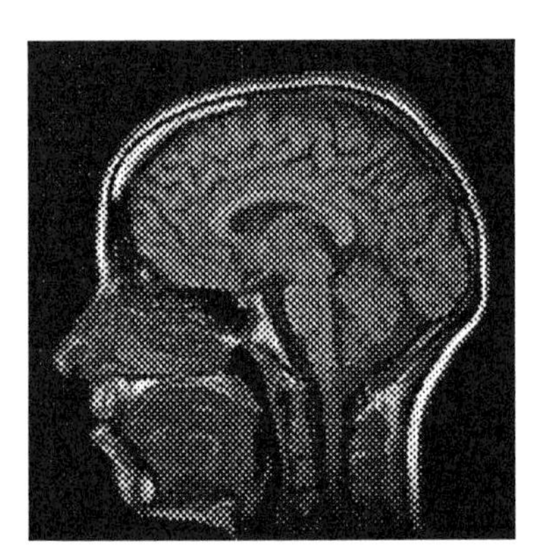 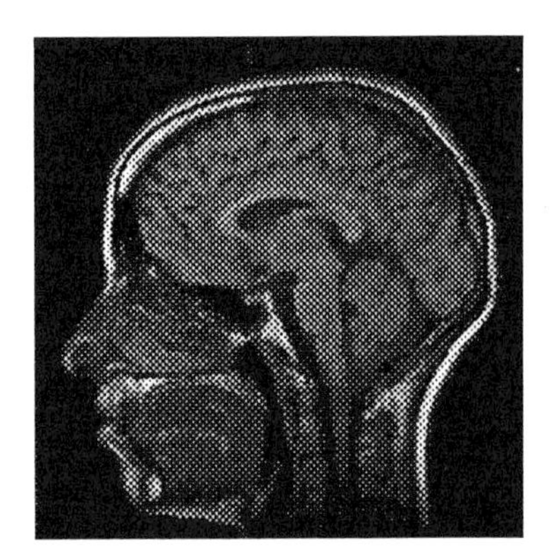

Figure 8: Original Image and Predictive PTSVQ

The simple prediction clearly improves the image significantly, but the compressed image at .75 bpp is clearly inferior to the original. This raises the fundamental issue of how many bits are needed to get a "good enough" reproduction, and that clearly depends on the application and image modality. One can compare the photographs for a variety of images and algorithms in [10], but we here mention a quantitative approach currently being pursued.

In a study designed to quantify the diagnostic accuracy of compressed images using predictive PTSVQ [20], three Stanford radiologists evaluated chest CT images with two common radiologic findings: pulmonary nodules and mediastinal adenopathy. Twenty-four images with lung nodules or abnormal nodes or no clinical abnormalities were viewed in randomized experiments at various levels of compression. Diagnoses were made in a typical fashion and the results analyzed for the effects of compression on sensitivity (the probability of correctly detecting a lesion) and predictive value positive (the probability that a detected lesion is in fact a lesion). Although detailed statistical analyses are currently in progress, preliminary results indicate that predictive value positive is largely unaffected by compression down to the lowest bit rates considered (approximately .5 bpp from an 11 bpp original) and that sensitivity is fairly constant down to rates near 1 bpp.

We mention this experiment as we believe that such clinical evaluations done cooperatively by engineers, biostatisticians, and radiologists will be required before compression (or other signal processing) will be acceptable to the medical community. Similar results quantitatively verifying differences or lack thereof between compressed and uncompressed images will also be necessary in other applications where users are convinced that no loss is acceptable but face the alternative of minimized loss or no data at all.

8 Histogram Equalized VQ

One of the potential advantages of VQ is the ability to combine compression with other signal processing by incorporating the processing into the VQ codebook. One can process the codewords off-line and make the processed codebook available to the decoder. This gives the decoder the option of producing the same reproduction as selected by the encoder or a processed reproduction without further computation. An example of such an application is histogram equalization, a contrast enhancement technique in which each pixel is remapped to an intensity proportional to its rank among surrounding pixels [15]. By transforming each pixel according to the inverse of a cumulative distribution function, the histogram or empirical distribution of the pixel intensities can be altered. If the cumulative distribution function is the empirical distribution of the image, the result is to form a more uniform distribution of pixel intensities, effectively widening the perceived dynamic range of the image. Histogram equalization is a competitor of interactive intensity windowing, which is the established contrast enhancement technique for medical images.

In global histogram equalization, one calculates the intensity histogram for the entire image and then remaps each pixel's intensity proportional to its rank among all the pixel intensities. Instead of performing the decoding and equalizing operations sequentially, one can perform them simultaneously by equalizing the decoder's codebook off-line [7]. This way, the decoder's reconstruction of the image and the histogram equalization would be performed in the same time required by the decompression alone. For example, one can construct a global histogram containing all pixels that composed the training images, and each pixel of each codeword can be equalized using this global histogram. Thus each pixel of each terminal node will be remapped to a new intensity that is proportional to its rank in the global histogram. These new codewords can be stored at the decoder, along with the original codewords. The resulting system is diagramed in Figure 9. The encoder is unchanged. The decoder takes the same set of indices and puts them through the same tree, but, upon reaching a terminal node of the tree, the decoder now has the option of outputting either

Y_i, the compressed reproduction, or $\tilde{Y}_i$, the compressed and histogram equalized reproduction. The radiologist thus has the option of looking at either the equalized or the unequalized series of compressed scans and either way requires the same amount of time to reconstruct the image. The idea of using VQ to perform image processing was first suggested by Read *et al.* at Brigham Young University [23, 24].

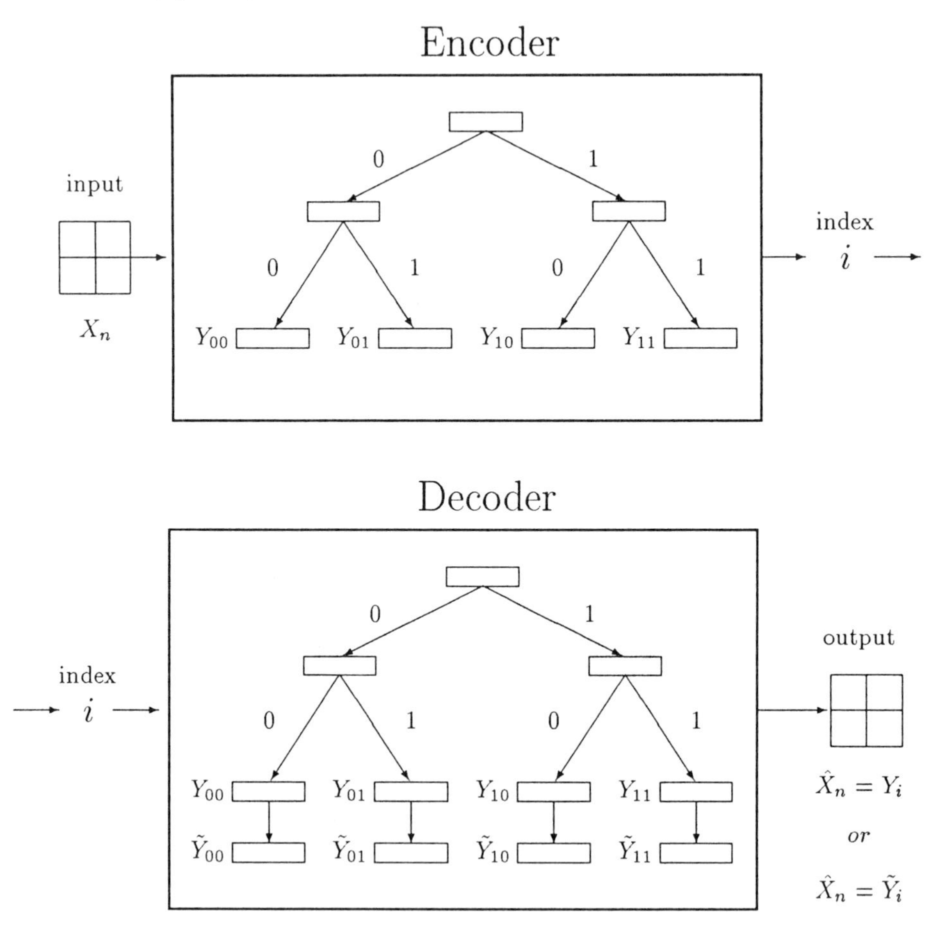

Figure 9: A vector quantizer with both equalized and unequalized codebooks

To demonstrate the technique, the MRI PTSVQ of the previous section was modified so that the codebook was equalized according the global histogram of the learning set. Application of the code to the same MR test image of Figure 7 yields the histogram equalized compressed image at 1.7 bpp along with the unequalized compressed image in Figure 10. The image quality of the equalized compressed image is very high, and its contrast is enhanced, e.g., the invaginations of the cortex are more obvious, and the vertebrae are more clearly differentiated from the interstitial spaces between them. The quality of this enhancement is essentially indistinguishable from the case in which the

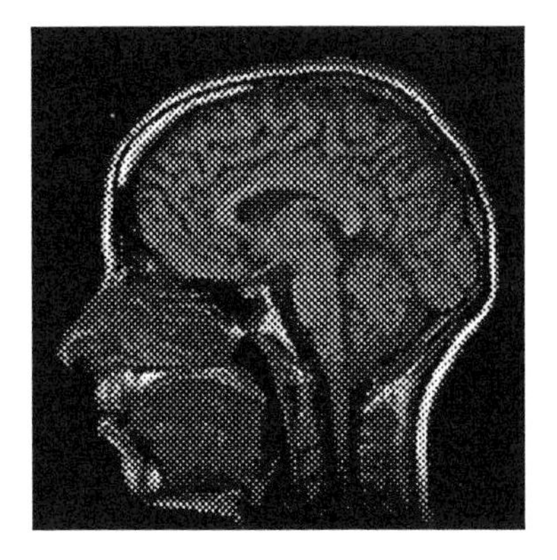 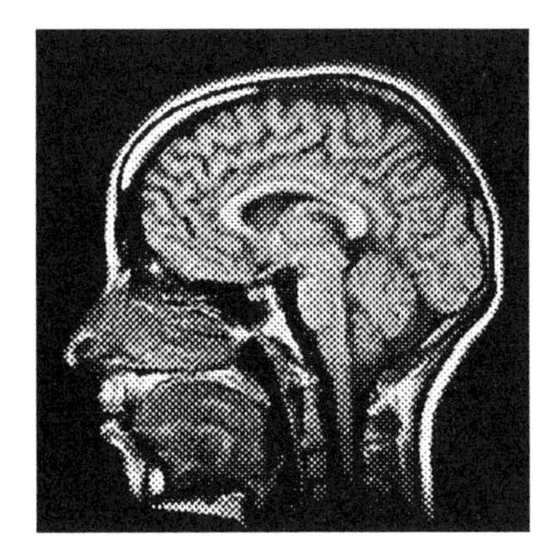

Figure 10: Compressed images: regular and equalized

decoding and equalization are done *sequentially*, despite the fact that sequential operations allow one to use the histogram of the decoded image, instead of the histogram of the training images.

9 Input-Weighted Distortion Measures

So far the only distortion measure considered has been the traditional mean squared error

$$d(X, \hat{X}) = (X - \hat{X})^t (X - \hat{X}).$$

We now consider a subset of the family of input-weighted distortion measures defined in (1). Consider distortion measures of the form

$$d(X, \hat{X}) = w_X (X - \hat{X})^t (X - \hat{X}),$$

where w_X is a positive real scalar depending on the input vector X. In other words, the weighting matrix has the form $B_X = w_X I$, where I is the identity matrix. The distortion measure is used in ordinary VQ in three distinct ways:

1. The minimum distortion selection rule.
2. The computation of a centroid corresponding to a binary channel codeword.
3. The splitting criterion in a greedily grown PTSVQ.

The last attribute means that by weighting a distortion measure according to input behavior, the tree can be forced to throw more bits at important inputs. It has the advantage that the tree splitting is done entirely off-line. The centroid computation is also performed off-line and could use the weighted distortion measure, but this has not always helped (and sometimes hurt perceptually). The nearest neighbor selection is done without the weighting for two reasons. First, the weighting w_X can be difficult to compute (e.g., it is based on an expert classification of the training set). Second, and more importantly, it makes no difference; since the weighting depends only on the input, the weighted squared error and squared error will be minimized by the same code vector. (This would not necessarily be the case if the weighting were a more general positive definite matrix.)

We consider three examples of weighted distortion measures: weighting proportional to image intensity, weighting dependent on automatic classification as textured or not, and weighting determined by hand-labeling a training set for important features.

Intensity Weighting

A classification based on intensity, in which bright training vectors are weighted more heavily than dark ones, is appropriate for MR brain scans because the bright parts of the image correspond typically to what is medically most important in the image. Certainly the dark background of the image is of no importance, and a high distortion can be tolerated there. The training sequence consisting of 8 MR brain scans was blocked into 2×2 pixel blocks, and each vector was assigned a weight proportional to its energy:

$$w_x = \left\lceil \frac{1}{20} \sqrt{\sum_{i=1}^{4} {x_i}^2} \right\rceil . \tag{2}$$

The images were 8-bit, and thus the range of possible weights was from 1 to 26. A tree was grown using the weighted distortion measure for splitting. The tree was grown to 2 bpp, pruned back to 0.75 bpp, and evaluated on images not in the training sequence. The same training sequence without weights assigned was used to grow an unweighted unbalanced tree according to the original greedy growing algorithm. The original test image of Figure 7 was used. An example of the compressed images at 0.75 bpp produced by the two different trees is given in Figure 11. The image made from the weighted tree looks better in the bright regions (e.g., the cortex and cerebellum) which generally correspond to the diagnostically important part of these images. According to the usual unweighted squared error distortion measure, the image made from the classified tree has less distortion in the regions of the cortex and cerebellum, but more in other parts, resulting in a higher distortion overall.

Texture Weighting

Due to pattern masking, the human visual system is generally less sensitive to noise in highly active or textured regions of an image. The greedy growing algorithm can be used to accomplish such a redistribution of quantization noise into regions of more texture by using a texture-based classification scheme. It is not appropriate to use MR scans for this purpose, because the most highly textured part of the image is the cortex, which is often the most important medically. Hence noise should not be placed in textured regions of MR scans. Instead we considered the USC database images for this application. The training sequence consisted of six USC database images, blocked into 4×4 vectors. The weights were assigned to the training vectors as follows: for each 4×4 block, 24

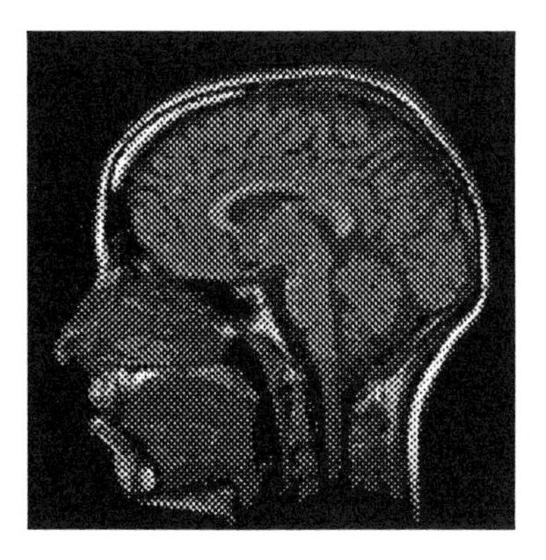
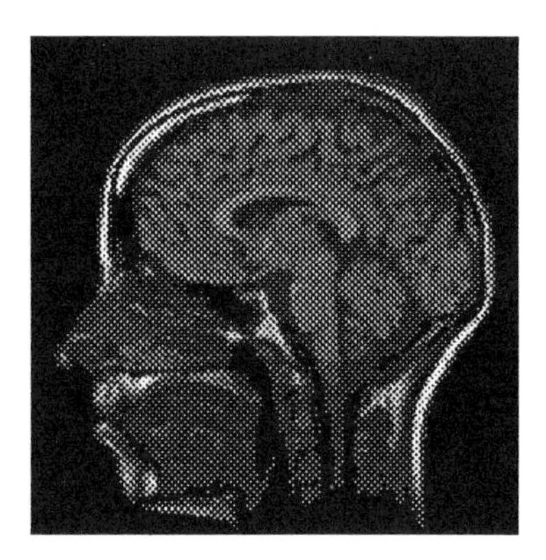

Figure 11: Compressed images at 0.75 bpp: regular (left) and weighted distortion measure (right)

possible pairs of adjacent pixels were examined to see if the difference between them exceeded some threshold. Thus highly textured vectors, which have many pairs exceeding the threshold, are assigned low weights. Highly homogeneous vectors are assigned large weights.

A tree was grown to 2 bpp using the weighted distortion for the splitting criterion, and then was pruned back to 0.54 bpp. We have not shown the resulting images here as good quality photographs are needed to see the differences. The reader is referred to [6] for the photos. The compressed image from the weighted tree has less distortion in the cloud regions, where distortion is most noticeable, and it has more in the areas of trees, where the high texture masks the noise. Nontextured regions misclassified as textured have clearly inferior quality and hence the performance of the technique is quite sensitive to the classifier.

Hand-Labeled Classification

In this section the emphasis is on classification rather than compression. The classification of the training set is done by a person labeling by hand those features that are to be recognized in subsequent images. The VQ encoder described here classifies image subblocks while compressing the image. In our example aerial photographs are hand-labeled as regions of man-made and natural objects and are used to construct a VQ. Class information generated from the encoding process is used to highlight the reconstructed image to aid image analysis by human observers.

The new idea is that the classification is done simultaneously as the image is encoded. Hence, one tree-structured search is sufficient to both encode a particular subblock and to classify it into one of the previously determined categories. Stored with each codeword in the codebook is a class label representing the best class prediction for image subblocks that will be represented by that particular codeword. Thus, once the encoder selects the most appropriate codeword, the preliminary classification of the subblock at hand is simply a matter of memory lookup; no other computations are required. In effect, we are getting this classification knowledge "for free."

The training set consisted of 5 images provided by ESL, Inc. The images were 512×512 pixels of 8 bit grayscale consisting of aerial photography of the San Francisco Bay area. Each 16×16 pixel subblock in the training set was assigned to be either man-made or natural based on the perceptions of a human observer. While the codebook construction and image encoding were carried out using 4×4 pixel subblocks as vectors, the training vectors were classified in 16×16 subblocks to simplify the task of the human classifier (even using the 16×16 subblocks, over 5000 decisions had to be made.) Using a training set classified with finer resolution might improve classification ability.

The original test image to be compressed and classified is shown in Figure 12.

Figure 12: Original aerial image

The best predictive class for a given codeword in a VQ codebook was determined by a majority vote of the *a priori* class assignments of the training vectors represented by that codeword.

Codebooks were generated using three different splitting criteria:

Criterion 1 Ignore the classification information and split the node with the largest $\lambda_1 = |\Delta D / \Delta R|$, where ΔD is the change in squared error resulting from the split and ΔR is the change in average rate. This design yields an ordinary VQ for comparison.

Criterion 2 Split the node that has the greatest percentage of misclassified training vectors, i.e. split the node with the largest value of

$$\lambda_2 = \frac{\text{(Number of misclassified vectors in node)}}{\text{(Total number of vectors in node)}}. \quad (3)$$

Here the encoder nearest neighbor mapping and the centroid reproduction levels are chosen to minimize squared error, but the tree is grown to reduce classification error.

Criterion 3 Split the node that has the greatest number of misclassified training vectors.

Sets of codebooks having either the same rate or same number of nodes were constructed to allow for comparison. In general, the first splitting criterion provided the lowest mean squared error in the encoded image at the expense of relatively poor classification ability. The latter two splitting methods provided poorer encoded images (much more blocky in appearance) but better classification ability. Criterion 3 classified more vectors as man-made than criterion 2. For a given number of nodes, criterion 2 produced a higher rate code than 1, and 3 higher than 2. Likewise, for a given rate the tree structure produced by criterion 1 has substantially more nodes than criterion 2, and 2 more nodes than 3. Choosing the splitting criterion involves a tradeoff between compression and classification quality; some splits serve one purpose better than the other.

Images outside the training sequence were encoded with the resulting codebooks. The compressed images are best viewed on a color monitor so that classification information can be indicated by color superimposed on the grayscale compressed image. For example, the classification information can be viewed by highlighting subblocks that were deemed man-made by the encoder with a red tint and subblocks deemed natural with a green tint. Such a contrast can make the natural and man-made features of the image easier for a human viewer to differentiate. Although this information is not as amenable to grayscale display, the images shown here reflect the accuracy of the classification encoder. The classification ability was modest; at 0.5 bpp the best classification encoder still had 25% misclassification error on the training sequence. However, this large error was partly due to the quality of the training sequence. The hand-labeling was affected by the human observer's resolution and consistency limitations.

Experimental data is shown for images encoded and classified with a codebook grown using criterion 2. Figure 13 shows the image after compression at 0.46 bpp. Figure 14 shows the subblocks in the compressed image that the encoder classified as natural, the man-made subblocks are replaced by solid white subblocks.

10 Tiling Shapes

Virtually all published applications of vector quantization to image compression use square or rectangular vectors. Indeed, we have followed this approach thus far. Square and rectangular blocks are a natural and convenient means

Figure 13: Image compressed at 0.46 bpp using compression/classification encoder.

Figure 14: Natural sublocks using compression/classification encoder. Man-made subblocks are replaced by solid white subblocks.

of tiling an image raster; that is, of covering the image with nonoverlapping replicates of a single pattern or tile. Alternative shapes possess potential advantages, however, including the reduction of unwanted visual artifacts such as the blockiness perceived with rectangular dimensional vectors. We describe two alternative tiling strategies and some preliminary results on their quantitative and qualitative performance. A wider variety of tiles may be found in [12]. We have chosen two of the better ones as illustrative of the potential gains: the 17-dimensional # tile and the 23-dimensional hexagonal tile. The shading transitions are much improved in the alternative tiles. The sawtooth nature of edges is subdued with the alternative tiles, but not eliminated. It is perceptually the least with the hexagonal tile, but a speckling artifact not present with the square tile appears there.

Dimension 17 # Tile

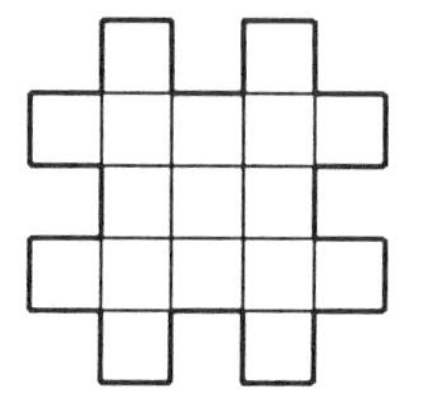

Figure 15: Test Image with 17-Dimensional Sharp Tile

The interlocking pixels around the border were intended to break the blocking effect of the usual square tile. The PSNR was 30.20 dB at 0.4937 bpp.

Dimension 23 Hexagonal Tile

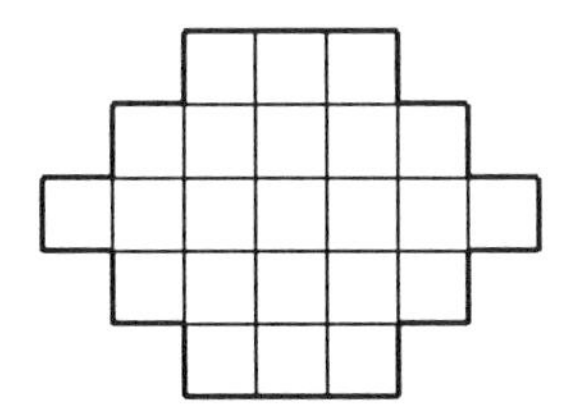

Figure 16: Test Image with 23-Dimensional Hexagonal Tile

Here the idea was to make the edges lie along the diagonal since the human eye is known to be less sensitive to diagonal edges than to horizontal and vertical edges, the so-called "oblique effect" of vision.

11 Parting Comments

Our aim has been twofold: We have tried to survey the fundamental ideas underlying image compression from the point of view of vector quantization and have sketched the design algorithms and relative advantages of a particular family of vector quantizers: pruned tree-structured vector quantizers with and without prediction. We have tried to show by example that in addition to good compression at small complexity, these codes are capable of side-benefits that have only begun to be explored. By modifying decoder codebooks, one can incorporate additional signal processing with virtually no computational cost if one has the memory. By modifying the distortion measure using an input-dependent weighting, one can force the code to work harder for inputs that are more important. Importance can be conferred by automatically classifying the input or by hand-labeling the training set. This opens the door for the design of trees that combine their traditional role of classification with their newer role of low complexity quantization and suggests a variety of ways of simultaneously compressing and improving an image for a specific application.

References

[1] H. Abut, editor. *Vector Quantization.* IEEE Reprint Collection. IEEE Press, Piscataway, New Jersey, May 1990.

[2] L. Breiman, J. H. Friedman, R. A. Olshen, and C. J. Stone. *Classification and Regression Trees.* Wadsworth, Belmont,California, 1984.

[3] P. J. Burt and E. H. Adelson. The Laplacian pyramid as a compact image code. *IEEE Trans. Comm.*, COM-31:552–540, April 1983.

[4] P. A. Chou, T. Lookabaugh, and R. M. Gray. Optimal pruning with applications to tree-structured source coding and modeling. *IEEE Transactions on Information Theory*, 35(2):299 - 315, March 1989.

[5] R. J. Clarke. *Transform Coding of Images.* Academic Press, Orlando, Fla., 1985.

[6] P. C. Cosman, K. L. Oehler, A. A. Heaton, and R. M. Gray. Tree-structured vector quantization with input-weighted distortion measures. In *Proceedings Visual Communications and Image Processing '91*, Boston, Massachusetts, November 1991. SPIE- The International Society for Optical Enginering.

[7] P. C. Cosman, E. A. Riskin, and R. M. Gray. Combining vector quantization and histogram equalization. In J. A. Storer and J. H. Reif, editors, *Proceedings Data Compression Conference*, pages 113–118, Snowbird, Utah, April 1991. IEEE Computer Society Press.

[8] W. H. Equitz. A new vector quantization clustering algorithm. *IEEE Trans. Acoust. Speech Signal Process.*, pages 1568–1575, October 1989.

[9] R. G. Gallager. Variations on a theme by Huffman. *IEEE Trans. Inform. Theory*, IT-24:668–674, Nov. 1978.

[10] A. Gersho and R. M. Gray. *Vector Quantization and Signal Compression.* Kluwer Academic Publishers, Boston, 1992.

[11] R. M. Gray and E. Karnin. Multiple local optima in vector quantizers. *IEEE Trans. Inform. Theory*, IT-28:256–261, March 1982.

[12] R. M. Gray, S. J. Park, and B. Andrews. Tiling shapes for image vector quantization. In *Proceedings Third International Conference on Advances in Communication and Control Systems*, Victoria, British Columbia, October 1991.

[13] R. M. Gray and E. A. Riskin. Variable rate pruned tree-structured vector quantizers for medical image coding. In *Proceedings 1988 ComCon (Advances in Communication and Control)*, Baton Rouge, LA, October 1988.

[14] D. A. Huffman. A method for the construction of minimum redundancy codes. *Proceedings of the IRE*, 40:1098–1101, 1952.

[15] R. Hummel. Image eartnhancement by histogram transformation. *Computer graphics and image processing*, 6:184–195, 1977.

[16] N. S. Jayant and P. Noll. *Digital Coding of Waveforms*. Prentice-Hall, Englewood Cliffs,New Jersey, 1984.

[17] M. Kunt, M. Bénard, and R. Leonardi. Recent results in high compression image coding. *IEEE Trans. Circuits and Systems*, CAS-34(11):1306–1336, November 1987.

[18] J. Makhoul, S. Roucos, and H. Gish. Vector quantization in speech coding. *Proc. IEEE*, 73. No. 11:1551–1587, November 1985.

[19] K. L. Oehler, P. C. Cosman, R. M. Gray, and J. May. Classification using vector quantization. In *Conference Record of the Twenty-Fifth Asilomar Conference on Signals, Systems and Computers*, Pacific Grove, California, November 1991.

[20] R. A. Olshen, P. C. Cosman, C. Tseng, C. Davidson, L. Moses, R. M. Gray, and C. Bergin. Evaluating compressed medical images. In *Proceedings Third International Conference on Advances in Communication and Control Systems*, Victoria, British Columbia, October 1991.

[21] R. Pasco. *Source coding algorithms for fast data compression*. Ph. D. Dissertation, Stanford University, 1976.

[22] M. Rabbani and P. W. Jones. *Digital Image Compression Techniques*, volume TT7 of *Tutorial Texts in Optical Engineering*. SPIE Optical Engineering Press, Bellingham, Washington, 1991.

[23] C. J. Read, D. V. Arnold, D. M. Chabries, P. L. Jackson, and R. W. Christiansen. Synthetic aperture radar image formation from compressed data using a new computation technique. *IEEE Aerospace and Electronic Systems Magazine*, 3(10):3 – 10, October 1988.

[24] C. J. Read, D. M. Chabries, R. W. Christiansen, and J. K. Flanagan. A method for computing the DFT of vector quantized data. In *Proceedings of ICASSP*, pages 1015–1018. IEEE Acoustics Speech and Signal Processing Society, 1989.

[25] E. A. Riskin. *Variable Rate Vector Quantization of Images*. Ph. D. Dissertation, Stanford University, 1990.

[26] E. A. Riskin and R. M. Gray. A greedy tree growing algorithm for the design of variable rate vector quantizers. *IEEE Trans. Signal Process.*, November 1991. To appear.

[27] E. A. Riskin, T. Lookabaugh, P. A. Chou, and R. M. Gray. Variable rate vector quantization for medical image compression. *IEEE Transactions on Medical Imaging*, 9:290–298, September 1990.

[28] J. Rissanen. Generalized Kraft inequality and arithmetic coding. *IBM J. Res. Develop.*, 20:198–203, 1976.

[29] J. Storer. *Data Compression*. Computer Science Press, Rockville, Maryland, 1988.

[30] T. A. Welch. A technique for high-performance data compression. *Computer*, pages 8–18, 1984.

[31] P. A. Wintz. Transform picture coding. In *Proc. IEEE*, volume 60, pages 809–820, July 1972.

[32] John W. Woods, editor. *Subband Image Coding*. Kluwer Academic Publishers, Boston, 1991.

[33] X. Wu and K. Zhang. A better tree-structured vector quantizer. In J. A. Storer and J. H. Reif, editors, *Proceedings Data Compression Conference*, pages 392–401, Snowbird, Utah, April 1991. IEEE Computer Society Press.

[34] J. Ziv and A. Lempel. A universal algorithm for sequential data compression. *IEEE Trans. Inform. Theory*, IT-23:337–343, 1977.

[35] J. Ziv and A. Lempel. Compression of individual sequences via variable-rate coding. *IEEE Trans. Inform. Theory*, IT-24:530–536, 1978.

2

Fractal Image Compression Using Iterated Transforms

Y. Fisher, E.W. Jacobs, and R.D. Boss

Naval Ocean Systems Center, San Diego, CA 92152-5000

Abstract: This article presents background, theory, and specific implementation notes for an image compression scheme based on fractal transforms. Results from various implementations are presented and compared to standard image compression techniques.

Section 1. Introduction.

The image compression scheme presented in this article derives its "fractal" forename from the fact that the method used to encode an image shares many features in common with simple fractal generating algorithms. In fact the decoded images generated by the scheme can have fractal characteristics; by "zooming" into a decoded image detail may be seen at finer and finer resolution. The basis for the encoding scheme is completely different from traditional compression methods: an image is partitioned into parts that can be approximated by other parts after some scaling operations. The result of the encoding process is a set of transformations, which, when iterated upon *any* initial image, possess a fixed point which approximates the target image. The figure below shows the iterative decoding process for an encoding of Lena. Starting with the cross hatch image, the transformations (i.e., the encoding of Lena) are applied until their fixed point is reached. This article presents some background into this methodology, followed by a theoretical motivation, notes on implementing the algorithms, and results.

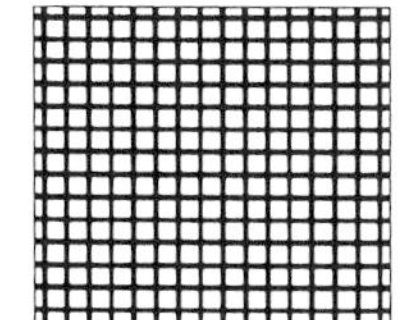
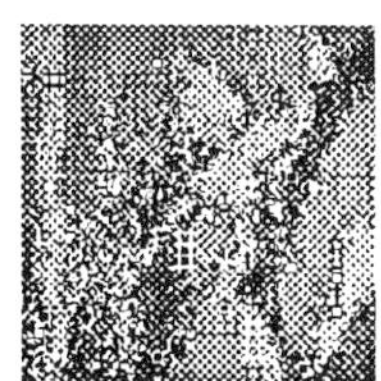
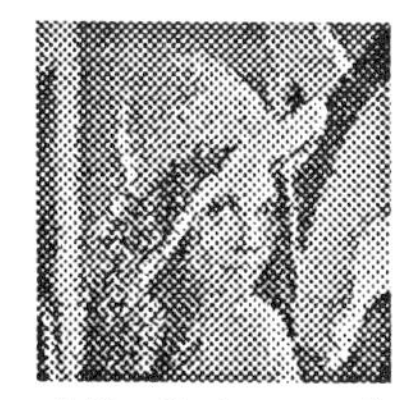
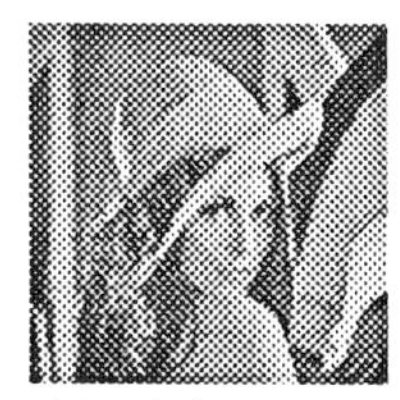

Figure 1. An arbitrary initial image, and the first, second, and tenth iterates of the transformations for Lena.

One of the first applications of fractals to image generation was Mandelbrot's publication of R.F. Voss's fractal generated mountain scenes.[1] Voss used fractal algorithms to generate realistic images of landscapes. Soon after, M.F. Barnsley suggested using fractals to encode natural images. Whereas Voss used fractals to generate a natural looking image, Barnsley suggested that given an image, fractal algorithms approximating the image should be sought. Since simple fractal algorithms can typically generate very complex images, he suggested that only relevant parameters of an algorithm need to be stored, resulting in reduced memory requirements for the storage of images. Finding good fractal algorithms and parameterizing them well has become known as the "inverse problem". The first published automated scheme to solve the inverse problem was presented by Jacquin in his PhD dissertation.[2]

§ 1.1. Iterated Function Systems (IFS).

Barnsley's suggested algorithms for the solution of the inverse problem rested on a paper by J. Hutchinson[3] describing a simple scheme for generating a variety of self-similar fractals. Barnsley coined the term *iterated function system* to describe the collection $(F, \delta, w_1, \ldots, w_n)$, where F is a complete metric space with metric δ, and $w_1, \ldots, w_n$ is a collection of contractive maps.[4]

Definition. *A mapping $w : F \to F$, from a metric space F with metric δ to itself is called contractive if there exists a positive real number $s < 1$ such that*

$$\delta(w(x), w(y)) < s\delta(x, y) \tag{1}$$

for every $x, y \in F$.

Typically, F is taken to be the space of compact subsets of the plane R^2, and δ is taken to be the Hausdorff metric. A detailed explanation of these definitions is beyond the scope of this article. It is sufficient to think of δ as measuring the extent to which two subsets of the plane differ.

In this article, the term *contractivity* will mean the infimum over values of s satisfying equation 1, even if this value is greater than 1.

§ 1.1.1. A Simple Example.

The example in this section serves as a simple illustration of the concepts involved in the image encoding scheme presented later. Although iterated function systems and the fractal encoding scheme to be discussed later have several features in common, the reader should be forewarned that they are not the same. The central concept is that the image of a set (a Sierpinski gasket, in this case) can be reconstructed from a set of transformations which may take less memory to store than the original image of the set.

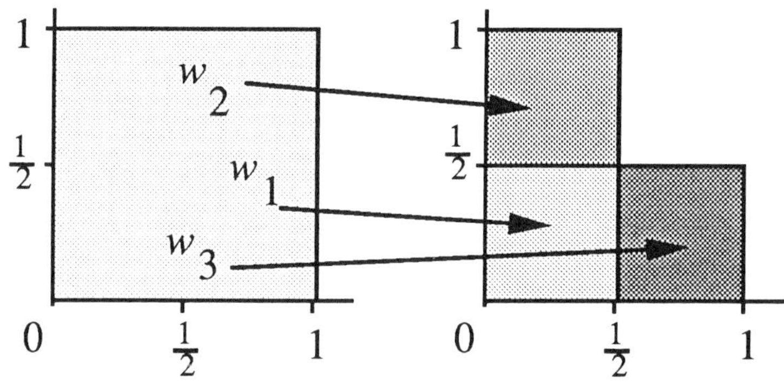

Figure 2. Three affine transformations in the plane.

Consider the three transformations shown in figure 2. They are

$$w_1\begin{bmatrix} x \\ y \end{bmatrix} = \begin{bmatrix} \frac{1}{2} & 0 \\ 0 & \frac{1}{2} \end{bmatrix}\begin{bmatrix} x \\ y \end{bmatrix} + \begin{bmatrix} 0 \\ 0 \end{bmatrix}$$
$$w_2\begin{bmatrix} x \\ y \end{bmatrix} = \begin{bmatrix} \frac{1}{2} & 0 \\ 0 & \frac{1}{2} \end{bmatrix}\begin{bmatrix} x \\ y \end{bmatrix} + \begin{bmatrix} 0 \\ \frac{1}{2} \end{bmatrix}$$
$$w_3\begin{bmatrix} x \\ y \end{bmatrix} = \begin{bmatrix} \frac{1}{2} & 0 \\ 0 & \frac{1}{2} \end{bmatrix}\begin{bmatrix} x \\ y \end{bmatrix} + \begin{bmatrix} \frac{1}{2} \\ 0 \end{bmatrix}.$$

For any set S, let

$$W(S) = \bigcup_{i=1}^{3} w_i(S).$$

Denote the n-fold composition of W with itself by $W^{\circ n}$ and let $I = [0,1]$. Now define $A_0 = \{(x,y) : 0 \leq x \leq 1, 0 \leq y \leq 1\} = I^2$ and $A_n = W(A_{n-1})$. Then, as $n \to \infty$, the set A_n converges (in the Hausdorff metric, and certainly visually) to a limit set A_∞. In fact, for any compact set $S \subset \mathrm{R}^2$, $W^{\circ n}(S) \to A_\infty$ as $n \to \infty$. Figure 3 shows A_1, A_2, A_3 and A_4. Figure 4 shows the limit set A_∞.

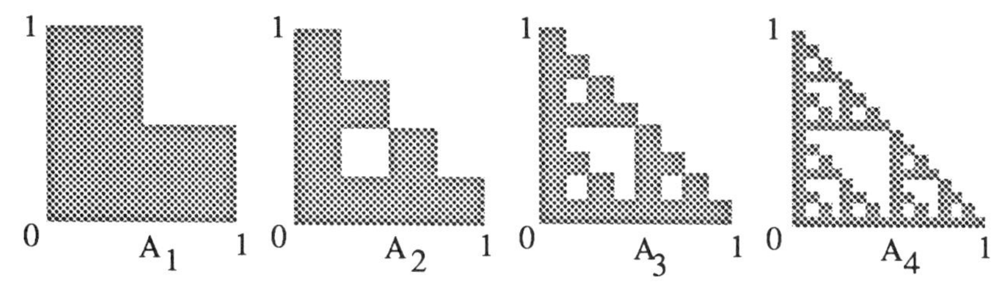

Figure 3. $A_1 = W(A_0)$ and its images A_2, A_3, and A_4.

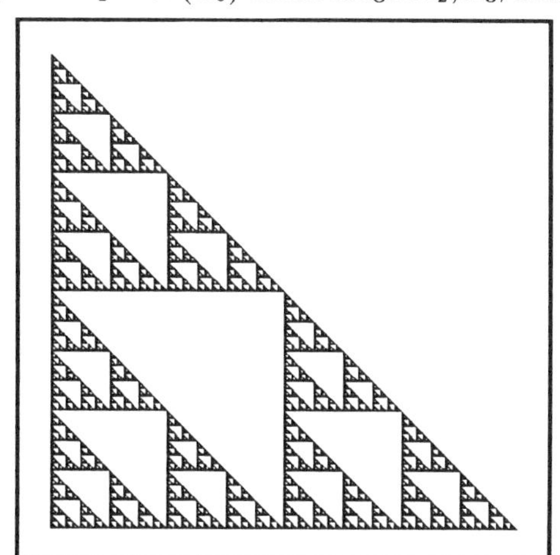

Figure 4. The limit set $A_\infty = \lim_{n\to\infty} W^{\circ n}(A_0)$.

That all compact initial sets converge under iteration to A_∞ is important – it means that the set A_∞ is defined by the w_i only. Also, it is not difficult to see why such a thing is true. The w_i are contractive; they halve the diameter of any set to which they are applied. Thus, any initial A_0 will shrink to a point in the limit as the w_i are repeatedly applied.

Each w_i is determined by 6 real values, so that for this example, 18 floating point numbers are required to define the image. In single precision, this is 72 bytes. The memory required to store an image of the set depends on the resolution; figure 4 requires $256 \times 256 \times 1$ bit $= 8192$ bytes of memory. The resulting compression ratio in this particular example is thus $113.7 : 1$.

Barnsley noticed that IFS's could be used to store more natural looking sets than Sierpinskii gaskets. For example, figure 5 shows 4 transformations which generate a limit set that looks like a fern.[4]

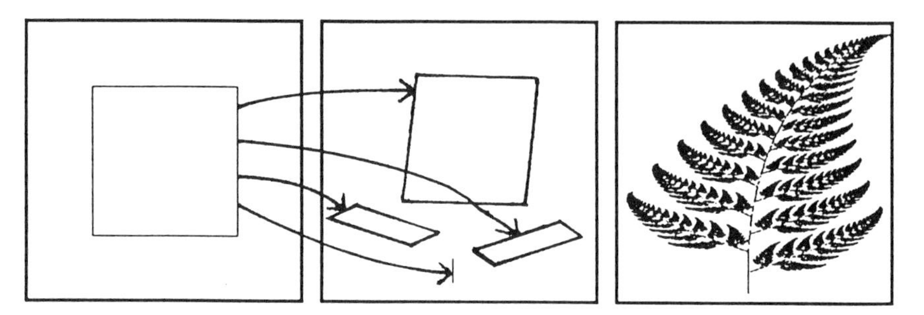

Figure 5. Transformations for the fern and their limit set.

It is difficult to extend this scheme to encode general images. First, it is inherently difficult to find a set of transformations that encode an arbitrary image. Second, how can color or a monochrome level be encoded? These points are addressed in the following sections.

§ 1.1.2. IFS Theory.

This section contains several theoretical results that are shared between IFS's and the encoding schemes to be present later.

IFS's are typically taken to have contractive affine transformations of the form $w(x) = Ax + b$, with $b, x \in \mathrm{R}^2$ and A a 2×2 matrix with norm less than 1. That is, $w(x)$ is contractive. Such affine transformations can compress, rotate, translate, and skew a set. It is not hard to show that a collection $w_1, \ldots, w_n$ of such transformations determines a map $W = \cup_{i=1}^{n} w_i$ that is contractive using the Hausdorff metric in the space of compact subsets of the plane.[4]

The following theorem then implies that the mapping $W : F \to F$ has a unique fixed point, denoted $|W|$.

Contractive Mapping Fixed Point Theorem. *If F is a complete metric space, and $W : F \to F$ is a contractive transformation, then there exists a unique fixed point $|W| = \lim_{n \to \infty} W^{\circ n}(A_0)$, for any $A_0 \in F$.*

Since the limit set is in fact a fixed point,

$$|W| = W(|W|) = w_1(|W|) \cup \cdots \cup w_n(|W|). \tag{2}$$

This formula suggests how one would seek the transformations $w_1, \ldots, w_n$ which encode a given fractal. The transformations should satisfy equation 2, which means that the fractal $|W|$ is built up of parts that are the result of compressing, rotating, translating, and skewing the fractal itself. The parts $w_1(|W|), \ldots, w_n(|W|)$ are said to *cover* the fractal $|W|$.

Given an arbitrary set f, it is not possible in general to exactly cover it with a finite number of transformations of itself. The obvious question is then: what happens if the covering $W(f)$ is approximate? A corollary of the contractive mapping fixed point theorem, which Barnsley calls the collage theorem, gives a relation between the degree to which a set can cover itself and the degree to which the resulting fixed point resembles the original set.

Collage Theorem ([4]). *Let $W : F \to F$ be a contraction with contractivity s and let $f \in F$, then $\delta(|W|, f) \leq (1-s)^{-1}\delta(W(f), f)$.*

This says that the closer the covering $W(f)$ is to the original set f, the closer the fixed point $|W|$ will be to f, and that this is especially true if the transformations composing W are very contractive.

§ 1.2. Partitioned Iterated Function Systems (PIFS).

An extension of the iterated function system concept (due to Barnsley and Jacquin, and which they call a *recurrent iterated function system*) is the *partitioned iterated function system*, in which the maps $w_1, \ldots, w_n$ are not applied to the whole plane but rather have restricted domains.[5] Such a system allows for the simple encoding of more general shapes. For example, it is not obvious how to encode the "bowtie" in figure 6 using an IFS, however when the image is partitioned (and then covered with a PIFS) the process becomes trivial. For example, four of the transformations in the figure are restricted to the left half of the set, and four to the right half, and they exactly cover the bowtie.

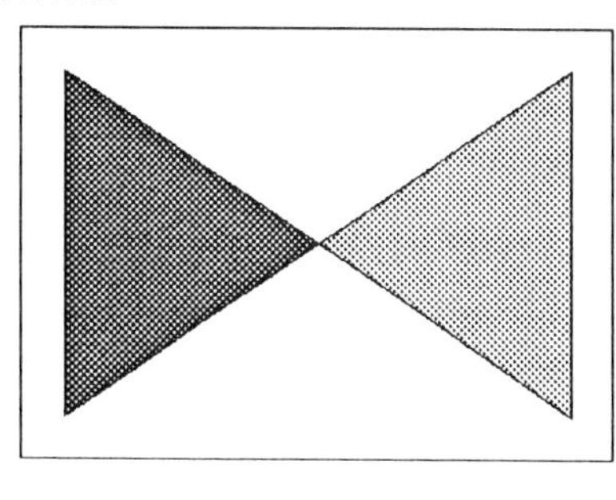

(a)

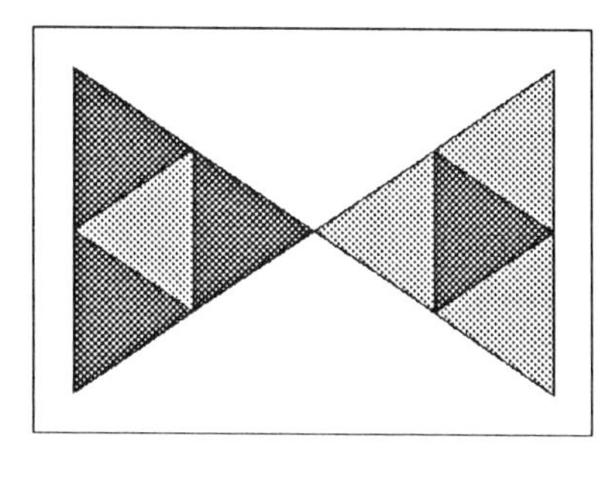

(b)

Figure 6. a) A "bowtie" with a partition indicated by the shading, and b) eight transformations (indicated by the shading) forming an exact collage.

In all PIFS, a transformation w_i is specified not only by an affine map, but also by the domain to which w_i is applied. The contractive mapping fixed point theorem and the collage theorem still hold, but there is an important difference. For an IFS with an expansive w_i, the iterates $W^{\circ n}$ will not converge to a compact fixed point; the expansive part will cause the limit set to be infinitely stretched along some direction. This is not necessarily true for a

PIFS, which can contain expansive transformations and still have a bounded attractor. Further discussion of this point can be found in section 2.2.

§ 1.2.1. Compressing Contours using PIFS.

Barnsley and Jacquin discussed the compression of contours in a paper[5] in which they applied the technique to cloud boundary data. In this scheme, contour data is partitioned (in some unspecified way) into large scale and small scale contours. Affine maps of the plane $w_1, \ldots, w_n$ are found which map large scale partitions $P_1, \ldots P_m$ onto smaller scale partitions $p_1, \ldots, p_m$ in such a way that the distance between $w_i(P_j)$ and p_j is minimized. The resulting map $W = \cup_i w_i$ has a fixed point which approximates the original contour.

In their paper, Barnsley and Jacquin described the concepts involved, but they did not detail the partitioning method, nor demonstrate a fully automated system. In reference 6, these concepts were applied to an automated contour encoding system. The contours were segmented and classified based on their curvature at different scales. A search for longer segments (domains) that best covered shorter segments was then performed. During the encoding process, end points of the available domains were adjusted such that the resulting map conformed to the necessary requirements of a contractive PIFS. Results from applying this scheme to geographic contour maps showed moderate compression with good fidelity.

§ 1.2.2. Compressing Grey Scale Images using PIFS.

Partitioned iterated function systems lend themselves to compressing gray scale images. In this case the maps w_i are no longer plane affine transformations, but rather they operate in R^3, with the new dimension encoding the gray level.

Iterated Systems Inc. uses proprietary algorithms to automatically encode gray scale images.[7] A patented scheme of theirs, using a variation of the PIFS idea, is an algorithm in which an image is partitioned, and an IFS (in 3 dimensions) is found for each partition.[8] The details of automating such a scheme are unclear.

The first published automated PIFS scheme was presented by Jacquin,[2,9] who called it *iterated transformation* based image encoding. He presented a theory based on iterated Markov operators on measure spaces, and demonstrated the encoding of gray scale images. The next sections discuss theory and methods for encoding gray scale images using iterated transforms.

Section 2. Theory of Fractal Image Encoding.

The goal of fractal image encoding is to store an image as the fixed point of a map $W : F \to F$ from a complete metric space of images F, to itself. The space F can be taken to be any reasonable image model, such as the space of bounded measurable functions on the unit square, etc. In this model, $f(x, y)$ represents the gray level at the point (x, y) in the image, and thus $f \in F$ is an image with infinite resolution.

In order to assure that a fixed point of W exists, the transformation is constructed so that W or $W^{\circ m}$ is contractive. The contractive mapping fixed point theorem then ensures convergence to a fixed point upon iteration *from any initial image*. Insisting that $W^{\circ m}$ be contractive is less restrictive than requiring that W be contractive. The goal is to construct the mapping W with fixed point "close" (based on a properly chosen metric $\delta(f,g)$ for $f,g \in F$) to a given image that is to be encoded, and such that W can be stored compactly.

A simple metric that can be used is

$$\delta_{\sup}(f,g) = \sup_{(x,y)\in I^2} |f(x,y) - g(x,y)|.$$

With this metric, a map is contractive if it contracts in the z-direction. Other metrics, such as the rms-metric

$$\delta_{\text{rms}}(f,g) = \left[\int_{I^2} (f(x,y) - g(x,y))^2\right]^{\frac{1}{2}}$$

have more complicated contractivity requirements. The rms metric is particularly useful since it can be easily minimized by performing standard regression of the algorithmic parameters. In the remainder of this article the metric will be specified when relevant.

§ 2.1. Constructing the Map W.

Let F be the space consisting of all graphs of real Lebesgue measurable functions $z = f(x,y)$ with $(x,y,f(x,y)) \in I^3$. The map W is constructed from local maps w_i, defined in the following way.

Let $D_1,\ldots,D_n$ and $R_1,\ldots,R_n$ be subsets of I^2 (called *domains* and *ranges*, respectively, even though they will not exactly be domains and ranges), and let $v_1,\ldots,v_n : I^3 \to I^3$ be some collection of maps. Define w_i as the restriction

$$w_i = v_i|_{D_i \times I}.$$

The maps $w_1,\ldots,w_n$ will form the PIFS.

The maps $w_1,\ldots,w_n$ are said to *tile* I^2 if for all $f \in F$, $\bigcup_{i=1}^n w_i(f) \in F$. This means the following: for any image $f \in F$, each D_i defines a part of the image $f \cap (D_i \times I)$ to which w_i is restricted. When w_i is applied to this part, the result must be a graph of a function over R_i, and $I^2 = \cup_{i=1}^n R_i$. This is illustrated in figure 7. By implication, the union $\cup_{i=1}^n w_i(f)$ yields a graph of a function over I^2, and the R_i's are disjoint.

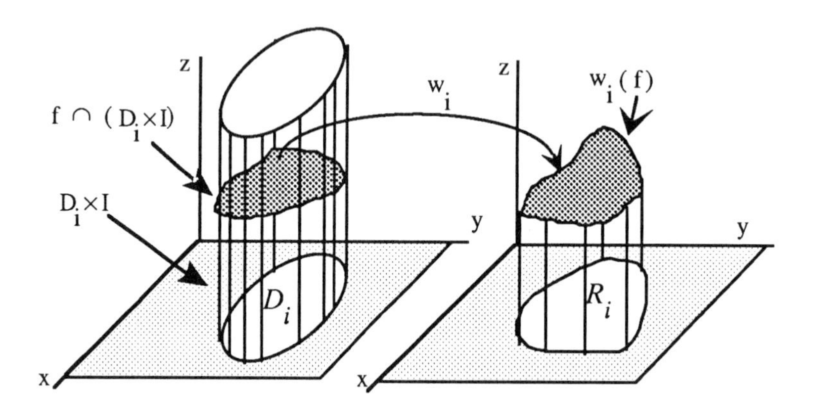

Figure 7. Parts of the tiling of an image.

The map W is defined as

$$W = \bigcup_{i=1}^{n} w_i. \tag{3}$$

Given W, it is easy to find the image that it encodes – begin with any image f_0 and successively compute $W(f_0), W(W(f_0)), \ldots$ until the images converge to $|W|$. The converse is considerably more difficult: given an image f, how can a mapping W be found such that $|W| = f$? A general non-trivial solution to this problem remains elusive. Instead an image $f' \in F$ is sought such that $\delta(f, f')$ is minimal with $f' = |W|$. Since $|W| = W(|W|) = \bigcup_{i=1}^{n} w_i(|W|)$, it is reasonable to seek domains $D_1, \ldots, D_n$ and corresponding transformations $w_1, \ldots, w_n$ such that

$$f \approx W(f) = \bigcup_{i=1}^{n} w_i(f). \tag{4}$$

This equation says: cover f with parts of itself; the parts are defined by the D_i and the way those parts cover f is determined by the w_i. Equality in equation (4) would imply that $f = |W|$. Being able to exactly cover f with parts of itself is not likely, so the best possible cover is sought with the hope that $|W|$ and f will not look too different, i.e., that $\delta(|W|, f)$ is small. The hope for this comes from the collage theorem.[4]

Thus, the encoding process is: Partition I^2 by a set of ranges R_i. For each R_i, a $D_i \subset I^2$ and $w_i : D_i \times I \to I^3$ is sought such that $w_i(f)$ is as δ close to $f \cap (R_i \times I)$ as possible; that is,

$$\delta(f \cap (R_i \times I), w_i(f)) \tag{5}$$

is minimized. The map W is specified by specifying the maps w_i. The w_i's must be chosen such that W or $W^{\circ m}$ is contractive. Specific partitioning methods and ways of storing W compactly are presented in section 3.

§ 2.2. Eventually Contractive Maps.

Those initiated to IFS theory may find it surprising that when the transformations w_i are constructed, it is not necessary to impose any contractivity conditions on the individual transforms. A sufficient contractivity requirement is that W be *eventually contractive.*[10] A map $W : F \to F$ is eventually contractive if there exists a positive integer m called the *exponent of eventual contractivity* such that $W^{\circ m}$ is contractive. Note that m is not unique. All contractive maps are eventually contractive, but not vice versa.

A brief explanation of how a transformation $W : F \to F$ can be eventually contractive but not contractive is in order. The map W is composed of a union of maps w_i operating on disjoint parts of an image. If the w_i are chosen so that

$$|w_i(x, y, z_1) - w_i(x, y, z_2)| < s_i |z_1 - z_2|$$

and $(x', y', z') = w_i(x, y, z)$ has x' and y' independent of z, then W will be δ_{sup}-contractive if and only if every w_i is δ_{sup}-contractive, i.e., when $s_i < 1$ for all i. If any $s_i \geq 1$, then W will not be δ_{sup}-contractive. The iterated transform $W^{\circ m}$ is composed of a union of compositions of the form

$$w_{i_1} \circ w_{i_2} \circ \cdots w_{i_m}.$$

Since the product of the contractivities bounds the contractivity of the compositions, the compositions may be contractive if each contains sufficiently contractive w_{i_j}. Thus W will be eventually contractive (in the sup metric) if it contains sufficient "mixing" so that the contractive w_i eventually dominate the expansive ones. In practice, this condition is relatively simple to check (see section 3.3).

The collage theorem can be generalized for the case of eventually contractive mappings. Let s be the contractivity of W, and let σ be the contractivity of $W^{\circ m}$. Here, it is assumed that $W^{\circ m}$ is contractive ($\sigma < 1$), but s may be greater than 1.

Generalized Collage Theorem ([10]). *For $f \in F$ and $W : F \to F$ eventually contractive,*

$$\delta(|W|, f) \leq \frac{1}{1-\sigma} \frac{1-s^m}{1-s} \delta(W(f), f).$$

With some extra notation, it is possible to improve this estimate, but this will not be done for the following reason. Both the collage theorem and the generalized collage theorem serve only as motivation; they do not provide useful bounds. Empirical results demonstrating this will be given in sections 3.1 and 4.1.

It is important to note that a mapping W which is contractive for δ_{sup} may only be eventually contractive for δ_{rms}. Unlike the sup metric, the condition $s_i < 1.0$ is not sufficient to ensure contractivity for the rms metric. However, this condition is sufficient to ensure eventual contractivity for the rms metric.

Section 3. Implementation.

This section contains a simple example illustrating the encoding process described at the end of section 2.1, a discussion of relevant considerations in achieving an encoding with good compression and fidelity, and details on some specific implementations.

§ 3.1. A Simple Example.

Let f be the image shown in figure 8a, thought of as an integer valued function on the integer lattice in $[0,255] \times [0,255]$ with values in $[0,255]$. It is a 256×256 pixel image with 256 monochrome levels. Let $\mathbf{R} = \{R_1, \ldots, R_{1024}\}$ contain the 1024 non-intersecting 8×8 subsquares of $[0,255] \times [0,255]$. Let $\mathbf{D}$ be the collection of all 16×16 subsquares of $[0,255] \times [0,255]$. For each R_i find the best domain in $\mathbf{D}$, denoted D_i, and a w_i of the form

$$w_i \begin{bmatrix} x \\ y \\ z \end{bmatrix} = \begin{bmatrix} a_i & b_i & 0 \\ c_i & d_i & 0 \\ 0 & 0 & s_i \end{bmatrix} \begin{bmatrix} x \\ y \\ z \end{bmatrix} + \begin{bmatrix} e_i \\ f_i \\ o_i \end{bmatrix} \tag{6}$$

such that $w_i(f)$ takes values over R_i and such that expression (5) is minimized. The first condition determines a_i, b_i, c_i, d_i, e_i and f_i in equation (6) up to composition with a symmetry of the square, and the second condition determines s_i, o_i, and the symmetry operation. In this example, D_i and w_i were chosen by searching through all possible choices. Every D_i was tested in 8 possible orientations (corresponding to the elements of the symmetry group of the square). The rms metric was used to minimize expression (5), and s_i and o_i were computed by least squares regression. To ensure that $\lim_{n\to\infty} W^{\circ n}$ exists, W must be contractive or eventually contractive.

The largest allowable s_i in equation (6) is denoted

$$s_{max} = \max_i \{s_i\}.$$

The result of applying the above algorithm with $s_{max} = 1.5$ yields the $|W|$ in figure 8b. Table 1 summarizes some results for this example, all at compression 16:1.

Table 1

s_{max}	$\delta_{\text{rms}}(W(f), f)$	$\delta_{\text{rms}}(\lvert W\rvert, f)$
0.7	20.04	21.48
0.9	19.77	21.16
1.2	19.42	21.16
1.5	19.42	20.97

The collage theorem and generalized collage theorem both suggest that unless $\delta(W(f), f)$ markedly improves it would be disadvantageous to allow for less contractive mappings. However the results given in table 1 show that (for the rms metric) even though the covering accuracy improves only slightly, the resulting fixed point improves while s_{max} increases. Furthermore, the δ_{sup}

bounds given by the collage theorem and generalized collage theorem are much larger than the maximum possible value of 255 (for a 256 grey level image). From these results (and results in Section 4.1) it is evident that the bounds in the collage theorems should be viewed as motivation rather than useful bounds.

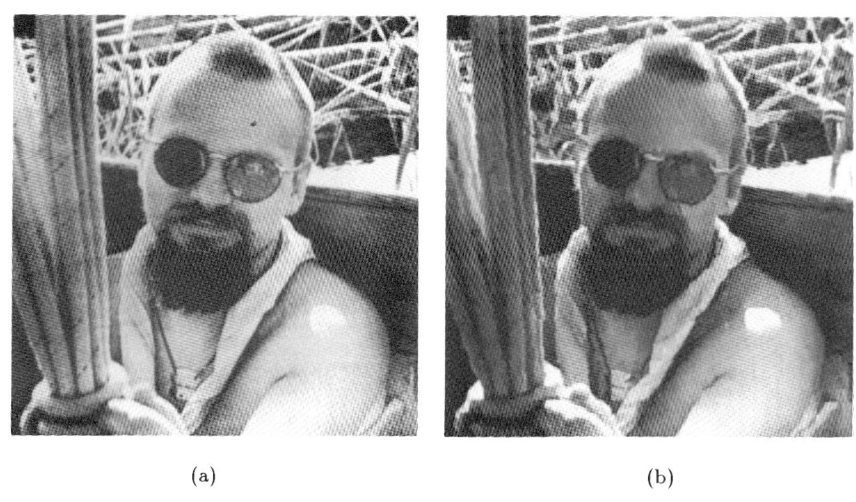

(a) (b)

Figure 8. a) Original Mara, and b) decoded Mara with s_{max}=1.5.

§ 3.2. Considerations in Compressing an Image.

Clearly, the selection of the domains, ranges, and maps w_i is one of the most important parts of an automated fractal encoding scheme. The notation **D** and **R** will be used to denote the set of all *potential* domains and ranges, respectively.

Since the goal of the encoding is compression, a second concern is the compact specification of the map w_i. To limit memory, only w_i of the form of equation (6) are used. The values of s_i and o_i are computed by least squares regression to minimize expression (5) using the rms metric, and they are stored using a fixed number of bits. One could compute the optimal s_i and o_i and then discretize them for storage. However, a *significant* improvement in fidelity can be obtained if only discretized s_i and o_i values are used when computing the error during encoding. Since specifying w_i requires specifying a domain D_i, the D_i are restricted to be geometrically simple. In the implementations described later, **D** and **R** will always be collections of rectangles.

Another concern is encoding time, which can be significantly reduced by employing a classification scheme on the ranges and domains. Both ranges and domains are classified using some criteria such as their edge like nature, or the orientation of bright spots, etc. Considerable time savings result from only using domains in the same class as a given range when seeking a cover, the rationale being that domains in the same class as a range should cover it best.

In the example of section 3.1, the number of transformations is fixed. In contrast, the algorithms described below are adaptive in the sense that they utilize a range size which varies depending on the local image complexity. For a given image, more transformations lead to better fidelity but worse compression. This trade-off between compression and fidelity leads to two different approaches to encoding an image f: one targeting fidelity and one targeting compression. These approaches are outlined in the pseudo-code in table 2. In the table, size(R_i) refers to the side length of the range; in the case of rectangles, size(R_i) is the length of the longest side.

A generalization of the simple example of section 3.1 (which can be implemented following the pseudo-code examples) is to select ranges from a quad-tree partitioning of an image. A quad-tree partition is one in which every square (starting with I^2) may be subdivided into four smaller squares with one fourth the area of the original square.

A generalization of the quad-tree scheme (called HV-partitioning) is to recursively partition the image along horizontal or vertical lines. Both ranges and domains are taken from the same partition, which is chosen in such a way that the ranges and domains share properties which allow domains to cover ranges well. For example, the image can be partitioned so that edge like features in the image tend to run diagonally through partitions. Schemes such as this allow the encoding process to be particularly adaptive to patterns in and content of individual images. Figure 9a shows a typical quadtree partition of an image and figure 9b shows an HV-partition.

While varying the number of ranges directly affects the compression, varying the number of possible domains has a more complicated effect. It may seem that using more domains should decrease the compression - since more information is required to specify the particular domain used. Likewise, it may seem that the encoding time must be increased - since more domains must be searched. But with more domains, a range which would otherwise be partitioned may be covered, so that fewer ranges need to be covered - resulting in time and memory saving. Similarly, using more domains does not guarantee a better fidelity, since an otherwise partitioned range may not be partitioned and thus covered more poorly.

Sections 3.4 and 3.5 contain specific implementation notes for image encoding schemes based on iterated transforms.

Table 2. Two pseudo-codes for an adaptive encoding algorithm

- Choose a tolerance level e_c.
- Set $R_1 = I^2$ and mark it uncovered.
- While there are uncovered ranges R_i do {
  - Out of the possible domains $\mathbf{D}$, find the domain D_i and the corresponding w_i which best covers R_i (i.e., which minimizes expression (5)).
  - If $\delta(f \cap (R_i \times I), w_i(f)) < e_c$ or size$(R_i) \leq r_{min}$ then
    - Mark R_i as covered, and write out the transformation w_i;
  - else
    - Partition R_i into smaller ranges which are marked as uncovered, and remove R_i from the list of uncovered ranges.

  }

a. Pseudo-code targeting a fidelity e_c.

- Choose a target number of ranges N_r.
- Set a list to contain $R_1 = I^2$ and mark it covered.
- While there are less than N_r ranges in the list do {
  - Out of the list of ranges, find the range R_j with size$(R_j) > r_{min}$ which has the largest

    $$\delta(f \cap (R_j \times I), w_j(f))$$

    (i.e., which is covered worst).
  - Partition R_j into smaller ranges which are added to the list and marked as uncovered.
  - Remove R_j, w_j and D_j from the list.
  - For each uncovered range in the list, find and store the domain $D_i \in \mathbf{D}$ and map w_i which covers it best.

  }
- Write out all the w_i in the list.

b. Pseudo-code targeting a compression having N_r transformations.

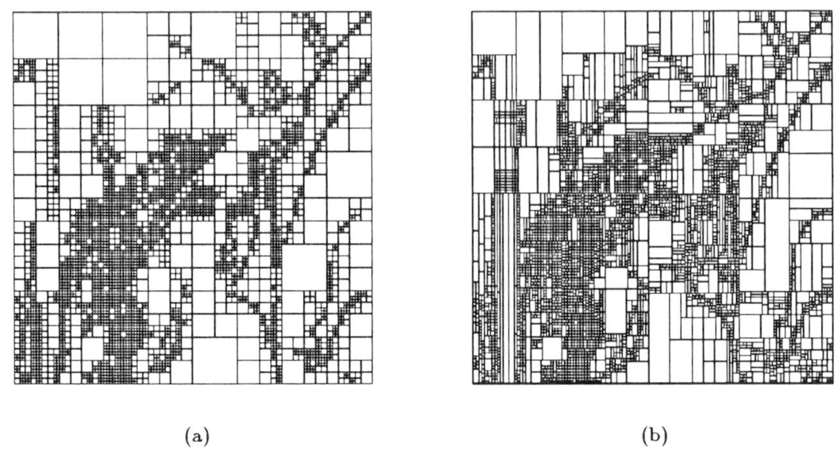

(a) (b)

Figure 9. A quadtree and hv-partition of Lena with equal number of partitions

§ 3.3. Choosing w_i.

The selection of the transformations w_i affects whether W is contractive or not. In fact, the choice of metric in which contractivity is measured is important too. For δ_{sup}, requiring that w_i be contractive in the z direction; that is, $s_i < 1$ in equation (6), is sufficient to ensure that W is contractive. If some w_i is expanding in the z direction, then W is exanding, but may be eventually contractive.

One way to check eventual contractivity for the metric δ_{sup} is to begin with an image f such that $f(x,y) = 1.0$. Define w_i' as w_i with $o_i = 0$, and $W' = \cup_i w_i'$. Then

$$\sup_{(x,y)\in I^2} \{W'^{\circ n}(f)(x,y)\}$$

will be the contractivity σ of $W'^{\circ n}$. To check if a map is eventually contractive, iterate W' until $\sigma < 1$. A similar procedure to determine eventual contractivity for the rms metric is not known to the authors. These tests only determine eventual contractivity after an encoding has been made. In practice, it is necessary to have a sufficient number of contractive maps so that encodings that are not eventually contractive do not occur with any significant probability.

The w_i can be taken to be contractive in the x and y directions, though this is not necessary when $s_{max} < 1$. Contractivity in the x and y directions increases the "mixing" of the transformations, thus promoting eventually contractive maps. The issue of the x and y contractivity when s_{max} is not restricted remains to be fully resolved.

Finally, how the w_i transform the domain pixel values is important. A pixel value can either be the average of all transformed domain pixel values that map to the pixel, or it can be chosen as one of the transformed domain pixel values. The former leads to better fidelity, at some computational cost.

§ 3.4. Quad-tree Partitioning.

References 2 and 9 contain a description of an adaptive partitioning scheme. In this scheme an image is divided into four quadrants that are independently encoded. If an image possesses local self-similarity in some general sense, then the quality of the encoding should not be significantly degraded by limiting **D** to be chosen from a quadrant of the image. Furthermore, the reduced cardinality of **D** decreases the encoding time. A two level partition is used to take advantage of the local characteristics of the image.

In the examples given in reference 2, 256 × 256 6-bpp images were divided into 128 × 128 subimages, each of which was independently encoded with 16 × 16 and 8 × 8 domain squares covering 8 × 8 and 4 × 4 range squares, respectively. The error in the 4 quadrants of each 8 × 8 range was required to meet a predetermined error criteria based on the δ_{rms} and δ_{sup} error. If the predetermined criteria was not met in a quadrant, that quadrant was encoded as a 4 × 4 range. The domains and ranges were classified, and domains covered only ranges in the same class. The allowed values for the scale s and offset o were class-dependent, with at most eight values for s and 128 values for o. This encoding scheme uses a relatively large number of transforms, each requiring a relatively small amount of memory storage.

In references 10 and 11, 8-bpp images were encoded with an implementation that, relative to the implementation summarized in the previous paragraph, used fewer transformations with a larger storage requirement per transformation. The algorithm was set up such that several encoding parameters could be varied, allowing for a general study of the implementation's performance. So that the discussion of results in the following section is clear, a somewhat detailed description of the implementation follows.

Encodings were performed with four different choices $\mathbf{D}^1, \mathbf{D}^{1/2}, \mathbf{D}^{1/4}$, and $\mathbf{D}^{1/8}$ for the set **D**. The domain squares had variable size, and were restricted to have corners on a lattice with a fixed vertical and horizontal spacing. Two of the domain sets included domains with sides slanted at 45 degree angles from the edges of the image. The combinations of square sizes, spacings, and orientation for the domain sets are summarized in table 3 below.

Table 3.

Set	size	Lattice Spacing	Includes 45° rotation
$\mathbf{D}^1$	8,16,32,64	size/2	Yes
$\mathbf{D}^{1/2}$	8,16,32,64	size/2	No
$\mathbf{D}^{1/4}$	8,16,32,64	size	Yes
$\mathbf{D}^{1/8}$	8,16,32,64	size	No

The encoding process followed the pseudo-code in Table 2a. However, a classification scheme was used to classify all possible domains into 72 classes. The classes were defined by the brightness and edge like character of the quadrants of each square. Choosing a classification scheme that ordered the quadrants of a domain by their brightness determined a symmetry operation which limited the search for a good cover, since only one of eight possible orientations was checked. For a more detailed description of the classification scheme, see reference 10. When covering a range, domains (with side length twice that of the the range square) from a fixed number n_c of classes were considered; those classes were chosen to be the "most" similar to the classification of the range.

The number of bits used to discretize and store s_i and o_i in equation (6) was adjustable, as was the maximum allowable s_i. These parameters are denoted n_s, n_o, and s_{max}, respectively. The values for s_i were restricted to the range $s_{max} \geq |s_i| \geq (s_{max}/10)$ and $s_i = 0$. The other parameters were $\mathbf{D}$, e_c, r_{min}, and the number of domain classes to search n_c. Results of varying these parameters are given in section 4.1.

§ 3.5. HV-Partitioning.

The HV-partitioning scheme derives its name from the horizontal and vertical partitions it makes. Unlike the quad-tree partition, which divides a square in a fixed way, HV-partitioning allows variable partitioning of rectangles. Each partitioning step (following the pseudo-code in table 2a) consists of subdividing a rectangle (initially the whole image) either horizontally or vertically into two rectangles. More detail can be found in reference 12. Partitioning methods similar to this have been applied to the image compression problem,[13] but the goal here is to use the adaptive partition as part of an iterated transform algorithm.

When a rectangle that contains a vertical or horizontal edge is partitioned, the partition occurs along the strongest such edge. If a rectangle contains other edges, the partition should occur in such a way that the edge intersects one of the generated rectangles at its corner, with the other rectangle not containing this edge. This results in rectangles which either contain no edges, or have edges which run diagonally.

The set $\mathbf{D}$ is dynamic, and changes for each covering step of the algorithm. It consists of all *previously* partitioned rectangles. Since the partition attempts to create rectangles with edges that run diagonally, it is reasonable to expect that this choice of $\mathbf{D}$ would work well. In fact, the number of domains searched using the HV-partitioning scheme is less than the number of domains used in the quad-tree method (by a factor of about 60), but the encoding results are comparable.

To find a horizontal partition, the pixel values of the rectangle are summed vertically. Edges are then detected by computing the successive difference of these sums. These differences are weighted with a bump function so that partitions would be preferentially chosen near the center of a rectangle. This is necessary to avoid many long, narrow partitions. Vertical partitions are sought in an analogous fashion.

Storing the partition compactly is simple, but requires more memory than the quad-tree scheme. The partition is stored as a sequence of labels denoting the orientation of the partition, and an offset value that determines the position of the partition at each step. As the rectangles get smaller, fewer bits are required to store the offset, so that the memory cost decreases as more rectangles are added. More rectangles also increase the domain pool. For these two reasons, the HV-partitioning scheme tends to work better at low compressions.

Mapping domains onto ranges is more complex in this scheme than in the quad-tree method. In the quad-tree scheme, domains were chosen to be larger by an integer multiple, so that averaging transformed domain pixels onto a range pixel is easy. In the HV-partitioning scheme, proper averaging is computationally expensive. One alternative is to simply choose a representative domain pixel for each range pixel. Another alternative, leading to better results at some computation time expense, is to average those domain pixels that map wholly within a range pixel, and to ignore those that contribute a fraction of their value.

Section 4.2 presents results from this partitioning scheme.

Section 4. Results.

The peak-signal-to-noise ratio (PSNR) was used to determine image fidelity. PSNR is defined as,

$$\mathrm{PSNR} = -20\log_{10}\left(\frac{rms}{2^n - 1}\right),$$

where rms is the root mean square difference of the reconstructed image with the original image, and n is the number of bits per pixel in the image.

§ 4.1. Results of Quad-tree Implementation.

Although the optimal values of n_s and n_o are image dependent, a thorough empirical study of the effect of these values on image compression and fidelity for many images suggests that, for this implementation, $n_s = 5$ and $n_o = 7$ are good choices for 8-bpp images.[11]

Figures 10 and 11 present results arising from varying s_{max}. Figure 10 shows the distribution of the scale factors used in three different encodings of 256×256 pixel resolution, 8-bpp Lena. Figure 10b ($s_{max} = 1.0$) shows that the larger scale factors are used more often, suggesting that a distribution of s_i's with more large values and less small values might yield improved encodings. However, experiments with linear and exponential distributions (opposed to the logarithmic distribution used here) and with truncation of small values resulted in no improvement.

The distribution of s_i values in figures 10b and c show a disproportionate number of values lying near the extremes of the allowed range, suggesting that increasing s_{max} may result in improved encodings. Figure 11, showing PSNR versus compression for 256×256 8-bpp Lena and San Francisco images, shows this to be the case. The original San Francisco image, and the decoded image for $s_{max} = 1.2$ are shown in figures 12a and b, respectively. Results for a variety of other images and encoding parameters indicate that $s_{max}=1.2$ or 1.5 usually yields the best PSNR versus compression results. As in section 3.1, this result indicates that relaxing the contractivity results in improved encodings.

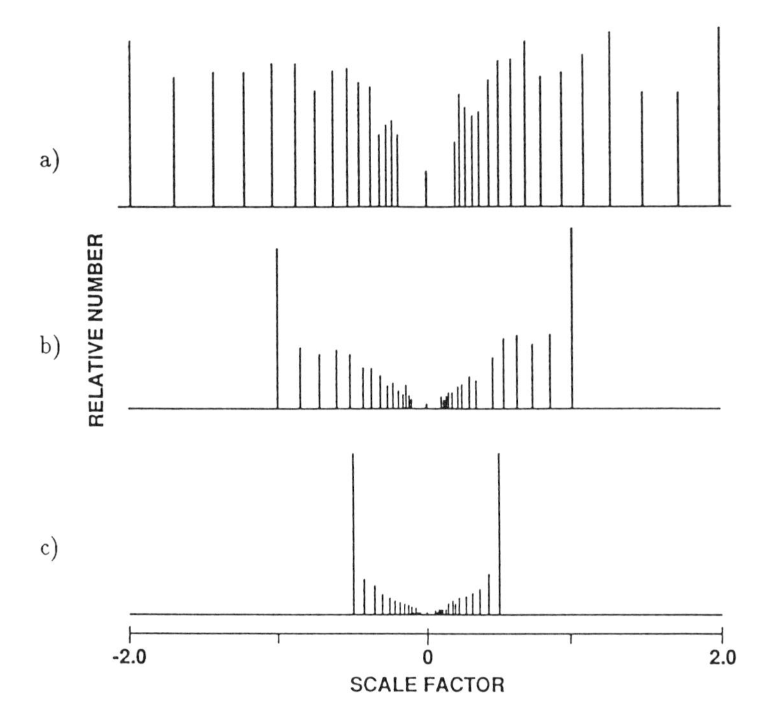

Figure 10. Relative number of s_i's at each allowed value for three encodings of 256 × 256 Lena. The value of s_{max} for 2a, b, and c were 2.0, 1.0, and 0.5 respectively. The domain set used was $\mathbf{D}^1$ with other parameters $(n_s, n_o, r_{min}, e_c, n_c)$ equal to (5, 7, 4, 8.0, 4) respectively.

It is of interest to note that for cases with $s_{max} \geq 1.0$, every one of the encodings performed (numerous images and several hundred separate encodings) resulted in a mapping that converged to a fixed point. In a few cases (with $s_{max} = 2.0$) a mapping took more than ten iterations to converge to its fixed point, but in all cases with $s_{max} \leq 1.2$, ten iterations were sufficient.

Figure 13 shows PSNR versus compression for various domain sets for 512×512 8-bpp Lena. The data indicate that in all cases increasing the number of domains improved both compression and fidelity. Continuing studies are underway to determine the point at which increasing the number of domains will begin to reduce the compression/fidelity results. It is important to note that number of domains more directly effects the encoding time than the other parameters which can be adjusted.

Figure 14 presents PSNR versus compression data of 512×512 Lena. The compression was varied by adjusting e_c and r_{min}. Data is also shown for different values of n_c. The value of e_c separates the data into three widely spaced clusters, with smaller e_c leading to higher PSNR. The original image of Lena and the decoded image with $r_{min} = 8, e_c = 8$, and $n_c = 72$ are shown in figure 15. For comparison, results are also shown for an ADCT algorithm similar to that described in reference 14. The ADCT data is comparable to the JPEG standard, being slightly better at higher compression.[15]

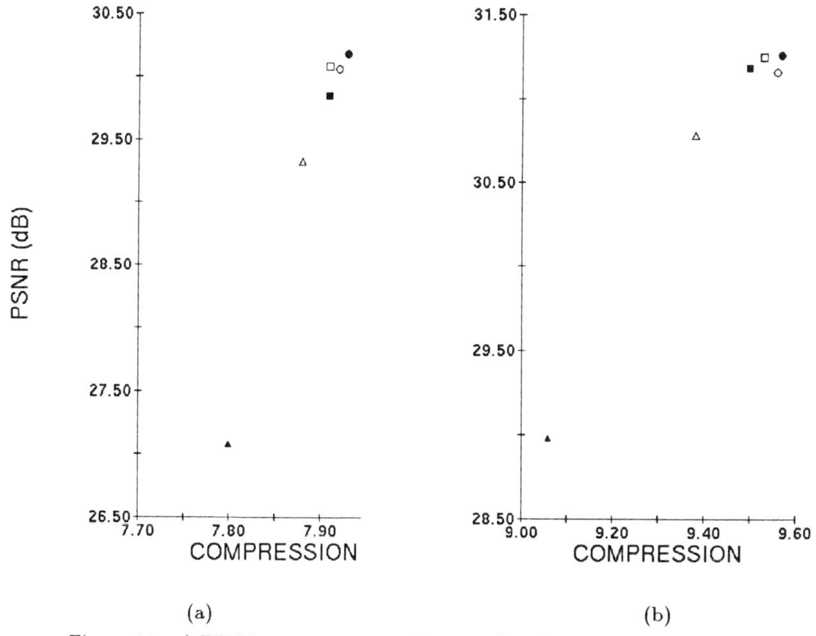

(a) (b)

Figure 11. a) PSNR versus compression as a function of s_{max} for 256 × 256 San Francisco, and b) 256 × 256 Lena. The values of s_{max}= 0.5, 0.8, 1.0, 1.2 1.5, and 2.0 are denoted by ▲, △, ▮, ▯, •, and ◦ respectively. Other parameters are the same as figure 10.

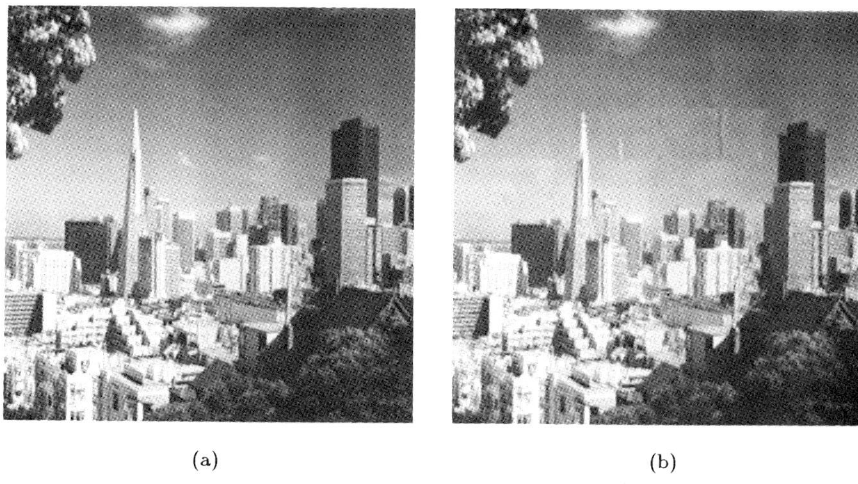

(a) (b)

Figure 12. a) Original San Francisco, and b) decoded image with s_{max}=1.2. For this image the compression is 7.9:1 and the PSNR is 30.08 dB.

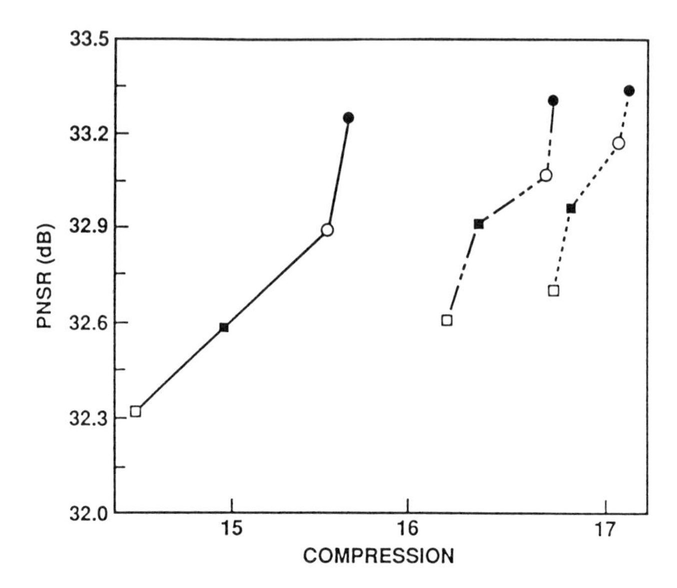

Figure 13. PSNR versus compression as function of number of domains for encodings of 512 × 512 Lena. The sets $\mathbf{D}^1$, $\mathbf{D}^{1/2}$, $\mathbf{D}^{1/4}$, and $\mathbf{D}^{1/8}$ are denoted by •, ○, ▮, and ▯ respectively. Data for n_c=1, 4, and 12 appear as the solid, dashed, and dotted curves respectively. Other parameters were $s_{max} = 1.0$ with other parameters the same as figure 10.

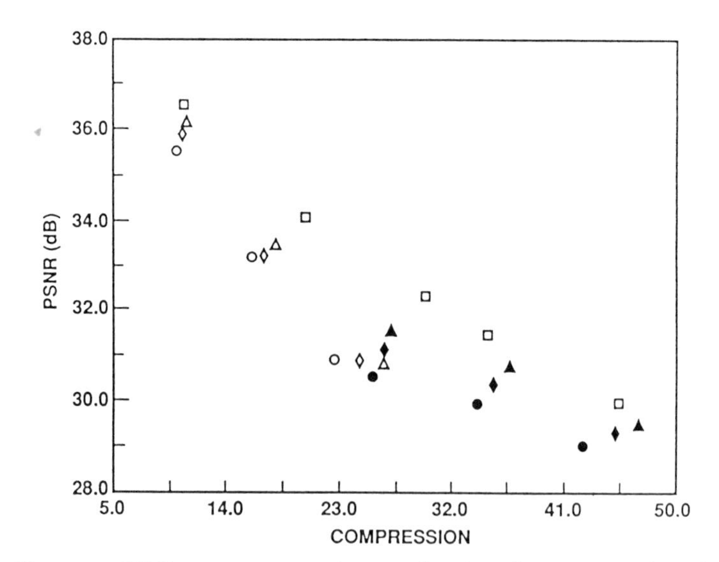

Figure 14. PSNR versus compression as a function of e_c, r_{min}, and n_c for 512 × 512 Lena. Open symbols represent $r_{min} = 4$ and solid symbols $r_{min} = 8$. Data for e_c = 5.0, 8.0, and 11.0 appear in clusters with increasing compression. Within each cluster, n_c= 1 (○), 4 (◇), and 72 (△). ADCT data (▯) is shown for comparison.

(a)

(b)

Figure 15. a) 512 × 512 original image of Lena, and b). Decoded image with compression is 36.78:1 and the PSNR is 30.71 dB.

In table 4, the relative encoding times for several encodings are presented. The data indicate the relative encoding time as a function of n_c, e_c, r_{min}, and image size. On an HP-Apollo 400t workstation, the relative time units in the table are approximately 1170 cpu seconds. The programs have not been optimized for speed.

Table 4. Relative time to encode Lena for various parameters.

s_{max}	n_c	r_{min}	e_c	size	time(rel)	Compression	PSNR(dB)
1.2	1	4	8	512	1	15.95:1	33.13
1.2	4	4	8	512	3.1	17.04:1	33.19
1.2	72	4	8	512	35.5	17.87:1	33.40
1.0	4	4	8	512	3.1	16.74:1	33.30
1.2	4	4	5	512	5.3	10.49:1	35.92
1.2	4	4	11	512	2.0	24.62:1	30.85
1.2	72	8	8	512	7.5	36.78:1	30.71
1.2	1	4	8	256	0.14	9.09:1	30.63
1.2	72	4	8	256	4.5	9.97:1	31.53
1.2	72	4	10	256	3.7	11.85:1	30.58

Finally, figure 16 shows the distribution of distances between R_i and D_i. The data shown is for an encoding of Lena, and is typical for the images that have been studied. If this distribution indicates that a disproportional number of domains are relatively near to their ranges, then the image has some local self similarity. The solid curve is the theoretical distribution of the distance between two randomly chosen points in I^2. The figure shows that there is no significant local self similarity; the slight shift between the theoretical and experimental distribution is due in part to the distance being measured between small but not infinitesimal squares. It should be noted that this result does not necessarily indicate that limiting the domains to localized areas degrades performance. For a fixed encoding time and compression, localizing the potential domains allows for an increased density of domains. This was the approach taken in reference 2.

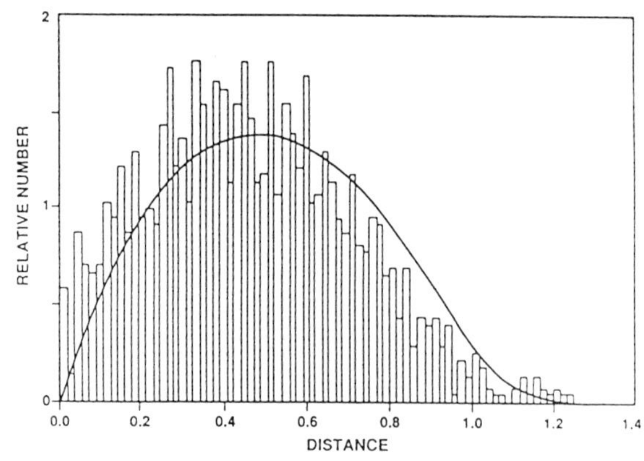

Figure 16. Theoretical distribution for random domain-range distance and the actual distribution for an encoding of Lena.

This implementation differs from that of reference 2 in several ways. There $s_i \leq 1.0$; a δ_{sup} as well as rms criteria was used during encoding; a smaller set of values for s were allowed; and the set of allowed domains was localized and small (about the same in number as $\mathbf{D}^{1/8}$). The implementation presented here results (last encoding in table 4) in an increase of about 1.4 dB at the same bit rate.

§ 4.2. Results of HV Partitioning Implementation.

Results for the HV-partitioning are presented for those algorithmic parameters that are specifically relevant to this scheme. These are the size r_{min} of the smallest rectangles that were allowed to be partitioned, and whether averaging was used in mapping domains onto ranges. Except where indicated, the figures show data using $n_s = 5$, $n_o = 7$, $r_{min} = 2$, and $s_{max} = 1$.

Figure 17 shows comparative results for the quad-tree scheme (using a 4 class search) and the HV-partitioning scheme with and without pixel averaging for the San Francisco image. Averaging typically improves the results by about 1 dB. A 4 class search in the quad-tree scheme is roughly comparable in computational requirement as the search performed by the HV-partitioning scheme. The compression was changed by varying e_c. The results for this image are better, but this is not uniformly true for the method; for example, Lena at size 512 encodes at 36.04dB with 8.7:1 compression and 32.74dB with 16.9:1 compression.

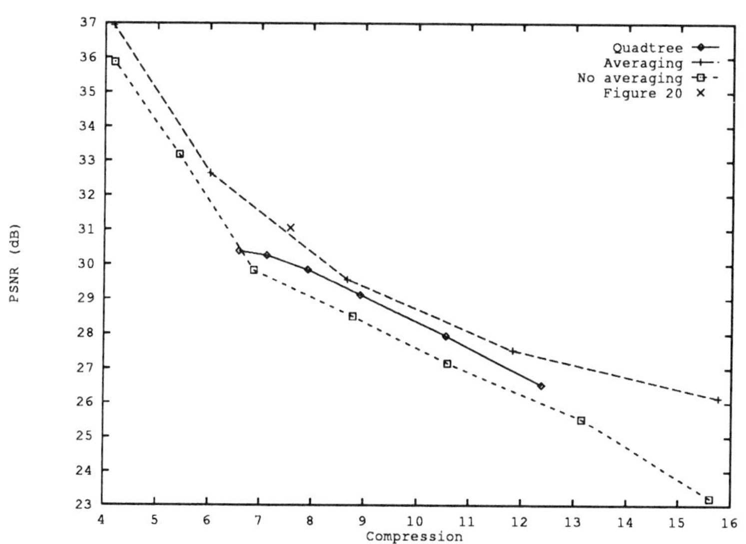

Figure 17. The effect of averaging pixels on the HV-partition scheme as compared with the quad-tree partition.

Figure 18 shows the effect of varying the size of the smallest partitionable rectangle for the 512 Lena image. It is not surprising that larger partitions yield better results at higher compressions, but it is surprising that allowing 2×2 pixel rectangles does not worsen the results. Taking $r_{min} = 2$ gives only a very slight compression (since transformations require about 30 bits on average). This means that the algorithm used some coverings that required nearly as much memory as simply storing the pixel value, but introduced some error in the process. This result suggests that the quad-tree algorithm may yield better results at low compression using smaller r_{min} than the results presented in section 4.1.

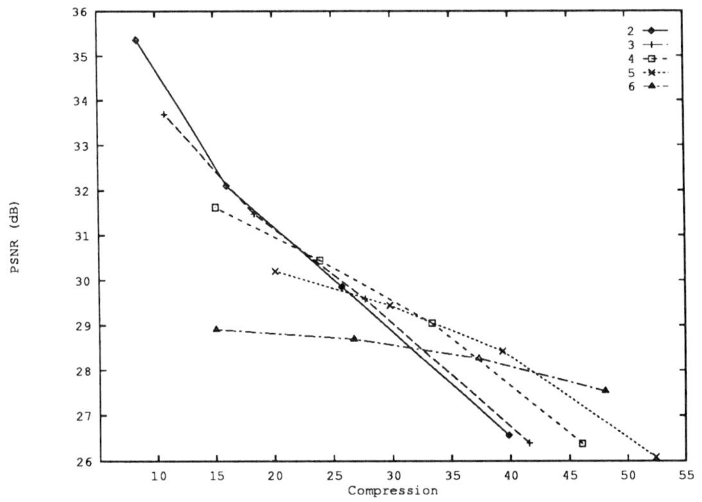

Figure 18. The effect of varying r_{min} on compression. The value of r_{min} is indicated by the legend.

Finally, preliminary results from varying other parameters suggest that these results can be improved. Domains used to cover ranges are restricted to be at least ρ times bigger in the x and y direction. Data presented above use $\rho = 1$, but $\rho = 1.5$ gives better results. Figure 19 is a 7.6:1 compression of San Francisco at 31.06 dB PSNR, using $\rho = 1.5$, $n_s = 4$, $n_o = 6$, $s_{max} = 1.0$ and $r_{min} = 2$.

Figure 19. HV-partitioned encoding of San Francisco at 7.6:1 compression and 31.06 dB.

Section 5. Other Work.

This section contains a brief summary of other work for which results are not included in this article but which nevertheless warrant attention. The discussion touches on the encoding of color images, postprocessing to eliminate artifacts, alternative encoding and decoding methods, and different image partitioning schemes.

A standard approach for encoding 24-bpp color images is to transform the images from the RGB representation to YIQ luminance and chrominance representation. The luminance is compressed as before, and the chrominance is spatially averaged and then compressed. Using this method, preliminary results indicate that performance is consistent (relative to ADCT methods) with the results for 8-bpp gray scale images presented here. Another approach attempts to encode one of R, G, or B signals, and then encode the remaining two components using (wherever possible) transformations similar to those used in the encoding of the first component. For some images this method works well, but in general the results are not so good as the luminance-chrominance approach.

As with other compression techniques, postprocessing can be performed to remove artifacts from a decoded image. A simple approach that averages between range boundaries to remove artifact edges worked well; even increasing the signal to noise ratio in some cases. None of the results presented in section 4 contained any postprocessing.

An alternative encoding method which showed few artifacts, but has not been fully evaluated, covered each range square with a linear combination of domains. This increases both the domain pool, the search time to find an optimal cover, and the storage requirement per transformation. Alternative decoding methods focus on decreasing the (already fast) decoding time. One approach is to decode an image at a smaller size until the fixed point is well approximated. Then, only a couple of iterations of W are required to reach the fixed point at the full image size.

Finally, current research is concentrating on finding alternative partitioning schemes. For example, a recursive triangle based scheme generalizing the HV-partitioning scheme shows promise because its memory requirements are the same (per range) as the HV-partitioning scheme, but the domains and ranges are considerably more versatile; allowing w_i to map with arbitrary orientation rather than at rotations of $90°$.

Section 6. Conclusion.

The results presented illustrate several interesting points. By adjusting appropriate parameters, encodings with reasonable compression/fidelity performance can be generated over a large range of compressions. Allowing non-contractive transformations can result in improved encodings. This is contrary to what the collage theorem suggests. The bound predicted by the collage theorem is much larger than the error found empirically from encoding images. Therefore, the collage and generalized collage theorems provide motivation that a good encoding can be found, not a useful bound on the error. Finally, based strictly on compression versus PSNR performance, current implementations do not outperform an ADCT method similar to the JPEG standard.

Although iterated transform image encoding is computationally intensive, the decoding process is computationally simple compared to ADCT methods. Because the iterated transform scheme is based on a self referential coding of an image as opposed to a fixed code book approach, it is well suited for the encoding of a more general class of images than vector quantization algorithms. Therefore, iterated transforms may be best applied in applications for which a large variety of pre-encoded images (possibly supplied by a central data base) need to be decoded quickly with software on off-the-shelf computer platforms. Future improvements in the method may lead to more general application.

In implementing the iterated transform method, the selection of classification method and scale and offset parameters are important, but it is the choices of the partition (**R**) and **D** which are least obvious. Very simple schemes work well for the same reason as vector quantization methods, i.e., the information in each transformation is relatively small, so that a large number of transformations can be used while maintaining reasonable compression. More elaborate partitions which might better exploit the self similar properties of iterated transform encodings must be stored as part of the encoding, and therefore become burdened by their own complexity. The data indicate that the improvements attained from generalizing a simple approach (such as the example in section 3.1) to a more elaborate approach (such as the HV partitioning discussed in section 3.5) are not negligible, but small. When devising an iterated transform encoder, the issue of optimal algorithmic complexity may be the most interesting problem.

Acknowledgments

The authors thank the Office of Naval Research and NOSC Independent Exploratory Development for their support; and John Hubbard for his useful comments. YF would like to thank ASEE for Financial support; and Hassan Aref and the San Diego Supercomputing Facility.

References

1 B.B. Mandelbrot, B. **The Fractal Geometry of Nature**. W.H. Freedman and Company, New York 1977.

2 A. Jacquin, A Fractal Theory of Iterated Markov Operators with Applications to Digital Image Coding. Doctoral Thesis. Georgia Institute of Technology.1989.

3 J.E. Hutchinson, *Fractals and Self Similarity*. Indiana University Mathamatics Journal, Vol. 35, No. 5. 1981.

4 M.F. Barnsley, **Fractals Everywhere**. Academic Press. San Diego, 1989.

5 M.F. Barnsley and A. Jacquin, "Application of Recurrent Iterated Function Systems to Images," SPIE vol. 1001, Vis. Comm. and Image Processing 1988, pp. 122-131.

6 E.W. Jacobs, R.D. Boss, and Y. Fisher, "Fractal-Based Image Compression, II" NOSC TR-1362, Naval Ocean Systems Center, San Diego, CA., 1990.

7 Scientific American, March 1990, p 77.

8 M.F. Barnsley and A.D. Sloan, "Methods and Apparatus for Image Compression by Iterated Function Systems," U.S. Patent #4941193, 1990.

9 A. Jacquin, "A Novel Fractal Block-Coding Technique for Digital Images," 1990 IEEE ICASSP Proc. Vol. 4, p.2225.

10 Y. Fisher, E.W. Jacobs, and R.D. Boss, "Iterated Transform Image Compression," NOSC TR-1408, Naval Ocean Systems Center, San Diego, CA., 1991.

11 E.W. Jacobs, Y. Fisher, and R.D. Boss, "Image Compression: A Study of the Iterated Transform Method," Submitted to Signal Processing, 1991.

12 R.D. Boss, E.W. Jacobs, and Y. Fisher, "Fractal Image Compression Via Adaptive Image Partitioning ," manuscript in preparation.

13 X. Wu and C. Yao, "Image Coding by Adaptive Tree Structured Segmentation", **Data Compression Conference Proceedings**, Snowbird Utah, IEEE Computer Society Press, Los Alamitos, California, 1991. pp 73-82.

14 W. Chen and W.K. Pratt, "Scene Adaptive Coder," IEEE Trans. Comm. 32, p. 225, (1984).

15 M. Rabbani, P.W. Jones, **Digital Image Compression Techniques**, SPIE Optical Engineering Press, Bellingham, WA, (1991).

3

Optical Techniques for Image Compression*

John H. Reif Akitoshi Yoshida

Department of Computer Science
Duke University
Durham, NC 27706.
Email: reif@cs.duke.edu, ay@cs.duke.edu

Abstract

Optical computing has recently become a very active research field. The advantage of optics is its capability of providing highly parallel operations in a three dimensional space. Image compression suffers from large computational requirements. We propose optical architectures to execute various image compression techniques, utilizing the inherent massive parallelism of optics.

We optically implement the following compression and corresponding decompression techniques:

- transform coding
- vector quantization
- interframe coding for video

We show many generally used transform coding methods, for example, the cosine transform, can be implemented by a simple optical system. The transform coding can be carried out in constant time.

Most of this paper is concerned with an innovative optical system for vector quantization using holographic associative matching. Limitations of conventional vector quantization schemes are caused by a large number of sequential searches through a large vector space. Holographic associative matching provided by multiple exposure holograms can offer advantageous techniques for vector quantization based compression schemes. Photorefractive crystals, which provide high density recording in real time, are used as our holographic media. The reconstruction alphabet can be dynamically constructed through training or stored in the photorefractive crystal in advance. Encoding a new vector can be carried out by holographic associative matching in constant time.

An extension to interframe coding is also discussed.

*Research supported in part by DARPA/ISTO Contracts N00014-88-K-0458, N00014-91-C-0114 and N00014-91-J-1985, NASA subcontract 550-63 of prime contract NAS5-30428, and US-Israel Binational NSF Grant 88-00282/2.

1 Introduction

1.1 Image Compression

Image compression is crucial for many applications [1] [2]. The objective of image compression is to reduce the bit rate for signal transmission or storage while maintaining an acceptable image quality for various purposes. In video signal transmission, one can compress original images so that the high quality images which would require high bandwidth can be transmitted through a medium with relatively low bandwidth. In medical imaging, one can use image compression techniques to store a large number of x-ray pictures that are routinely produced at hospitals.

Compression techniques generally exploit the redundancy in the image. The transform coding such as the fourier transform coding or the cosine transform coding decomposes an input image into its spectral components. One can achieve compression by appropriately coding the spectral components which are above a certain threshold value. Unfortunately, the difficulties of implementing the transform coding arise from the complexity of transforming algorithms, which require $O(n \log n)$ time for an n-point transformation.

A vector quantizer is a system that maps a set of continuous or discrete vectors into a finite set of discrete vectors that are suitable for transmission or storage. The vector quantization techniques for compressing image and speech signals have been extensively investigated by many researchers [3] [4]. The basic algorithm using full search in the vector space requires an extremely large number of computations. Although the algorithm is guaranteed to find a locally optimal quantizer, its computational complexity can be prohibitive. A similar approach using tree search has less computational complexity but requires larger storage and needs still a quite large number of computations.

The problem is caused by sequential execution of the algorithm on conventional electronic computers. Although, some parallelism can be obtained by standard parallel processing hardware, electrically implemented interconnection may not provide enough bandwidth to handle a large number of computations.

1.2 Power of Optical Computing

Optical computing has recently become a very active research field [5] [6] [7] [8]. The obvious advantage of optics is its freedom in space. A set of light beams can establish communication links among optical logic gates in a three dimensional space, whereas the VLSI model must confine electrical wires on a two dimensional plane.

From a theoretical computational point of view, for a given problem, there is a lower bound on the circuit area and its computational time. One such lower bound on the planar VLSI model called "AT^2 bounds" states that $AT^2 = \Omega(I^2)$, where A is the circuit area, T is the time used by the circuit, and I is information content[1] of the problem (See Ullman [9]). In a three dimensional electro-optical model described by Barakat and Reif called VLSIO, the similar lower bound can be expressed as $VT^{3/2} = \Omega(I^{3/2})$ [10]. This implies that as the information content becomes larger, the VLSI circuit requires a larger and larger area to solve the problem in a fixed amount of time. Using three dimensional optical systems as in the VLSIO model, we can overcome this interconnection problem by utilizing space in a volume. As an example, a $n \times n$ two-dimensional fourier transform can be computed by a simple optical system of volume $O(n^{3/2})$ in constant time. A n-point fourier transform has $AT^2 = \Omega(n^2 \log^2 n)$ in the VLSI model, whereas in the VLSIO model it has $VT^{3/2} = \Omega(n^{3/2} \log^{3/2} n)$. In fact, there is an optical system which implements a n-point fourier transform in constant time using volume $n^{3/2}$ [11]. This is optimal within a polylog factor.

1.3 Optics for Image Compression

Although the advantages of optics are well known, the realization of a general purpose optical computer is yet to come. On the other hand, special purpose optical computers have been implemented for areas such as image and signal processing [12] [13], associative memory [14], and neural networks [15] [16]. These systems exploit the advantages of optics such as the ability to perform a matrix-vector multiplication and a two dimensional fourier transform in constant time or to implement associative memory using holograms.

A lens can compute the fourier transform of an input image [17]. A transparency whose transmittance represents the input image is placed at the front focal plane of the lens. The amplitude of the light passing the transparency is modulated by the transmittance. The complex amplitude of the light at the back focal plane represents the fourier transform of the input image.

Holograms have been used for associative matching [18] [19] [20] [21] [22]. One can record multiple images on a single holographic medium using distinct reference beams as their associative keys. Later, the stored image can be reconstructed by using its corresponding reference beam

[1] Information content is the number of bits that must cross a boundary in order to solve the problem. The boundary separates the circuit into two sides, each of which holds approximately half the input bits.

as a key associated to the image. Recently, dynamically modifiable holographic media such as photorefractive crystals (i.e., iron doped $LiNbO_3$) have been widely investigated. The refractive index of these media can be optically changed to store holograms. The thickness of the media allows the superposition of many holograms in a common volume in the crystal. A large number of holograms can be stored in a volume by using a recording reference beam that has a distinct angle for each hologram. Later, each hologram can be read out using its corresponding reference beam. Photorefractive crystals are particularly attractive as holographic media, since they provide high density recording in real time.

In this paper, we consider compression techniques using volume holograms in photorefractive crystals. We discuss their advantages and limitations. We also discuss an extension to interframe video compression.

Our results are:

- Cosine transform of a $n^{1/2} \times n^{1/2}$ image with n pixels can be carried out in constant time with the VLSIO model in volume $O(n^{3/2})$.
- Encoding and decoding a vector by an N-level k-dimensional vector quantizer can be carried out in constant time with the VLSIO model in volume $O(N^{3/2} + k^{3/2})$.
- Vector quantization of a $n^{1/2} \times n^{1/2}$ image with n pixels by an N-level k-dimensional quantizer can be carried out in constant time with the VLSIO model in volume $O((N^{1/2} + k^{1/2})n)$.

In the following, section 2 gives optical implementation of the cosine transform coding. Section 3 starts with a background in vector quantization and holographic associative matching, and describes our holographic vector quantizer. Section 4 discuss its extension to interframe coding. Section 5 concludes this paper.

2 Optical Implementation of Cosine Transform Coding

The cosine transform [23] [24] is a member of sinusoidal transforms. The fourier transform decomposes a spatially encoded input image into its spectral components. For an image with high inter-pixel correlation, high spatial frequency components are often small and negligible. Thus, appropriately encoding only significant components leads to compression of data.

The drawback of the discrete fourier transform is that it produces coefficients with real and imaginary components. An alternative approach

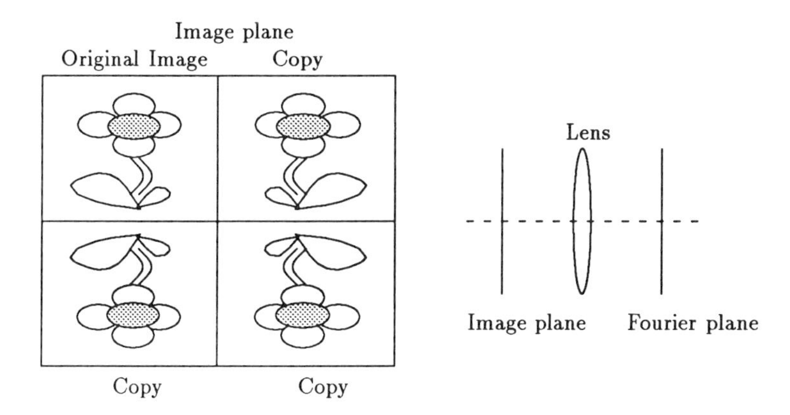

Figure 1: Optical cosine transform.

is the discrete cosine transform that generates only real coefficients. Suppose we have a vector of n sampled data. We then form a symmetric vector of size $2n$ by concatenating the vector and its flipped copy. The resulting vector is a real even version of the original vector. To obtain its discrete cosine transform, we apply a discrete fourier transform of length $2n$ to this vector. Since the vector is real and even, only the coefficients of cosine terms are non-zero.

An optical implementation of a two-dimensional cosine transform is quite simple, as depicted in Figure 1. The input image is copied and flipped three times to form a symmetric image. A two-dimensional fourier transform can be computed optically in constant time using a lens. The symmetry of the input image allows its cosine transform to be obtained at the focal plane. The amplitude of each spectral component represents its corresponding fourier coefficient. Unfortunately, if the intensity is directly measured, we obtain the square of each coefficient. To determine the sign of each coefficient, we first detect the intensity of each spectral component. Then, we interfere the spectral components with a reference beam of a known phase and detect the intensity of interfered spectral components. We can determine the sign of each coefficient from the difference in intensity.

Encoding

Our optical system is used to compute the two dimensional cosine transform of an input image in constant time. Threshold operations are applied to each spectral component to cut off some components. The resulting components are appropriately quantized and encoded by a computer.

Decoding
The same optical system used for encoding can be used to decode the spectral components.

Theorem 2.1 *Cosine transform of a $n^{1/2} \times n^{1/2}$ image with n pixels can be carried out in constant time with the VLSIO model in volume $O(n^{3/2})$.*

Proof: Let d_i and d_o represent the spacings between input pixels and detector pixels, respectively. The highest spatial frequency can be written as $1/d_i$. At the fourier plane, this spectral component is displayed at $\lambda f/d_i$, where λ is the wavelength, and f is the focal length of the lens. Since the detector array is of size $n^{1/2} \times n^{1/2}$, the spacing between detector pixels, d_o, becomes $2\lambda f/d_i n^{1/2}$. Assuming $d_i = d_o$, we obtain $f = O(n^{1/2})$. Thus, we can place the whole optical system in volume of size $O((n^{1/2})^3)$

3 Holographic Vector Quantization

3.1 Vector Quantizer

Extensive studies of vector quantizers have been made by many researchers [3] [4]. Lloyd proposed an iterative nonvariational method for optimal PCM design algorithm for scalar variables. Linde, Buzo, and Gray generalized Lloyd's approach to the general vector quantizer design and gave a method now called LBG algorithm [25].

Representation
An N-level k-dimensional vector quantizer can be defined as a mapping of a k-dimensional input vector $x = (x_1, \ldots, x_k)$ into a reproduction vector $\hat{x} = q(x)$, which is an element of a finite reproduction alphabet $\hat{A} = \{y_i | i = 1, \ldots, N\}$.

It can be viewed as two mappings: an encoder assigning to each input vector x a channel symbol from some channel symbol set, and a decoder assigning to each channel symbol a reproduction vector $\hat{x}$ drawn from the reproduction alphabet. The channel symbol set is typically the set of N binary vectors, each of size $\log N$.

Distortion Measure
The goal of quantization is to produce the best possible reproduction vectors for a set of input vectors under some given constrains. The performance of a quantizer is determined by using a distortion measure.

One simple distortion measure is the squared error distortion measure defined by

$$d(x, \hat{x}) = ||x - \hat{x}||^2 = \sum_{j=1}^{k}(x_j - \hat{x}_j)^2$$

The good quantizer is the one which minimizes the expected distortion for a given input sequence. The above algorithm and its variations are based on this minimization principle.

3.2 Holographic Associative Matching

Consider two mutually coherent beams interfering with each other on a photosensitive plate. After the plate is developed, the resultant intensity distribution is stored as a spatial modulation of the transmittance of the plate. When the plate is later illuminated by one of these beams, the incident beam will be diffracted by the modulation of the transmittance. Then, the resulting wavefront from the plate will be observed as if there are both beams incident on the plate.

Principle of Holography

A simple explanation for a plane recording medium is as follows. Two mutually coherent beams, the *object beam* illuminating the object, and the *reference beam* illuminating the recording medium, are used in recording. Here for simplicity we consider plane absorption holograms to discuss the basic aspects of holographic associative matching. Let O represent the complex amplitude of the object beam arriving at the recording plane. Similarly, let R represent the complex amplitude of the reference beam. The interference of the two beams is given by

$$I = |O + R|^2 = (O + R)(O + R)^*$$

where Z^* denotes the complex conjugate of Z.

The transmittance T of the medium is changed by $\Delta T = -\beta I$, where β describes the sensitivity of the medium. If the medium is illuminated by a reconstruction beam that is identical to the reference beam R, the distribution of light transmitted by the medium is

$$\begin{aligned} F &= (T + \Delta T)R \\ &= (T - \beta(|O|^2 + |R|^2))R - \beta|R|^2 O - \beta R^2 O^* \end{aligned}$$

Assume the wave intensity $|R|^2$ are constant over the recording medium. The first term corresponds to the reconstruction beam itself. The second term corresponds to the reconstructed object beam. The third term is the conjugate of the object beam with some phase shift caused by

R^2. Thus, one can create a replica of the object beam by illuminating the hologram using its associated reference beam. For detail, see the textbook by Collier, Burckhardt and Lin [26].

Angular Selectivity

Holograms recorded in volume media are considered as a spatial modulation of the dielectric constant (phase gratings) or the absorption constant (absorption gratings). When two plane waves are used in recording, the interference of the two beams causes either the dielectric constant or the absorption constant to vary sinusoidally in the volume. A set of equally spaced planes that are loci of the constant dielectric or absorption constant are drawn in Figure 2. Two waves represented by E_i and E_j enter the holographic medium. The wavelength in the medium is λ. Two vectors, $\vec{k}_i$ and $\vec{k}_j$, are called the wave vectors of E_i and E_j, respectively. The length of a wave vector is $2\pi/\lambda$, and its direction is parallel to the direction of the wave propagation in the medium. A vector, $\vec{K}_{i,j}$ is called the grating vector between wave E_i and wave E_j, and normal to the fringe planes. The length of the grating vector is defined as $2\pi/d$, where d is a distance between the fringe planes. Later, when the hologram is illuminated by a reconstruction beam, the incident beam is scattered by the set of planes. Only the reconstruction beam that satisfies the Bragg condition can constructively interfere with its scattered waves to maximize its diffraction efficiency, which is defined as the ratio of the incident intensity to the diffracted intensity. This is shown in Figure 2. Thus, the Bragg condition gives volume holograms strong angular selectivity upon reconstruction.

Volume media are suited for multiple exposure holograms because of their strong angler selectivity. A large number of holograms have been recorded in a volume of dynamic media such as $LiNbO_3$ [27] [28] [29].

3.3 Design of Holographic Compression

3.3.1 General Configuration

For a given image, we partition the image into a set of blocks (vectors). Let the size of each block be $\sqrt{k} \times \sqrt{k}$. We use two arrays of liquid crystal light valve (LCLV) [30] spatial light modulators (SLM) at the input plane: one for the input image array, S_1, and the other for the label array, S_2. The first array, S_1, is of size $2\sqrt{k} \times \sqrt{k}$ and is used to represent each vector of the input image. Each pixel of the input vector is represented in S_1 by two pixels of complementary intensity values. A pixel with an intensity value c is encoded into two pixels: one with an intensity value c and the other with an intensity value $\bar{c}$, which is equal to the maximum intensity value minus value c. This encoding scheme keeps

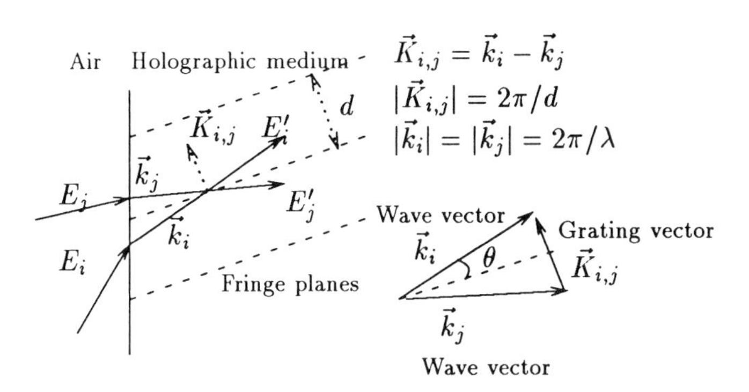

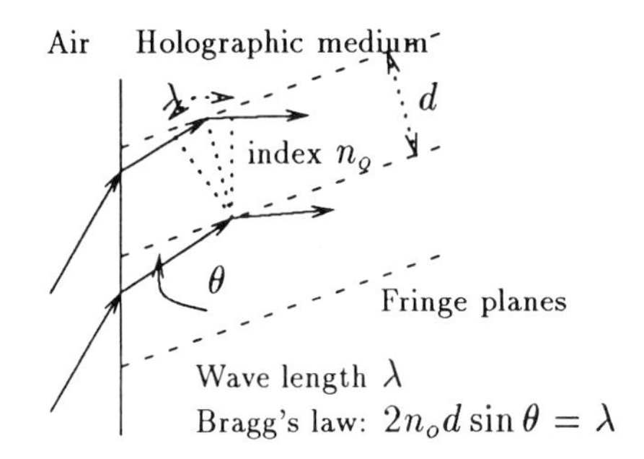

Figure 2: Formation of Holograms and the Bragg condition.

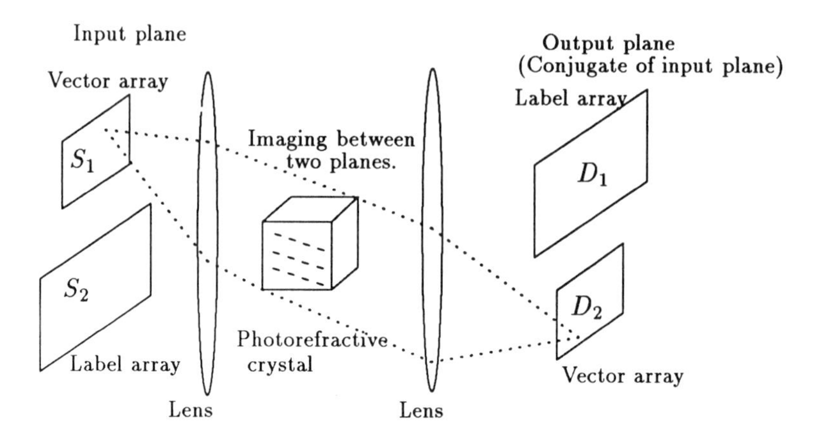

Figure 3: Holographic Vector Quantizer.

each encoded block at a constant intensity level, maintaining an equal energy incident on the holographic medium for every input vector. The other array, S_2, is of size $\sqrt{N} \times \sqrt{N}$, and provides one of N label beams to the holographic medium. We have two arrays of photodetectors at the output plane: one for the reconstruction vector array, D_1, and the other for the label array, D_2. The detector array, D_1, is of size $2\sqrt{k} \times \sqrt{k}$ and is used for reconstructing the decoded vector. The other array, D_2, is of size $\sqrt{N} \times \sqrt{N}$ and is used to detect the best matching label for the input vector.

Figure 3 shows a configuration of the system. Two mutually coherent beams illuminate array S_1 and array S_2. Two fourier lenses are used to form fourier transform holograms in the photorefractive crystal placed between the lenses.

We consider each pixel in the input plane as a point source. Each point source produces a plane wave after passing the first lens. There are $2k$ pixels in S_1 and N pixels in S_2. At the output plane, there are $2k$ pixels in D_1 and N pixels in D_2. We give a global index number to each pixel of the vector arrays and the label arrays. The pixels of the vector arrays are indexed from 1 to $2k$, and those of the label arrays are indexed from $2k + 1$ to $2k + N$. The image of the i-th pixel in the input plane is formed at the position of the i-th pixel in the output plane via two fourier lenses.

In photorefractive crystals, holograms are recorded via the modulation of the index of refraction by the space charge field induced by a spatially varying optical intensity distribution. Interaction between light waves and multiple gratings in the volume can be treated as linear if the diffraction efficiencies of the gratings are very small [31]. In this case, each incident plane wave interacts with every grating in the volume independently. The diffracted light for an arbitrary configuration of the input pixels is thus computed as a sum of each diffracted wave via each grating. Each grating formed by recording a pair of two plane waves is capable of connecting the pair of pixels.

Let η_{ij} be the diffraction efficiency of the grating connecting the i-th pixel and the j-th pixel. The intensity of light at the j-th pixel due to a light wave from the i-th pixel with intensity I_i is given by [32]

$$I_j = I_i \eta_{ij} + \sum_{p \neq i} \sum_{q \neq j} I_i \eta_{pq}(i,j)$$

where $\eta_{pq}(i,j)$ is the diffraction efficiency of light from the i-th pixel to the j-th pixel due to a grating connecting the p-th pixel and the q-th pixel. The crosstalk diffraction efficiency $\eta_{pq}(i,j)$ is given approximately by

$$\eta_{pq}(i,j) = \begin{cases} \eta_{pq} \ \mathrm{sinc}^2(K_{pq}(i,j)L/2) & \text{if } \vec{k}_i - \vec{k}_j = \vec{K}_{p,q}|\vec{k}_i - \vec{k}_j|/|\vec{K}_{p,q}| \\ 0 & \text{otherwise} \end{cases}$$

where $\vec{k}_i$ and $\vec{k}_j$ denote the wave vectors of light from pixels, i and j, respectively. $\vec{K}_{p,q}$ denotes the grating vector of the grating connecting the p-th pixel and the q-th pixel. L is the thickness of the crystal. $K_{pq}(i,j)$ denotes the phase mismatch of the wave from pixel i to pixel j due to the grating connecting the p-th pixel and the q-th pixel. This is given by

$$K_{pq}(i,j) = |(\vec{k}_i - \vec{k}_j) - \vec{K}_{p,q}|$$

In $LiNbO_3$, the diffraction efficiencies η_{ij} are proportional to the square of the component of the space charge density in the crystal at spatial frequency $\vec{K}_{ij} = \vec{k}_i - \vec{k}_j$.

The phase mismatched crosstalk term can be eliminated under several conditions [31] [32]. One condition is satisfied when the phase mismatch term $K_{pq}(i,j)$ is larger than $2\pi/L$. Another condition requires that the diffracted light waves do not propagate to any one of the output pixels. Then, recording a hologram connecting a pair of pixels increases the connection between the pair without seriously affecting the connections between other pairs. If we assume the phase mismatched crosstalk term is negligible, we can consider the photorefractive crystal as a linear mapping from input plane vector P_{in} to output plane vector P_{out}. P_{in} is a vector of length $2k + N$ whose elements correspond to the intensity

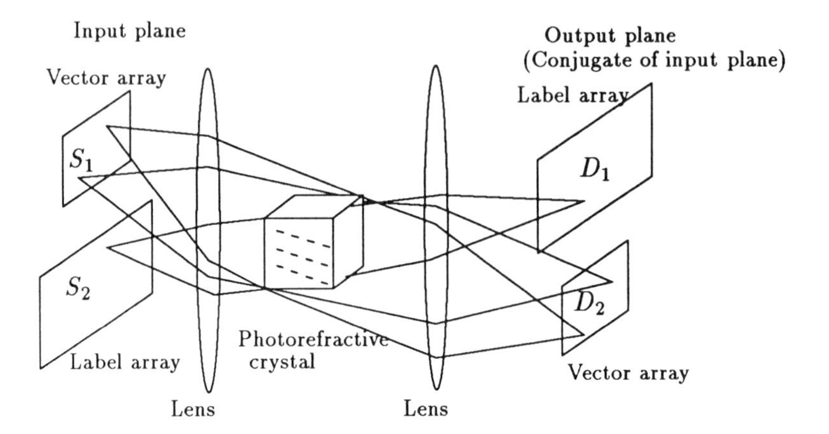

Figure 4: Recording vector and its associated label.

values of pixels at the input plane. Similarly, P_{out} is a vector of length $2k+N$ whose elements correspond to the intensity values of pixels at the output plane. Then, the mapping from the input plane to the output plane can be written as a matrix vector multiplication form.

$$P_{out} = H P_{in}$$

where H is a $(2k+N) \times (2k+N)$ matrix whose (i,j)-th entry is η_{ij}.

In this paper, we assume the above model to characterize photorefractive crystals. However, the nonlinear dynamics of multiple exposure holograms make it extremely difficult to characterize analytically the recorded holograms. It is not possible to record multiple holograms independently. The exposure of each new hologram partially erases previously recorded holograms. For more complete analysis, one may have to rely on experiments or computer simulations.

3.3.2 Encoding

Static Code Book

First, we consider a case where the reconstruction alphabet of size N is already stored in the photorefractive crystal. The photorefractive crystal is initialized by a uniform plane light wave. To record the reconstruction alphabet, each vector in the reconstruction alphabet is loaded into array S_1 one at a time. Array S_2 can produce one of the N label beams as in Figure 4.

The exposure of a pair of each vector wave and its associated label beam increases the connection strength by an amount proportional to the intensity of each pixel. The photorefractive crystal stores each η_{ij} as an element in matrix H. [16] A reproduction alphabet $\hat{A} = \{y_i | i = 1, \ldots, N\}$ is represented by a set of new vectors $\{y_i, \bar{y}_i | i = 1, \ldots, N\}$, where $\bar{y}_i$ denotes the complement of vector y_i. We denote this alphabet $\{z_i | i = 1, \ldots, N\}$. Each vector z_i is of length $2k$ and is represented with intensity. Each diffraction efficiency η_{ij} is proportional to the product of the intensity values of the i-th pixel and the j-th pixel. Thus, after recording each vector in this alphabet, we have

$$H \propto \begin{pmatrix} H_{11} & z_1 \quad z_2 \quad \cdots \quad z_N \\ z_1^t & \\ z_2^t & I \\ \cdots & \\ z_N^t & \end{pmatrix}$$

where z_i^t denotes the transposed vector of z_i, and H_{11} is the sum of every outer product of z_i, which is given as

$$H_{11} = \sum_{i=1}^{N} z_i z_i^t$$

Consider encoding a new vector z. The input plane vector P_{in} can be written as

$$P_{in} = (\underbrace{z^t}_{2k}, \underbrace{0, \ldots, 0}_{N})$$

Thus, the output plane vector P_{out} can be written as

$$P_{out} \propto (\underbrace{(H_{11} z^t)^t}_{2k}, \underbrace{z_1^t z, z_2^t z, \ldots, z_N^t z}_{N})$$

The first $2k$ elements of P_{out} correspond to the intensity values at D_1 and are not of our interest. The rest of N elements correspond to the intensity values at label array D_2. Each pixel receives intensity proportional to the inner product of the input vector and the recorded vectors. In the LBG algorithm, the distortion measure of two vectors x and y_i is written as

$$d(x, y_i) = ||x - y_i||^2 = ||x||^2 + ||y_i||^2 - 2xy_i$$

Our complementary encoding scheme keeps the first two terms constant. Thus, minimizing the distortion measure in vector quantization corresponds to maximizing the inner product in our holographic matching.

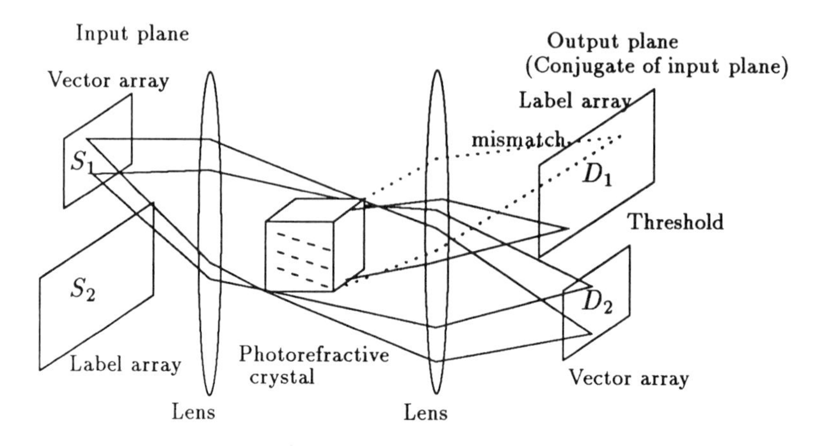

Figure 5: Encoding a new vector.

We choose the label of the pixel which has the maximum intensity value as an encoded symbol.

Thus, once the reconstruction alphabet is recorded, encoding a new vector is simple and fast, as depicted in Figure 5. There are no explicit computations or searches required to find the best vector in the reconstruction alphabet. Each new vector can be loaded into array S_1, which in turn illuminates the photorefractive crystal to retrieve its associated label. The retrieved light reaches some pixels at array D_2. Array D_2 selects the pixel with the largest intensity and transmits its label as an encoded symbol.

The speed of the system is only limited by the switching speed of the SLM.

3.3.3 Dynamic Code Book

Each vector from the training sequence is loaded into array S_1 one at a time. Array S_1 illuminates the photorefractive crystal. The diffracted waves from the crystal are detected by array D_2. If there is a pixel that has intensity greater than a threshold value t_r, we select the index of the pixel as a quantized label for the input vector. Otherwise, we arbitrarily select any unused label pixel that has intensity greater than a threshold value, t_u. The corresponding label is then loaded into array S_2 to create a new label beam for recording. Both S_1 and S_2 then illuminate the holographic medium, and the pair is recorded. After all the training vectors are presented to the system, the convergence of the system can be

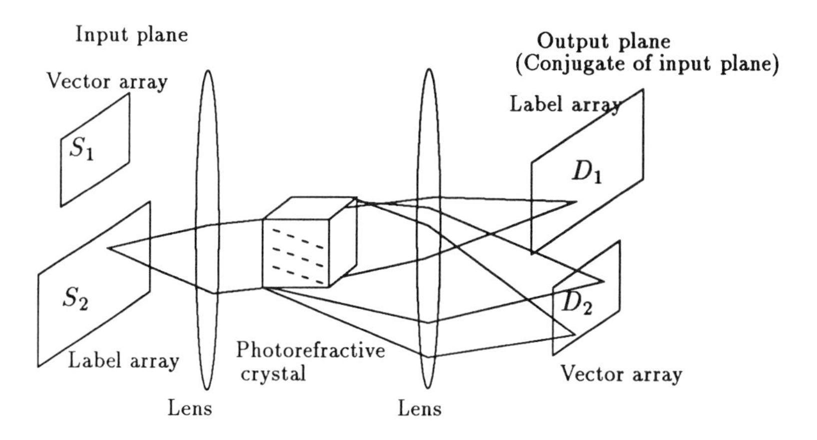

Figure 6: Decoding a label.

examined as follows. First, we illuminate the photorefractive crystal with a label beam. We then load its associated image vector that is obtained at array D_1 into array S_1. We then examine whether the illumination from array S_1 produces its associated label at array D_2. For each label beam, we repeat this process to see if there is any mismatch. If there is, we may repeat the recording phase.

Theorem 3.1 *Encoding a new vector by an N-level k-dimensional vector quantizer can be carried out in constant time with the VLSIO model in volume $O(N^{3/2} + k^{3/2})$.*

Proof: Using a similar argument as before, we have volume of size $O((N^{1/2} + k^{1/2})^3)$.

3.4 Decoding

For a given label, the reconstruction of its associated image vector is straight forward. The label is loaded into array S_2 that will illuminate the photorefractive crystal. By associative matching of the holograms, the image vector associated with the label will be obtained at array D_1 as shown in Figure 6.

Consider decoding a label s. The input plane vector P_{in} can be written as

$$P_{in} = (\underbrace{0, \ldots, 0}_{2k}, \underbrace{s}_{N})$$

Suppose s is the i-th label vector, thus s has zeros except at the i-th position, where it has a 1. Then, the output plane vector P_{out} can be written as

$$P_{out} \propto (\underbrace{z_i^t}_{2k}, \underbrace{\ldots, 0, 1, 0, \ldots}_{N})$$

The first $2k$ elements correspond to a reconstruction vector of the i-th vector in the alphabet.

3.5 Increased Parallelism: Coding a Frame in Parallel

If the technology permits, one may construct an array of our holographic vector quantizers to encode and decode a whole image frame in parallel. As we showed in our theorem, each encoder of an N-level k-dimensional vector quantizer occupies a volume of $O(N^{3/2}+k^{3/2})$. In order to encode a $n^{1/2} \times n^{1/2}$ image with n pixels, we need a $(n/k)^{1/2} \times (n/k)^{1/2}$ array of the quantizers. Thus, for a fixed ratio N/k, the total volume becomes $O((N^{1/2} + k^{1/2})n)$.

Theorem 3.2 *Vector quantization of a $n^{1/2} \times n^{1/2}$ image with n pixels by an N-level k-dimensional quantizer can be carried out in constant time with the VLSIO model in volume $O((N^{1/2} + k^{1/2})n)$.*

Proof: From the above argument, the theorem follows.

3.6 Advantages and Limitations

Advantages of our holographic vector quantizer are its speed and size. Encoding a vector is carried out in constant time. The size of the encoder is optimal within a polylog factor.

Limitations are caused by a maximum degree of multiple-exposures in a photorefractive crystal. Although $LiNbO_3$ has large dynamic range to allow a number of multiple exposures, a large number of exposures may reduce the dynamic range of the image.

4 Interframe Compression, Motion Detection

In this paper, we consider only coding of the lateral movement of each block in the image frame. In our further research, we investigate coding methods for the movement of multiple objects and for zooming.

We use dynamic cross-correlation filters to track the movement of the image. Optical correlators are divided into two types, the matched filtering correlator [33] and the joint transform correlator [34] [35] [36]. The matched filtering correlator requires a preparation of a fourier hologram of the model in advance. The joint transform correlator is well suited for real-time applications such as interframe coding, since it provides real-time pattern recognition without requiring preprocessed holograms.

We describe our interframe coding method using the joint transform correlator and our holographic quantizer. Suppose the first frame is already decomposed into blocks and encoded by the previous algorithm. The encoded labels for each block are stored in memory. Each block of the second frame is encoded as follows. First, its previous label is used to reconstruct the previous block. The fourier transform of the reconstructed block is formed on an SLM array that implements a cross correlation filter. The fourier transform of the block from the second frame is formed on the SLM. The light passing the filter forms the cross-correlation of the previous block and the current block at a detector array via another fourier lens. If there is a correlation peak at the detector array, the value of its displacement is transmitted as a movement of the block. If there is no correlation peak, the holographic vector quantizer is used to assign the block a new label.

5 Conclusion and Further Research

We considered several compression techniques using optical systems. Optics can offer an alternative approach to overcome the limitations of current compression schemes. We gave a simple optical system for the cosine transform. We designed a new optical vector quantizer system using holographic associative matching and discussed the issues concerning the system. We also discussed an extension of this optical system to interframe coding.

Further research will include computer simulation and the construction of a small prototype system.

References

[1] A. K. Jain, "Image Data Compression: A Review," *Proceedings of the IEEE* **69**, 349-389 (1981).

[2] —, *DCC' 91 Data Compression Conference,* J. A. Storer and J. H. Reif, ed. (IEEE Computer Society Press, Los Alamitos, California, 1991).

[3] R. M. Gray, "Vector Quantization," *IEEE ASSP Mag.* April, 4-29 (1984).

[4] N. M. Nasrabadi and R. A. King, "Image Coding Using Vector Quantization: A Review," *IEEE Trans. Commun.* **COM-36,** 957-971 (1988).

[5] J. W. Goodman, F. J. Leonberger, S. Kung, and R. A. Athale, "Optical interconnections for VLSI systems," *Proceedings of the IEEE* **72,** 850-866 (1984).

[6] A. A. Sawchuk and T. C. Strand, "Digital optical computing," Proceedings of the IEEE **72,** 758-779 (1984).

[7] T. E. Bell, "Optical Computing: A field in flux," *IEEE Spectrum* **23(8),** 34-57 (1986).

[8] D. Feitelson, *Optical Computing, A Survey for Computer Scientists,* (MIT Press, Cambridge, Mass., 1988).

[9] J. D. Ullman, *Computational Aspects of VLSI,* (Computer Science Press, Rockwille, Md., 1984).

[10] R. Barakat and J. H. Reif, "Lower Bounds on the Computational Efficiency of Optical Computing Systems," *Appl. Opt.* **26,** 1015-1018 (1987).

[11] J. H. Reif and A. Tyagi, "Efficient Parallel Algorithms for Optical Computing with the DFT Primitive," *10th Conference on Foundations of Software Technology and Theoretical Computer Science, Lecture Notes in Computer Science,* (Springer-Verlag, Bangalor, India 1990).

[12] K. Preston, *Coherent optical computers,* (McGraw-Hill, New York, 1972).

[13] F. T. S. Yu, *Optical information processing,* (Wiley, New York, 1983).

[14] T. Kohonen, *Self-Organization and Associative Memory,* 2nd ed. (Springer-Verlag, New York, 1988).

[15] N. Farhat, D. Psaltis, A. Prata, and E. Paek, "Optical implementation of the Hopfield model," *Appl. Opt.* **24,** 1469-1475(1985).

[16] D. Psaltis, D. Brady, and K. Wagner, "Adaptive optical networks using photorefractive crystals," *Appl. Opt.* **27,** 1752-1759 (1988).

[17] J. W. Goodman, *Introduction to Fourier Optics,* (McGraw-Hill, New York 1968).

[18] H. J. Caulfield, "Associative mapping by optical holography," *Opt. Commun.* **55**, 80-82 (1985).

[19] A. Yariv, S. Kwong, and K. Kyuma, "Demonstration of an all-optical associative holographic memory," *Appl. Phy. Lett.* **48**, 1114-1116 (1986).

[20] B. H. Soffer, G. J. Dunning, Y. Owechko, and E. Marom, "Associative holographic memory with feedback using phase conjugate mirrors," *Opt. Lett.* **11**, 118-120 (1986).

[21] Y. Owechko, G. J. Dunning, E. Marom, and B. H. Soffer, "Holographic associative memory with nonlinearities in the correlation domain," *Appl. Opt.* **26**, 1900-1910 (1987).

[22] H. Kang, C. X. Yang, G. G. Mu, and Z. K. Wu, "Real-time holographic associative memory using doped $LiNbO_3$ in a phase-conjugating resonator," *Opt. Lett.* **15**, 637-639 (1990).

[23] N. Ahmed, T. Natarajan, and K. R. Rao, "Discrete cosine transform," *IEEE Trans. Comput.* **C-23**, 90-93 (1974).

[24] R. J. Clarke, *Transform Coding of Images,* (Academic Press, London, 1985).

[25] Y. Linde, A. Buzo, and R. M. Gray, "An Algorithm for Vector Quantizer Design," *IEEE Trans. Commun.* **COM-28**, 84-95 (1980).

[26] R. J. Collier, C. B. Burckhardt, and L. H. Lin, *Optical Holography,* (Academic Press, Orlando, Florida, 1971).

[27] D. L. Staebler, W. J. Burke, W. Phillips, and J. J. Amodei, "Multiple storage and erasure of fixed holograms in Fe-doped $LiNbO_3$," *Appl. Phys. Lett.* **26**, 182-184 (1975).

[28] W. J. Burke, D. L. Staebler, W. Phillips, and G. A. Alphonse, "Volume Phase Holographic Storage in Ferroelectric Crystals," *Opt. Eng.* **17**, 308 (1978).

[29] A. C. Strasser, E. S. Maniloff, K. M. Johnson, and S. D. D. Goggin, "Procedure for recording multiple-exposure holograms with equal diffraction efficiency in photorefractive media," *Opt. Lett.* **14**, 6-8 (1989).

[30] W. P. Bleha, L. T. Lipton, E. W. Wiener-Avner, J. Grinberg, P. G. Reif, D. Casasent, H. B. Brown, and B. V. Markevitch, "Application of the Liquid Crystal Light Valve to Real-Time Optical Data Processing," *Opt. Eng.* **17**, 371-384 (1978).

[31] H. Lee, "Volume holographic global and local interconnecting patterns with maximal capacity and minimal first-order crosstalk," *Appl. Opt.* **28,** 5312-5316 (1989).

[32] H. Lee, X. Gu and D. Psaltis, "Volume holographic interconnections with maximal capacity and minimal cross talk," *J. Appl. Phys.* **65,** 2191-2194 (1989).

[33] A. VanderLugt, "Coherent optical processing," *Proceedings of the IEEE* **62,** 1300-1319 (1974).

[34] D. Casasent, "Coherent optical pattern recognition," *Proceedings of the IEEE* **67,** 813-825 (1979).

[35] B. Javidi and J. L. Horner, "Single spatial light modulator joint transform correlator," *Appl. Opt.* **28,** 1027-1032 (1989).

[36] T. D. Hudson and D. A. Gregory, "Joint transform correlation using an optically addressed ferroelectric LC spatial light modulator," *Appl. Opt.* **29,** 1064-1066 (1990).

Part 2: TEXT COMPRESSION

4

Practical Implementations of Arithmetic Coding

Paul G. Howard[1] and Jeffrey Scott Vitter[2]

Department of Computer Science
Brown University
Providence, R. I. 02912–1910

Abstract

We provide a tutorial on arithmetic coding, showing how it provides nearly optimal data compression and how it can be matched with almost any probabilistic model. We indicate the main disadvantage of arithmetic coding, its slowness, and give the basis of a fast, space-efficient, approximate arithmetic coder with only minimal loss of compression efficiency. Our coder is based on the replacement of arithmetic by table lookups coupled with a new deterministic probability estimation scheme.

Index terms: data compression, arithmetic coding, adaptive modeling, analysis of algorithms, data structures, low precision arithmetic.

1 Data Compression and Arithmetic Coding

Data can be compressed whenever some data symbols are more likely than others. Shannon [54] showed that for the best possible compression code (in the sense of minimum average code length), the output length contains a contribution of $-\lg p$ bits from the encoding of each symbol whose probability of occurrence is p. If we can provide an accurate model for the probability of occurrence of each possible symbol at every point in a file, we can use arithmetic coding to encode the symbols that actually occur; the number of bits used by arithmetic coding to encode a symbol with probability p is very nearly $-\lg p$, so the encoding is very nearly optimal for the given probability estimates.

In this article we show by theorems and examples how arithmetic coding achieves its performance. We also point out some of the drawbacks of arithmetic coding in practice, and propose a unified compression system for overcoming them. We begin by attempting to clear up some of the false impressions commonly held about arithmetic coding; it offers some genuine benefits, but it is not the solution to all data compression problems.

---

[1]Support was provided in part by NASA Graduate Student Researchers Program grant NGT–50420 and by a National Science Foundation Presidential Young Investigators Award grant with matching funds from IBM. Additional support was provided by a Universities Space Research Association/CESDIS associate membership.

[2]Support was provided in part by National Science Foundation Presidential Young Investigator Award CCR–9047466 with matching funds from IBM, by NSF research grant CCR–9007851, by Army Research Office grant DAAL03–91–G–0035, and by the Office of Naval Research and the Defense Advanced Research Projects Agency under contract N00014–91–J–4052, ARPA Order No. 8225.

The most important advantage of arithmetic coding is its flexibility: it can be used in conjunction with any model that can provide a sequence of event probabilities. This advantage is significant because large compression gains can be obtained only through the use of sophisticated models of the input data. Models used for arithmetic coding may be adaptive, and in fact a number of independent models may be used in succession in coding a single file. This great flexibility results from the sharp separation of the coder from the modeling process [47]. There is a cost associated with this flexibility: the interface between the model and the coder, while simple, places considerable time and space demands on the model's data structures, especially in the case of a multi-symbol input alphabet.

The other important advantage of arithmetic coding is its optimality. Arithmetic coding is optimal in theory and very nearly optimal in practice, in the sense of encoding using minimal average code length. This optimality is often less important than it might seem, since Huffman coding [25] is also very nearly optimal in most cases [8,9,18,39]. When the probability of some single symbol is close to 1, however, arithmetic coding does give considerably better compression than other methods. The case of highly unbalanced probabilities occurs naturally in bilevel (black and white) image coding, and it can also arise in the decomposition of a multi-symbol alphabet into a sequence of binary choices.

The main disadvantage of arithmetic coding is that it tends to be slow. We shall see that the full precision form of arithmetic coding requires at least one multiplication per event and in some implementations up to two multiplications and two divisions per event. In addition, the model lookup and update operations are slow because of the input requirements of the coder. Both Huffman coding and Ziv-Lempel [59,60] coding are faster because the model is represented directly in the data structures used for coding. (This reduces the coding efficiency of those methods by narrowing the range of possible models.) Much of the current research in arithmetic coding concerns finding approximations that increase coding speed without compromising compression efficiency. The most common method is to use an approximation to the multiplication operation [10,27,29,43]; in this article we present an alternative approach using table lookups and approximate probability estimation.

Another disadvantage of arithmetic coding is that it does not in general produce a prefix code. This precludes parallel coding with multiple processors. In addition, the potentially unbounded output delay makes real-time coding problematical in critical applications, but in practice the delay seldom exceeds a few symbols, so this is not a major problem. A minor disadvantage is the need to indicate the end of the file.

One final minor problem is that arithmetic codes have poor error resistance, especially when used with adaptive models [5]. A single bit error in the encoded file causes the decoder's internal state to be in error, making the remainder of the decoded file wrong. In fact this is a drawback of *all* adaptive codes, including Ziv-Lempel codes and adaptive Huffman codes [12,15,18,26,55,56]. In practice, the poor error resistance of adaptive coding is unimportant, since we can simply apply appropriate error correction coding to the encoded file. More complicated solutions appear in [5,20], in which errors are made easy to detect, and upon detection of an error, bits are changed until no errors are detected.

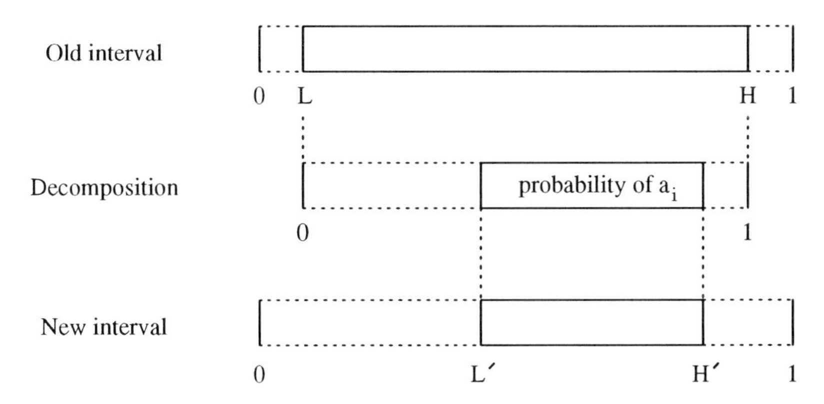

Figure 1: Subdivision of the current interval based on the probability of the input symbol a_i that occurs next.

Overview of this article. In Section 2 we give a tutorial on arithmetic coding. We include an introduction to modeling for text compression. We also restate several important theorems from [23] relating to the optimality of arithmetic coding in theory and in practice.

In Section 3 we present some of our current research into practical ways of improving the speed of arithmetic coding without sacrificing much compression efficiency. The center of this research is a reduced-precision arithmetic coder, supported by efficient data structures for text modeling.

2 Tutorial on Arithmetic Coding

In this section we explain how arithmetic coding works and give implementation details; our treatment is based on that of Witten, Neal, and Cleary [58]. We point out the usefulness of binary arithmetic coding (that is, coding with a 2-symbol alphabet), and discuss the modeling issue, particularly high-order Markov modeling for text compression. Our focus is on encoding, but the decoding process is similar.

2.1 Arithmetic coding and its implementation

Basic algorithm. The algorithm for encoding a file using arithmetic coding works conceptually as follows:

1. We begin with a "current interval" $[L, H)$ initialized to $[0, 1)$.
2. For each symbol of the file, we perform two steps (see Figure 1):
   (a) We subdivide the current interval into subintervals, one for each possible alphabet symbol. The size of a symbol's subinterval is proportional to the estimated probability that the symbol will be the next symbol in the file, according to the model of the input.

(b) We select the subinterval corresponding to the symbol that actually occurs next in the file, and make it the new current interval.

3. We output enough bits to distinguish the final current interval from all other possible final intervals.

The length of the final subinterval is clearly equal to the product of the probabilities of the individual symbols, which is the probability p of the particular sequence of symbols in the file. The final step uses almost exactly $-\lg p$ bits to distinguish the file from all other possible files. We need some mechanism to indicate the end of the file, either a special end-of-file symbol coded just once, or some external indication of the file's length.

In step 2, we need to compute only the subinterval corresponding to the symbol a_i that actually occurs. To do this we need two cumulative probabilities, $P_C = \sum_{k=1}^{i-1} p_k$ and $P_N = \sum_{k=1}^{i} p_k$. The new subinterval is $[L + P_C(H - L), L + P_N(H - L))$. The need to maintain and supply cumulative probabilities requires the model to have a complicated data structure; Moffat [35] investigates this problem, and concludes for a multi-symbol alphabet that binary search trees are about twice as fast as move-to-front lists.

Example 1: We illustrate a non-adaptive code, encoding the file containing the symbols **bbb** using arbitrary fixed probability estimates $p_{\mathrm{a}} = 0.4$, $p_{\mathrm{b}} = 0.5$, and $p_{\mathrm{EOF}} = 0.1$. Encoding proceeds as follows:

Current Interval	Action	Subintervals a	b	EOF	Input
$[0.000, 1.000)$	Subdivide	$[0.000, 0.400)$	$[0.400, 0.900)$	$[0.900, 1.000)$	b
$[0.400, 0.900)$	Subdivide	$[0.400, 0.600)$	$[0.600, 0.850)$	$[0.850, 0.900)$	b
$[0.600, 0.850)$	Subdivide	$[0.600, 0.700)$	$[0.700, 0.825)$	$[0.825, 0.850)$	b
$[0.700, 0.825)$	Subdivide	$[0.700, 0.750)$	$[0.750, 0.812)$	$[0.812, 0.825)$	EOF
$[0.812, 0.825)$					

The final interval (without rounding) is $[0.8125, 0.825)$, which in binary is approximately $[0.11010\ 00000, 0.11010\ 01100)$. We can uniquely identify this interval by outputting **1101000**. According to the fixed model, the probability p of this particular file is $(0.5)^3(0.1) = 0.0125$ (exactly the size of the final interval) and the code length (in bits) should be $-\lg p = 6.322$. In practice we have to output 7 bits. □

The idea of arithmetic coding originated with Shannon in his seminal 1948 paper on information theory [54]. It was rediscovered by Elias about 15 years later, as briefly mentioned in [1].

Implementation details. The basic implementation of arithmetic coding described above has two major difficulties: the shrinking current interval requires the use of high precision arithmetic, and no output is produced until the entire file has been read. The most straightforward solution to both of these problems is to output each leading bit as soon as it is known, and then to double the length of the current interval so that it reflects only the unknown part of the

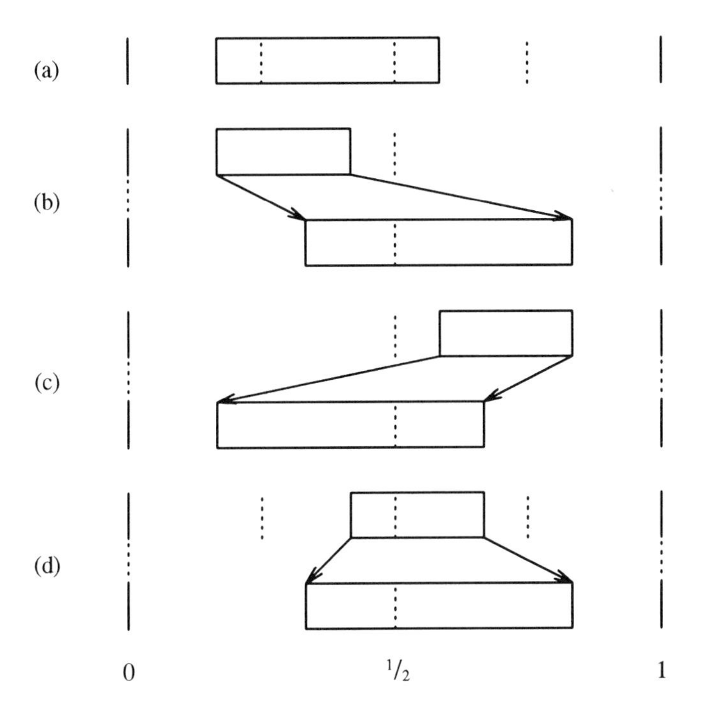

Figure 2: Interval expansion process. (a) No expansion. (b) Interval in $[0, 1/2)$. (c) Interval in $[1/2, 1)$. (d) Interval in $[1/4, 3/4)$ (follow-on case).

final interval. Witten, Neal, and Cleary [58] add a clever mechanism for preventing the current interval from shrinking too much when the endpoints are close to $1/2$ but straddle $1/2$. In that case we do not yet know the next output bit, but we do know that whatever it is, the *following* bit will have the opposite value; we merely keep track of that fact, and expand the current interval symmetrically about $1/2$. This follow-on procedure may be repeated any number of times, so the current interval size is always longer than $1/4$.

Mechanisms for incremental transmission and fixed precision arithmetic have been developed through the years by Pasco [40], Rissanen [48], Rubin [52], Rissanen and Langdon [49], Guazzo [19], and Witten, Neal, and Cleary [58]. The bit-stuffing idea of Langdon and others at IBM that limits the propagation of carries in the additions is roughly equivalent to the follow-on procedure described above.

We now describe in detail how the coding and interval expansion work. This process takes place immediately after the selection of the subinterval corresponding to an input symbol.

We repeat the following steps (illustrated schematically in Figure 2) as many times as possible:

a. If the new subinterval is not entirely within one of the intervals $[0, 1/2)$, $[1/4, 3/4)$, or $[1/2, 1)$, we stop iterating and return.

b. If the new subinterval lies entirely within $[0, 1/2)$, we output **0** and any **1**s left over from previous symbols; then we double the size of the interval $[0, 1/2)$, expanding toward the right.

c. If the new subinterval lies entirely within $[1/2, 1)$, we output **1** and any **0**s left over from previous symbols; then we double the size of the interval $[1/2, 1)$, expanding toward the left.

d. If the new subinterval lies entirely within $[1/4, 3/4)$, we keep track of this fact for future output; then we double the size of the interval $[1/4, 3/4)$, expanding in both directions away from the midpoint.

Example 2: We show the details of encoding the same file as in Example 1.

Current Interval	Action	Subintervals a	b	EOF	Input
$[0.00, 1.00)$	Subdivide	$[0.00, 0.40)$	$[0.40, 0.90)$	$[0.90, 1.00)$	b
$[0.40, 0.90)$	Subdivide	$[0.40, 0.60)$	$[0.60, 0.85)$	$[0.85, 0.90)$	b
$[0.60, 0.85)$	Output **1**				
	Expand $[1/2, 1)$				
$[0.20, 0.70)$	Subdivide	$[0.20, 0.40)$	$[0.40, 0.65)$	$[0.65, 0.70)$	b
$[0.40, 0.65)$	*follow*				
	Expand $[1/4, 3/4)$				
$[0.30, 0.80)$	Subdivide	$[0.30, 0.50)$	$[0.50, 0.75)$	$[0.75, 0.80)$	EOF
$[0.75, 0.80)$	Output **10**				
	Expand $[1/2, 1)$				
$[0.50, 0.60)$	Output **1**				
	Expand $[1/2, 1)$				
$[0.00, 0.20)$	Output **0**				
	Expand $[0, 1/2)$				
$[0.00, 0.40)$	Output **0**				
	Expand $[0, 1/2)$				
$[0.00, 0.80)$	Output **0**				

The "*follow*" output in the sixth line indicates the follow-on procedure: we keep track of our knowledge that the next output bit will be followed by its opposite; this "opposite" bit is the **0** output in the ninth line. The encoded file is **1101000**, as before. □

Clearly the current interval contains some information about the preceding inputs; this information has not yet been output, so we can think of it as the coder's state. If a is the length of the current interval, the state holds $-\lg a$ bits not yet output. In the basic method (illustrated by Example 1) the state contains *all* the information about the output, since nothing is output until the end. In the implementation illustrated by Example 2, the state always contains fewer than two bits of output information, since the length of the current interval is always more than $1/4$. The final state in Example 2 is $[0, 0.8)$, which contains $-\lg 0.8 \approx 0.322$ bits of information.

Use of integer arithmetic. In practice, the arithmetic can be done by storing the current interval in sufficiently long integers rather than in floating point or exact rational numbers. (We can think of Example 2 as using the integer interval $[0, 100)$ by omitting all the decimal points.) We also use integers for the frequency counts used to estimate symbol probabilities. The subdivision process involves selecting non-overlapping intervals (of length at least 1) with lengths approximately proportional to the counts. To encode symbol a_i we need two cumulative counts, $C = \sum_{k=1}^{i-1} c_k$ and $N = \sum_{k=1}^{i} c_k$, and the sum T of all counts, $T = \sum_{k=1}^{n} c_k$. (Here and elsewhere we denote the alphabet size by n.) The new subinterval is $[L + \lfloor \frac{C(H-L)}{T} \rfloor, L + \lfloor \frac{N(H-L)}{T} \rfloor)$. (In this discussion we continue to use half-open intervals as in the real arithmetic case. In implementations [58] it is more convenient to subtract 1 from the right endpoints and use closed intervals. Moffat [36] considers the calculation of cumulative frequency counts for large alphabets.)

Example 3: Suppose that at a certain point in the encoding we have symbol counts $c_{\mathrm{a}} = 4$, $c_{\mathrm{b}} = 5$, and $c_{\mathrm{EOF}} = 1$ and current interval $[25, 89)$ from the full interval $[0, 128)$. Let the next input symbol be **b**. The cumulative counts for **b** are $C = 4$ and $N = 9$, and $T = 10$, so the new interval is $[25 + \lfloor \frac{4(89-25)}{10} \rfloor, 25 + \lfloor \frac{9(89-25)}{10} \rfloor) = [50, 82)$; we then increment the follow-on count and expand the interval once about the midpoint 64, giving $[36, 100)$. It is possible to maintain higher precision, truncating (and adjusting to avoid overlapping subintervals) only when the expansion process is complete; this makes it possible to prove a tight analytical bound on the lost compression caused by the use of integer arithmetic, as we do in [23], restated as Theorem 1 below. In practice this refinement makes the coding more difficult without improving compression. □

Analysis. In [23] we prove a number of theorems about the code lengths of files coded with arithmetic coding. Most of the results involve the use of arithmetic coding in conjunction with various models of the input; these will be discussed in Section 2.3. Here we note two results that apply to implementations of the arithmetic coder. The first shows that using integer arithmetic has negligible effect on code length.

Theorem 1 *If we use integers from the range $[0, N)$ and use the high precision algorithm for scaling up the subrange, the code length is provably bounded by $4/(N \ln 2)$ bits per input symbol more than the ideal code length for the file.*

For a typical value $N = 65{,}536$, the excess code length is less than 10^{-4} bit per input symbol.

The second result shows that if we indicate end-of-file by encoding a special symbol just once for the entire file, the additional code length is negligible.

Theorem 2 *The use of a special end-of-file symbol when coding a file of length t using integers from the range $[0, N)$ results in additional code length of less than $8t/(N \ln 2) + \lg N + 7$ bits.*

Again the extra code length is negligible, less than 0.01 bit per input symbol for a typical 100,000 byte file.

Since we seldom know the exact probabilities of the process that generated an input file, we would like to know how errors in the estimated probabilities affect the code length. We can estimate the extra code length by a straightforward asymptotic analysis. The average code length L for symbols produced by a given model in a given state is given by

$$L = -\sum_{i=1}^{n} p_i \lg q_i,$$

where p_i is the actual probability of the ith alphabet symbol and q_i is its estimated probability. The optimal average code length for symbols in the state is the entropy of the state, given by

$$H = -\sum_{i=1}^{n} p_i \lg p_i.$$

The excess code length is $E = L - H$; if we let $d_i = q_i - p_i$ and expand asymptotically in d, we obtain

$$E = \sum_{i=1}^{n} \left(\frac{1}{2 \ln 2} \frac{d_i^2}{p_i} + O\left(\frac{d_i^3}{p_i^2} \right) \right). \tag{1}$$

(This corrects a similar derivation in [5], in which the factor of $1/\ln 2$ is omitted.) The vanishing of the linear terms means that small errors in the probabilities used by the coder lead to very small increases in code length. Because of this property, *any* coding method that uses approximately correct probabilities will achieve a code length close to the entropy of the underlying source. We use this fact in Section 3.1 to design a class of fast approximate arithmetic coders with small compression loss.

2.2 Binary arithmetic coding

The preceding discussion and analysis has focused on coding with a multi-symbol alphabet, although in principle it applies to a binary alphabet as well. It is useful to distinguish the two cases since both the coder and the interface to the model are simpler for a binary alphabet. The coding of bilevel images, an important problem with a natural two-symbol alphabet, often produces probabilities close to 1, indicating the use of arithmetic coding to obtain good compression. Historically, much of the arithmetic coding research by Rissanen, Langdon, and others at IBM has focused on bilevel images [29]. The Q-Coder [2,27,33,41,42,43] is a binary arithmetic coder; work by Rissanen and Mohiuddin [50] and Chevion *et al.* [10] extends some of the Q-Coder ideas to multi-symbol alphabets.

In most other text and image compression applications, a multi-symbol alphabet is more natural, but even then we can map the possible symbols to the leaves of a binary tree, and encode an event by traversing the tree and encoding a decision at each internal node. If we do this, the model no longer has to maintain and produce cumulative probabilities; a single probability suffices to encode each decision. Calculating the new current interval is also simplified, since just one endpoint changes after each decision. On the other hand, we now usually have to encode more than one event for each input symbol, and we have a new data structure problem, maintaining the coding trees efficiently without using excessive space. The smallest average number of events coded per input symbol occurs when the tree is a Huffman tree, since such trees have minimum average weighted path length; however, maintaining such trees dynamically is complicated and slow [12,26,55,56]. In Section 3.3 we present a new data structure, the *compressed tree*, suitable for binary encoding of multi-symbol alphabets.

2.3 Modeling for text compression

Arithmetic coding allows us to compress a file as well as possible for a given model of the process that generated the file. To obtain maximum compression of a file, we need both a good model and an efficient way of representing (or learning) the model. (Rissanen calls this principle the *minimum description length* principle; he has investigated it thoroughly from a theoretical point of view [44,45,46].) If we allow two passes over the file, we can identify a suitable model during the first pass, encode it, and use it for optimal coding during the second pass. An alternative approach is to allow the model to adapt to the characteristics of the file during a single pass, in effect learning the model. The adaptive approach has advantages in practice: there is no coding delay and no need to encode the model, since the decoder can maintain the same model as the encoder in a synchronized fashion.

In the following theorem from [23] we compare context-free coding using a two-pass method and a one-pass adaptive method. In the two-pass method, the exact symbol counts are encoded after the first pass; during the second pass each symbol's count is decremented whenever it occurs, so at each point the relative counts reflect the correct symbol probabilities for the remainder of the file (as in [34]). In the one-pass adaptive method, all symbols are given initial counts of 1; we add 1 to a symbol's count whenever it occurs.

Theorem 3 *For all input files, the adaptive code with initial 1-weights gives exactly the same code length as the semi-adaptive decrementing code in which the input model is encoded based on the assumption that all symbol distributions are equally likely.*

Hence we see that use of an adaptive code does not incur any extra overhead, but it does not eliminate the cost of describing the model.

Adaptive models. The simplest adaptive models do not rely on contexts for conditioning probabilities; a symbol's probability is just its relative frequency in the part of the file already coded. (We need a mechanism for encoding a symbol for the first time, when its frequency is 0; the easiest way [58] is to start all symbol counts at 1 instead of 0.) The average code length per input symbol of a file encoded using such a 0-order adaptive model is very close to the 0-order entropy of the file. We shall see that adaptive compression can be improved by taking advantage of locality of reference and especially by using higher order models.

Scaling. One problem with maintaining symbol counts is that the counts can become arbitrarily large, requiring increased precision arithmetic in the coder and more memory to store the counts themselves. By periodically reducing all symbol's counts by the same factor, we can keep the relative frequencies approximately the same while using only a fixed amount of storage for each count. This process is called *scaling*. It allows us to use lower precision arithmetic, possibly hurting compression because of the reduced accuracy of the model. On the other hand, it introduces a *locality of reference* (or *recency*) effect, which often improves compression. We now discuss and quantify the locality effect.

In most text files we find that most of the occurrences of at least some words are clustered in one part of the file. We can take advantage of this locality by assigning more weight to recent occurrences of a symbol in an adaptive model. In practice there are several ways to do this:

- Periodically restarting the model. This often discards too much information to be effective, although Cormack and Horspool find that it gives good results when growing large dynamic Markov models [11].
- Using a sliding window on the text [26]. This requires excessive computational resources.
- Recency rank coding [7,13,53]. This is simple but corresponds to a rather coarse model of recency.
- Exponential aging (giving exponentially increasing weights to successive symbols) [12,38]. This is moderately difficult to implement because of the changing weight increments, although our probability estimation method in Section 3.4 uses an approximate form of this technique.
- Periodic scaling [58]. This is simple to implement, fast and effective in operation, and amenable to analysis. It also has the computationally desirable property of keeping the symbol weights small. In effect, scaling is a practical version of exponential aging.

Analysis of scaling. In [23] we give a precise characterization of the effect of scaling on code length, in terms of an elegant notion we introduce called *weighted entropy*. The weighted entropy of a file at the end of the mth block, denoted by H_m, is the entropy implied by the probability distribution at that time, computed according to the scaling model described above.

We prove the following theorem for a file compressed using arithmetic coding and a zero-order adaptive model with scaling. All counts are halved and rounded up when the sum of the counts reaches $2B$; in effect, we divide the file into b blocks of length B.

Theorem 4 *Let L be the compressed length of a file. Then we have*

$$B\left(\left(\sum_{m=1}^{b} H_m\right) + H_b - H_0\right) - t\frac{k}{B}$$
$$< L < B\left(\left(\sum_{m=1}^{b} H_m\right) + H_b - H_0\right) + t\left(\frac{k}{B}\lg\left(\frac{B}{k_{\min}}\right) + O\left(\frac{k^2}{B^2}\right)\right),$$

where $H_0 = \lg n$ is the entropy of the initial model, H_m is the (weighted) entropy implied by the scaling model's probability distribution at the end of block m, k

Table 1: PPM escape probabilities (p_{esc}) and symbol probabilities (p_i). The number of symbols that have occurred j times is denoted by n_j.

	PPMA	PPMB	PPMC	PPMP	PPMX
p_{esc}	$\frac{1}{t+1}$	$\frac{k}{t}$	$\frac{k}{t+k}$	$\frac{n_1}{t} - \frac{n_2}{t^2} + \cdots$	$\frac{n_1}{t}$
p_i	$\frac{c_i}{t+1}$	$\frac{c_i - 1}{t}$	$\frac{c_i}{t+k}$		

is the number of different alphabet symbols that appear in the file, and k_{min} *is the smallest number of different symbols that occur in any block.*

When scaling is done, we must ensure that no symbol's count becomes 0; an easy way to do this is to round fractional counts to the next higher integer. We show in the following theorem from [23] that this roundup effect is negligible.

Theorem 5 *Rounding counts up to the next higher integer increases the code length for the file by no more than* $n/2B$ *bits per input symbol.*

When we compare code lengths with and without scaling, we find that the differences are small, both theoretically and in practice.

High order models. The only way to obtain substantial improvements in compression is to use more sophisticated models. For text files, the increased sophistication invariably takes the form of conditioning the symbol probabilities on contexts consisting of one or more symbols of preceding text. (Langdon [28] and Bell, Witten, Cleary, and Moffat [3,4,5] have proven that both Ziv-Lempel coding and the dynamic Markov coding method of Cormack and Horspool [11] can be reduced to finite context models, despite superficial indications to the contrary.)

One significant difficulty with using high-order models is that many contexts do not occur often enough to provide reliable symbol probability estimates. Cleary and Witten deal with this problem with a technique called *Prediction by Partial Matching* (*PPM*). In the PPM methods we maintain models of various context lengths, or *orders*. At each point we use the highest order model in which the symbol has occurred in the current context, with a special *escape* symbol indicating the need to drop to a lower order. Cleary and Witten specify two *ad hoc* methods, called PPMA and PPMB, for computing the probability of the escape symbol. Moffat [37] implements the algorithm and proposes a third method, PPMC, for computing the escape probability: he treats the escape event as a separate symbol; when a symbol occurs for the first time he adds 1 to both the escape count and the new symbol's count. In practice, PPMC compresses better than PPMA and PPMB. PPMP and PPMX appear in [57]; they are based on the assumption that the appearance of symbols for the first time in a file is approximately a Poisson process. See Table 1 for formulas for the probabilities used by the different methods, and see [5] or [6] for a detailed description of the PPM method. In Section 3.5 we indicate two methods that provide improved estimation of the escape probability.

2.4 Other applications of arithmetic coding

Because of its nearly optimal compression performance, arithmetic coding has been proposed as an enhancement to other compression methods and activities related to compression. The output values produced by Ziv-Lempel coding are not uniformly distributed, leading several researchers [21,32,51] to suggest using arithmetic coding to further compress the output. Compression is indeed improved, but at the cost of slowing down the algorithm and increasing its complexity.

Lossless image compression is often performed using predictive coding, and it is often found that the prediction errors follow a Laplace distribution. In [24] we present methods that use tables of the Laplace distribution precomputed for arithmetic coding to obtain excellent compression ratios of grayscale images. The distributions are chosen to guarantee that, for a given variance estimate, the resulting code length exceeds the ideal for the estimate by only a small fixed amount.

Especially when encoding model parameters, it is often necessary to encode arbitrarily large non-negative integers. Witten *et al.* [58] note that arithmetic coding can encode integers according to any given distribution. In the examples in Section 3.1 we show how some encodings of integers found in the literature can be derived as low-precision arithmetic codes.

We point out here that arithmetic coding can also be used to generate random variables from any desired distribution, as well as to produce nearly random bits from the output of any random process. In particular, it is easy to convert random numbers from one base to another, and to convert random bits with an unknown but fixed probability to bits with a probability of 1/2.

3 Fast Arithmetic Coding

In this section we present some of our current research into several aspects of arithmetic coding. We show the construction of a fast, reduced-precision binary arithmetic coder, and indicate a theoretical construct, called the *ϵ-partition*, that can assist in choosing a representative set of probabilities to be used by the coder. We introduce a data structure that we call the *compressed tree* for efficiently representing a multi-symbol alphabet as a binary tree. We give a deterministic algorithm for estimating probabilities of binary events and storing them in 8-bit locations. We give two improved ways of handling the zero-frequency problem (symbols occurring in context for the first time). Finally we show that we can use hashing to obtain fast access of contexts with only a small loss of compression efficiency. All these components can be combined into a fast, space-efficient text coder.

3.1 Reduced-precision arithmetic coding

We have noted earlier that the primary disadvantage of arithmetic coding is its slowness. We have also seen that small errors in probability estimates cause very small increases in code length, so we can expect that by introducing approximations into the arithmetic coding process in a controlled way we can improve coding speed without significantly degrading compression performance. In the Q-Coder work at IBM, the time-consuming multiplications are replaced by additions and shifts, and low-order bits are ignored.

In this section, we take a different approach to approximate arithmetic coding: recalling that the fractional bits characteristic of arithmetic coding are stored as state information in the coder, we reduce the number of possible states, and replace arithmetic operations by table lookups. Here we present a fast, reduced-precision binary arithmetic coder (which we refer to as *quasi-arithmetic coding* in a companion paper [22]) and develop it through a series of examples. It should be noted that the compression is still completely reversible; using reduced precision merely affects the average code length.

The number of possible states (after applying the interval expansion procedure) of an arithmetic coder using the integer interval $[0, N)$ is $3N^2/16$. If we can reduce the number of states to a more manageable level, we can precompute all state transitions and outputs and substitute table lookups for arithmetic in the coder. The obvious way to reduce the number of states is to reduce N. The value of N must be even; for computational convenience we prefer that it be a multiple of 4.

Example 4: The simplest non-trivial coders have $N = 4$, and have only three states. By applying the arithmetic coding algorithm in a straightforward way, we obtain the following coding table. A "*follow*" output indicates application of the follow-on procedure described in Section 2.1.

State	Prob{**0**}	**0** input Output	**0** input Next state	**1** input Output	**1** input Next state
$[0,4)$	$0 < p < 1-\alpha$	**00**	$[0,4)$	-	$[1,4)$
	$1-\alpha \le p \le \alpha$	**0**	$[0,4)$	**1**	$[0,4)$
	$\alpha < p < 1$	-	$[0,3)$	**11**	$[0,4)$
$[0,3)$	$0 < p < 1/2$	**00**	$[0,4)$	*follow*	$[0,4)$
	$1/2 \le p < 1$	**0**	$[0,4)$	**10**	$[0,4)$
$[1,4)$	$0 < p < 1/2$	**01**	$[0,4)$	**1**	$[0,4)$
	$1/2 \le p < 1$	*follow*	$[0,4)$	**11**	$[0,4)$

The value of the cutoff probability α in state $[0,4)$ is clearly between $1/2$ and $3/4$. If this were an exact coder, the subintervals of length 3 would correspond to $-\lg 3/4 \approx 0.415$ bits of output information stored in the state, and we would choose $\alpha = 1/\lg 3 \approx 0.631$ to minimize the extra code length. But because of the approximate arithmetic, the optimal value of α depends on the distribution of Prob{**0**}; if Prob{**0**} is uniformly distributed on (0, 1), we find analytically that the excess code length is minimized when $\alpha = (15 - \sqrt{97})/8 \approx 0.644$. Fortunately, the amount of excess code length is not very sensitive to the value of α; in the uniform distribution case any value from about 0.55 to 0.73 gives less than one percent extra code length. □

Arithmetic coding does not mandate any particular assignment of subintervals to input symbols; all that is required is that subinterval lengths be proportional to symbol probabilities and that the decoder make the same assignment as the encoder. In Example 4 we uniformly assigned the left subinterval to symbol **0**. By preventing the longer subinterval from straddling the midpoint whenever possible, we can sometimes obtain a simpler coder that never requires the follow-on procedure; it may also use fewer states.

Example 5: This coder assigns the *right* subinterval to **0** in lines 4 and 7 of Example 4, eliminating the need for using the follow-on procedure; otherwise it is the same as Example 4.

		0 input		**1** input	
State	Prob{**0**}	Output	Next state	Output	Next state
$[0,4)$	$0 < p < 1-\alpha$	**00**	$[0,4)$	-	$[1,4)$
	$1-\alpha \le p \le \alpha$	**0**	$[0,4)$	**1**	$[0,4)$
	$\alpha < p < 1$	-	$[0,3)$	**11**	$[0,4)$
$[0,3)$	$0 < p < 1/2$	**10**	$[0,4)$	**0**	$[0,4)$
	$1/2 \le p < 1$	**0**	$[0,4)$	**10**	$[0,4)$
$[1,4)$	$0 < p < 1/2$	**01**	$[0,4)$	**1**	$[0,4)$
	$1/2 \le p < 1$	**1**	$[0,4)$	**01**	$[0,4)$

□

Langdon and Rissanen [29] suggest identifying the symbols as the more probable symbol (MPS) and less probable symbol (LPS) rather than as **1** and **0**. By doing this we can often combine transitions and eliminate states.

Example 6: We modify Example 5 to use the MPS/LPS idea. We are able to reduce the coder to just two states.

		LPS input		MPS input	
State	Prob{MPS}	Output	Next state	Output	Next state
$[0,4)$	$1/2 \le p \le \alpha$	**0**	$[0,4)$	**1**	$[0,4)$
	$\alpha < p < 1$	**00**	$[0,4)$	-	$[1,4)$
$[1,4)$	$1/2 \le p < 1$	**01**	$[0,4)$	**1**	$[0,4)$

□

Another way of simplifying an arithmetic coder is to allow only a subset of the possible interval subdivisions. Using integer arithmetic has the effect of making the symbol probabilities approximate, especially as the integer range is made smaller; limiting the number of subdivisions simply makes them even less precise. Since the main benefit of arithmetic coding is its ability to code efficiently when probabilities are close to 1, we usually want to allow at least some pairs of unequal probabilities.

Example 7: If we know that one symbol occurs considerably more often than the other, we can eliminate the transitions in Example 6 for approximately equal probabilities. This makes it unnecessary for the coder to decide which transition pair to use in the $[0, 4)$ state, and gives a very simple reduced-precision arithmetic coder.

	LPS input		MPS input	
State	Output	Next state	Output	Next state
$[0, 4)$	**00**	$[0, 4)$	-	$[1, 4)$
$[1, 4)$	**01**	$[0, 4)$	**1**	$[0, 4)$

This simple code is quite useful, providing almost a 50 percent improvement on the unary code for representing non-negative integers. To encode n in unary, we output n **1**s and a **0**. Using the code just derived, we re-encode the unary coding, treating **1** as the MPS. The resulting code consists of $\lfloor n/2 \rfloor$ **1**s, followed by **00** if n is even and **01** if n is odd. We can do even better with slightly more complex codes, as we shall see in examples that follow. □

We now introduce the *maximally unbalanced subdivision* and show how it can be used to obtain excellent compression when Prob{MPS} ≈ 1. Suppose the current interval is $[L, H)$. If Prob{MPS} is very high we can subdivide the interval at $L+1$ or $H-1$, indicating Prob{LPS} $= 1/(H-L)$ and Prob{MPS} $= 1 - 1/(H-L)$. Since the length of the current interval $H - L$ is always more than $N/4$, such a subdivision always indicates a Prob{MPS} of more than $1 - 4/N$. By choosing a large value of N and always including the maximally unbalanced subdivision in our coder, we ensure that very likely symbols can always be given an appropriately high probability.

Example 8: Let $N = 8$ and let the MPS always be **1**. We obtain the following four-state code if we allow only the maximally unbalanced subdivision in each state.

	0 (LPS) input		**1** (MPS) input	
State	Output	Next state	Output	Next state
$[0, 8)$	**000**	$[0, 8)$	-	$[1, 8)$
$[1, 8)$	**001**	$[0, 8)$	-	$[2, 8)$
$[2, 8)$	**010**	$[0, 8)$	-	$[3, 8)$
$[3, 8)$	**011**	$[0, 8)$	**1**	$[0, 8)$

We can use this code to re-encode unary-coded non-negative integers with $\lfloor n/4 \rfloor + 3$ bits. In effect, we represent n in the form $4a + b$; we encode a in unary, then use two bits to encode b in binary. □

Whenever the current interval coincides with the full interval, we can switch to a different code.

Example 9: We can derive the Elias code for the positive integers [14] by using the maximally unbalanced subdivision technique of Example 8 and by doubling the full integer range whenever we see enough **1**s to output a bit and expand the current interval so that it coincides with the full range. This coder has an infinite number of states; no state is visited more than once. We use the notation $[L, H)/M$ to indicate the subinterval $[L, H)$ selected from the range $[0, M)$.

State	**0** (LPS) input Output	Next state	**1** (MPS) input Output	Next state
$[0,2)/2$	**0**	STOP	**1**	$[0,4)/4$
$[0,4)/4$	**00**	STOP	-	$[1,4)/4$
$[1,4)/4$	**01**	STOP	**1**	$[0,8)/8$
$[0,8)/8$	**000**	STOP	-	$[1,8)/8$
$[1,8)/8$	**001**	STOP	-	$[2,8)/8$
⋮	⋮	⋮	⋮	⋮

This code corresponds to encoding positive integers as follows:

n	Code
1	**0**
2	**100**
3	**101**
4	**11000**
5	**11001**
⋮	⋮

In effect we represent n in the form $2^a + b$; we encode a in unary, then use a bits to encode b in binary. This is essentially the Elias code; it requires $\lfloor 2 \lg n \rfloor + 1$ bits to encode n. □

If we design a coder with more states, we obtain a more fine-grained set of probabilities.

Example 10: We show a six-state coder, obtained by letting $N = 8$ and allowing all possible subdivisions. We indicate only the center probability for each range; in practice any reasonable division will give good results. Output symbol f indicates application of the follow-on procedure.

State	Approximate Prob{MPS}	LPS input Output	LPS input Next state	MPS input Output	MPS input Next state
$[0,8)$	$1/2$	**1**	$[0,8)$	**0**	$[0,8)$
	$5/8$	**1**	$[2,8)$	-	$[0,5)$
	$3/4$	**11**	$[0,8)$	-	$[0,6)$
	$7/8$	**111**	$[0,8)$	-	$[0,7)$
$[0,7)$	$4/7$	**1**	$[0,6)$	**0**	$[0,8)$
	$5/7$	**1**f	$[0,8)$	-	$[0,5)$
	$6/7$	**110**	$[0,8)$	-	$[0,6)$
$[0,6)$	$1/2$	f	$[2,8)$	**0**	$[0,6)$
	$2/3$	**10**	$[0,8)$	**0**	$[0,8)$
	$5/6$	**101**	$[0,8)$	-	$[0,5)$
$[2,8)$	$1/2$	f	$[0,6)$	**1**	$[2,8)$
	$2/3$	**01**	$[0,8)$	**1**	$[0,8)$
	$5/6$	**010**	$[0,8)$	-	$[3,8)$
$[0,5)$	$3/5$	ff	$[0,8)$	**0**	$[0,6)$
	$4/5$	**100**	$[0,8)$	**0**	$[0,8)$
$[3,8)$	$3/5$	ff	$[0,8)$	**1**	$[2,8)$
	$4/5$	**011**	$[0,8)$	**1**	$[0,8)$

This coder is easily programmed and extremely fast. Its only shortcoming is that on average high-probability symbols require 1/4 bit (corresponding to Prob{MPS} $= 2^{-1/4} \approx 0.841$) no matter how high the actual probability is. □

Design of a class of reduced-precision coders. We now present a very flexible yet simple coder design incorporating most of the features just discussed. We choose N to be any power of 2. All states in the coder are of the form $[k, N)$, so the number of states is only $N/2$. (Intervals with $k \geq N/2$ will produce output, and the interval will be expanded.) In every state $[k, N)$ we include the maximally unbalanced subdivision (at $k+1$), which corresponds to values of Prob{MPS} between $(N-2)/N$ and $(N-1)/N$. We include a nearly balanced subdivision so that we will not lose efficiency when Prob{MPS} $\approx 1/2$. In addition, we locate other subdivision points such that the subinterval expansion that follows each input symbol leaves the coder in a state of the form $[k, N)$, and we choose one or more of them to correspond to intermediate values of Prob{MPS}. For simplicity we denote state $[k, N)$ by k.

We always allow the interval $[k, N)$ to be divided at $k+1$; if the LPS occurs we output the $\lg N$ bits of k and move to state 0, while if the MPS occurs we simply move to state $k+1$, then if the new state is $N/2$ we output a **1** and move to state 0. The other permitted subdivisions are given in the following table. In some cases additional output and expansion may be possible. It may not be necessary to include all subdivisions in the coder.

Range of states k	Subdivision LPS	Subdivision MPS	LPS input Output	LPS input Next State	MPS input Output	MPS input Next State
$[0, \frac{N}{2})$	$[k, \frac{N}{2})$	$[\frac{N}{2}, N)$	**0**	$2k$	**1**	0
$[0, \frac{N}{4})$	$[k, \frac{N}{4})$	$[\frac{N}{4}, N)$	**00**	$4k$	-	$\frac{N}{4}$
$[\frac{N}{8}, \frac{N}{4})$	$[k, \frac{3N}{8})$	$[\frac{3N}{8}, N)$	**0**f	$4k - \frac{N}{2}$	-	$\frac{3N}{8}$
$[\frac{N}{4}, \frac{3N}{8})$	$[k, \frac{3N}{8})$	$[\frac{3N}{8}, N)$	**010**	$8k - 2N$	-	$\frac{3N}{8}$
$[\frac{3N}{8}, \frac{N}{2})$	$[k, \frac{5N}{8})$	$[\frac{5N}{8}, N)$	ff	$4k - \frac{3N}{2}$	**1**	$\frac{N}{4}$
$[\frac{7N}{16}, \frac{N}{2})$	$[k, \frac{9N}{16})$	$[\frac{9N}{16}, N)$	fff	$8k - \frac{7N}{2}$	**1**	$\frac{N}{8}$
$[\frac{N}{4}, \frac{N}{2})$	$[\frac{3N}{4}, N)$	$[k, \frac{3N}{4})$	**11**	0	f	$2k - \frac{N}{2}$

For example, the fifth line indicates that for all states k for which $3N/8 \leq k < N/2$, we may subdivide the interval at $5N/8$. If the LPS occurs, we perform the follow-on procedure twice, which leaves us with the interval $[4k - 3N/2, N)$; otherwise we output a **1** and expand the interval to $[N/4, N)$.

A coder constructed using this procedure will have a small number of states, but in every state it will allow us to use estimates of Prob{MPS} near 1, near 1/2, and in between. Thus we can choose a large N so that highly probable events require negligible code length, while keeping the number of states small enough to allow table lookups rather than arithmetic.

3.2 ϵ-partitions and ρ-partitions

In Section 3.1 we have shown that is is possible to design a binary arithmetic coder that admits only a small number of possible probabilities. In this section we give a theoretical basis for selecting the probabilities. Often there are practical considerations limiting our choices, but we can show that it is reasonable to expect that choosing only a few probabilities will give close to optimal compression.

For a binary alphabet, we can use Equation (1) to compute $E(p, q)$, the extra code length resulting from using estimates q and $1 - q$ for actual probabilities p and $1 - p$, respectively. For any desired maximum excess code length ϵ, we can partition the space of possible probabilities to guarantee that the use of approximate probabilities will never add more than ϵ to the code length of any event. We select partitioning probabilities $P_0, P_1, \ldots$ and estimated probabilities $Q_0, Q_1, \ldots$. Each probability Q_i is used to encode all events whose probability p is in the range $P_i < p \leq P_{i+1}$. We compute the partition, which we call an *ϵ-partition*, as follows:

1. Set $i := 0$ and $Q_0 := 1/2$.

2. Find the value of P_{i+1} (greater than Q_i) such that $E(P_{i+1}, Q_i) = \epsilon$. We will use Q_i as the estimated probability for all probabilities p such that

$Q_i < p \leq P_{i+1}$.

3. Find the value of Q_{i+1} (greater than P_{i+1}) such that $E(P_{i+1}, Q_{i+1}) = \epsilon$. After we compute P_{i+2} in step 2 of the next iteration, we will use Q_{i+1} as the estimate for all probabilities p such that $P_{i+1} < p \leq P_{i+2}$.

We increment i and repeat steps 2 and 3 until P_{i+1} or Q_{i+1} reaches 1. The values for $p < 1/2$ are symmetrical with those for $p > 1/2$.

Example 11: We show the ϵ-partition for $\epsilon = 0.05$ bit per binary input symbol.

Range of actual probabilities	Probability to use
$[0.0000, 0.0130)$	0.0003
$[0.0130, 0.1427)$	0.0676
$[0.1427, 0.3691)$	0.2501
$[0.3691, 0.6309)$	0.5000
$[0.6309, 0.8579)$	0.7499
$[0.8579, 0.9870)$	0.9324
$[0.9870, 1.0000)$	0.9997

Thus by using only 7 probabilities we can guarantee that the excess code length does not exceed 0.05 bit for each binary decision coded. □

We might wish to limit the *relative* error so that the code length can never exceed the optimal by more than a factor of $1 + \rho$. We can begin to compute these *ρ-partitions* using a procedure similar to that for ϵ-partitions, but unfortunately the process does not terminate, since ρ-partitions are not finite. As P approaches 1, the optimal average code length grows very small, so to obtain a small relative loss Q must be very close to P. Nevertheless, we can obtain a partial ρ-partition.

Example 12: We show part of the ρ-partition for $\rho = 0.05$; the maximum relative error is 5 percent.

Range of actual probabilities	Probability to use
⋮	⋮
$[0.0033, 0.0154)$	0.0069
$[0.0154, 0.0573)$	0.0291
$[0.0573, 0.1670)$	0.0982
$[0.1670, 0.3722)$	0.2555
$[0.3722, 0.6278)$	0.5000
$[0.6278, 0.8330)$	0.7445
$[0.8330, 0.9427)$	0.9018
$[0.9427, 0.9846)$	0.9709
$[0.9846, 0.9967)$	0.9931
⋮	⋮

□

In practice we will use an approximation to an ϵ-partition or a ρ-partition for values of Prob{MPS} up to the maximum probability representable by our coder.

3.3 Compressed trees

To use the reduced-precision arithmetic coder described in Section 3.1 for an n-symbol alphabet, we need an efficient data structure to map each of n symbols to a sequence of binary choices. We might consider Huffman trees, since they minimize the average number of binary events encoded per input symbol; however, a great deal of effort is required to keep the probabilities on all branches near 1/2. For arithmetic coding maintaining this balance condition is unnecessary and wastes time.

In this section we present the *compressed tree*, a space-efficient data structure based on the complete binary tree. Because arithmetic coding allows us to obtain nearly optimal compression of binary events even when the two probabilities are unequal, we are free to represent the probability distribution of an n-symbol alphabet by a complete binary tree with a probability at each internal node. The tree can be flattened (linearized) by breadth-first traversal, and we can save space by storing only one probability at each internal node, say, the probability of taking the left branch. (This probability can be stored to sufficient precision in just one byte, as we shall see in Section 3.4.)

In high-order text models, many longer contexts occur only a few times, and only a few different alphabet symbols occur in each context. In such cases even the linear representation is wasteful of space, requiring $n - 1$ nodes regardless of the number of alphabet symbols that actually occur. Including pointers in the nodes would at least double their size. In the compressed tree we collapse the breath-first linear representation of the complete binary tree by omitting nodes with zero probability. If k different symbols have non-zero probability, the compressed tree representation requires at most $k \lg(2n/k) - 1$ nodes.

Example 13: Suppose we have the following probability distribution for an 8-symbol alphabet:

Symbol	a	b	c	d	e	f	g	h
Probability	0	0	$1/8$	$1/4$	$1/8$	0	$1/8$	$3/8$

We can represent this distribution by the tree in Figure 3(a), rounding probabilities and expressing them as multiples of 0.01. We show the linear representation in Figure 3(b) and the compressed tree representation in Figure 3(c). □

Traversing the compressed tree is mainly a matter of keeping track of omitted nodes. We do not have to process each node of the tree: for the first $\lg n - 2$ levels we have to process each node; but when we reach the desired node in the next-to-lowest level we have enough information to directly index the desired node of the lowest level. The operations are very simple, involving only one test and one or two increment operations at each node, plus a few extra operations at each level. Including the capability of adding new symbols to the tree makes the algorithm only slightly more complicated.

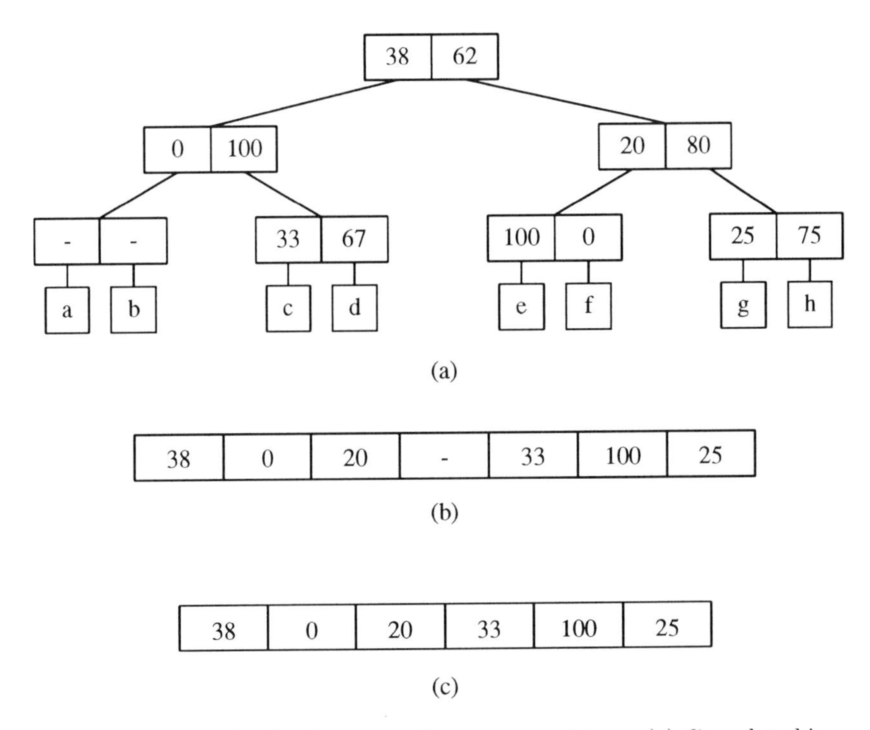

Figure 3: Steps in the development of a compressed tree. (a) Complete binary tree. (b) Linear representation. (c) Compressed tree.

3.4 Representing and estimating probabilities

In our binary coded representation of each context we wish to use only one byte for each probability, and we need the probability to limited precision. Therefore, we will represent the probability at a node as a state in a finite state automaton with about 256 states. Each state indicates a probability, and some of the states also indicate the size of the sample used to estimate the probability.

We need a method for estimating the probability at each node of the binary tree. Leighton and Rivest [30] and Pennebaker and Mitchell [41] describe probabilistic methods. Their estimators are also finite state automata, with each state corresponding to a probability. When a new symbol occurs, a transition to another state may occur, the probability of the transition depending on the current state and the new symbol. Generally, the transition probability is higher when the LPS occurs. In [30] transitions occur only between adjacent states. In [41] the LPS always causes a transition, possibly to a non-adjacent state; a transition after the MPS, when one occurs, is always to an adjacent state.

We give a deterministic estimator based on the same idea. In our estimator each input symbol causes a transition (unless the MPS occurs when the estimated probability is already at its maximum value). The probabilities represented by the states are so close together that transitions often occur between non-adjacent states. The transitions are selected so that we compute the new probability p_{new} of the left branch by

$$p_{\text{new}} \approx \begin{cases} f\,p_{\text{old}} + (1-f) & \text{if the left branch was taken} \\ f\,p_{\text{old}} & \text{if the right branch was taken,} \end{cases}$$

where f is a smoothing factor. This corresponds to exponential aging; hence the probability estimate can track changing probabilities and benefit from locality of reference, as discussed in Section 2.3.

In designing a probability estimator of this type we must choose both the scaling factor f and the set of probabilities represented by the states. We should be guided by the requirements of the coder and by our lack of *a priori* knowledge of the process generating the sequence of branches.

First we note that when the number of occurrences is small, our estimates cannot be very accurate. Laplace's law of succession, which gives the estimate

$$p = \frac{c+1}{t+2} \tag{2}$$

after c successes in t trials, offers a good balance between using all available information and allowing for random variation in the data; in effect, it gives the Bayesian estimate assuming a uniform *a priori* distribution for the true underlying probability P.

We recall that for values of P near $1/2$ we do not require a very accurate estimate, since any value will give about the same code length; hence we do not need many states in this probability region. When P is closer to 1, we would like our estimate to be more accurate, to allow the arithmetic coder to give near-optimal compression, so we assign states more densely for larger P. Unfortunately, in this case estimation by any means is difficult, because occurrences of the LPS are so infrequent. We also note that the underlying probability of any branch in the coding tree may change at any time, and we would like our estimate to adapt accordingly.

To handle the small-sample cases, we reserve a number of states simply to count occurrences when t is small, using Equation (2) to estimate the probabilities. We do the same for larger values of t when c is 0, 1, $t-1$, or t, to provide fast convergence to extreme values of P.

We can show that if the underlying probability P does not change, the expected value of the estimate p_k after k events is given by

$$E(p_k) = P + (p_0 - P)f^k,$$

which converges to P for all f, $0 \le f < 1$. The rapid convergence of $E(p_k)$ when $f = 0$ is misleading, since in that case the estimate is always 0 or 1, depending only on the preceding event. The expected value is clearly P, but the estimator is useless. A value of f near 1 provides resistance to random fluctuations in the input, but the estimate converges slowly, both initially and when the underlying P changes. A careful choice of f would depend on a detailed analysis like that performed by Flajolet for the related problem of approximate counting [16,17]. We make a more pragmatic decision. We know that periodic scaling is an approximation to exponential aging and we can show that a scaling factor of f corresponds to a scaling block size B of approximately $f \ln 2/(1-f)$. Since $B = 16$ works well for scaling [58], we choose $f = 0.96$.

Table 2: Comparsion of PPMC and PPMD. Compression figures are in bits per input symbol.

File	Text?	PPMC	PPMD	Improvement using PPMD
bib	Yes	2.11	2.09	0.02
book1	Yes	2.65	2.63	0.02
book2	Yes	2.37	2.35	0.02
news	Yes	2.91	2.90	0.01
paper1	Yes	2.48	2.46	0.02
paper2	Yes	2.45	2.42	0.03
paper3	Yes	2.70	2.68	0.02
paper4	Yes	2.93	2.91	0.02
paper5	Yes	3.01	3.00	0.01
paper6	Yes	2.52	2.50	0.02
progc	Yes	2.48	2.47	0.01
progl	Yes	1.87	1.85	0.02
progp	Yes	1.82	1.80	0.02
geo	No	5.11	5.10	0.01
obj1	No	3.68	3.70	-0.02
obj2	No	2.61	2.61	0.00
pic	No	0.95	0.94	0.01
trans	No	1.74	1.72	0.02

3.5 Improved modeling for text compression

To obtain good, fast text compression, we wish to use the multi-symbol extension of the reduced-precision arithmetic coder in conjunction with a good model. The PPM idea described in Section 2.3 has proven effective, but the *ad hoc* nature of the escape probability calculation is somewhat annoying. In this section we present yet another *ad hoc* method, which we call PPMD, and also a more complicated but more principled approach to the problem.

PPMD. Moffat's PPMC method [37] is widely considered to be the best method of estimating escape probabilities. In PPMC, each symbol's weight in a context is taken to be number of times it has occurred so far in the context. The escape "event," that is, the occurrence of a symbol for the first time in the context, is also treated as a "symbol," with its own count. When a letter occurs for the first time, its weight becomes 1; the escape count is incremented by 1, so the total weight increases by 2. At all other times the total weight increases by 1.

We have developed a new method, which we call PPMD, which is similar to PPMC except that it makes the treatment of new symbols more consistent by adding 1/2 instead of 1 to both the escape count and the new symbol's count when a new symbol occurs; hence the total weight always increases by 1.

We have compared PPMC and PPMD on the Bell-Cleary-Witten corpus [5] (including the four papers not described in the book). Table 2 shows that for text files PPMD compresses consistently about 0.02 bit per character better than PPMC. The compression results for PPMC differ from those reported in [5] because of implementation differences; we used versions of PPMC and PPMD that were identical except for the escape probability calculations. PPMD has the added advantage of making analysis more tractable by making the code length independent of the appearance order of symbols in the context.

Indirect probability estimation. Often we are faced with a situation where we have no theoretical basis for estimating the probability of an event, but where we know the factors that affect the probability. In such cases a logical and effective approach is to create conditioning classes based on the values of the factors, and to estimate the probability adaptively for each class. In the PPM method, we know that the number of occurrences of a state (t) and the number of different alphabet symbols that have occurred (k) are the factors affecting p_{esc}. We have done experiments, using all combinations of t and k as the conditioning classes (except that we group together all values of t greater than 48 and all values of k greater than 18). In our experiments we use a third-order model; when a symbol has not occurred previously in its context of length 3, we simply use 8 bits to indicate the ASCII value of the symbol. (The idea of skipping some shorter contexts for speed, space, and simplicity appears also in [31].) Even with this simplistic way of dropping to shorter contexts, the improved estimation of p_{esc} gives slightly better overall compression than PPMC for *book1*, the longest file in the Bell-Cleary-Witten corpus. We expect that using indirect probability estimation in conjunction with the full multi-order PPM mechanism will yield substantially improved compression.

3.6 Hashed high-order Markov models.

For finding contexts in the PPM method, Moffat [37] and Bell *et al.* [5] give complicated data structures called *backward trees* and *vine pointers*. For fast access and minimal memory usage we propose single hashing without collision resolution. One might expect that using the same bucket for accumulating statistics from unrelated contexts would significantly degrade compression performance, but we can show that often this is not the case.

Even in the worst case, when the symbols from the k colliding contexts in bucket b are mutually disjoint, the additional code length is only $H_b = H(p_1, p_2, p_3, \ldots, p_k)$, the entropy of the ensemble of probabilities of occurrence of the contexts. We show this by conceptually dividing the bucket into disjoint subtrees corresponding to the various contexts, and noting that the cost of identifying an individual symbol is just $L_C = -\lg p_i$, the cost of identifying the context that occurred, plus L_S, the cost of identifying the symbol in its own context. Hence the extra cost is just L_C, and the average extra cost is $\sum_{i=1}^{k} -p_i \lg p_i = H_b$. The maximum value of H_b is $\lg k$, so in buckets that contain data from only two contexts, the extra code length is at most 1 bit per input symbol.

In fact, when the number of colliding contexts in a bucket is large enough that H_b is significant, the symbols in the bucket, representing a combination of a number of contexts, will be a microcosm of the entire file; the bucket's average code length will approximately equal the 0-order entropy of the file. Lelewer and Hirschberg [31] apply hashing with collision resolution in a similar high-order scheme.

4 Conclusion

We have shown the details of an implementation of arithmetic coding and have pointed out its advantages (flexibility and near-optimality) and its main disadvantage (slowness). We have developed a fast coder, based on reduced-precision arithmetic coding, which gives only minimal loss of compression efficiency; we can use the concept of ϵ-partitions to find the probabilities to include in the coder to keep the compression loss small. In a companion paper [22], in which we refer to this fast coding method as *quasi-arithmetic coding*, we give implementation details and performance analysis for both binary and multi-symbol alphabets. We prove analytically that the loss in compression efficiency compared with exact arithmetic coding is negligible.

We introduce the compressed tree, a new data structure for efficiently representing a multi-symbol alphabet by a series of binary choices. Our new deterministic probability estimation scheme allows fast updating of the model stored in the compressed tree using only one byte for each node; the model can provide the reduced-precision coder with the probabilities it needs. Choosing one of our two new methods for computing the escape probability enables us to use the highly effective PPM algorithm, and use of a hashed Markov model keeps space and time requirements manageable even for a high-order model.

References

[1] N. Abramson, *Information Theory and Coding*, McGraw-Hill, New York, NY, 1963.

[2] R. B. Arps, T. K. Truong, D. J. Lu, R. C. Pasco, and T. D. Friedman, "A Multi-Purpose VLSI Chip for Adaptive Data Compression of Bilevel Images," *IBM J. Res. Develop.* 32 (Nov. 1988), 775–795.

[3] T. Bell, "A Unifying Theory and Improvements for Existing Approaches to Text Compression," Univ. of Canterbury, Ph.D. Thesis, 1986.

[4] T. Bell and A. M. Moffat, "A Note on the DMC Data Compression Scheme," *Computer Journal* 32 (1989), 16–20.

[5] T. C. Bell, J. G. Cleary, and I. H. Witten, *Text Compression*, Prentice-Hall, Englewood Cliffs, NJ, 1990.

[6] T. C. Bell, I. H. Witten, and J. G. Cleary, "Modeling for Text Compression," *Comput. Surveys* 21 (Dec. 1989), 557–591.

[7] J. L. Bentley, D. D. Sleator, R. E. Tarjan, and V. K. Wei, "A Locally Adaptive Data Compression Scheme," *Comm. ACM* 29 (Apr. 1986), 320–330.

[8] A. C. Blumer and R. J. McEliece, "The Rényi Redundancy of Generalized Huffman Codes," *IEEE Trans. Inform. Theory* IT–34 (Sept. 1988), 1242–1249.

[9] R. M. Capocelli, R. Giancarlo, and I. J. Taneja, "Bounds on the Redundancy of Huffman Codes," *IEEE Trans. Inform. Theory* IT–32 (Nov. 1986), 854–857.

[10] D. Chevion, E. D. Karnin, and E. Walach, "High Efficiency, Multiplication Free Approximation of Arithmetic Coding," in *Proc. Data Compression Conference*, J. A. Storer and J. H. Reif, eds., Snowbird, Utah, Apr. 8–11, 1991, 43–52.

[11] G. V. Cormack and R. N. Horspool, "Data Compression Using Dynamic Markov Modelling," *Computer Journal* 30 (Dec. 1987), 541–550.

[12] G. V. Cormack and R. N. Horspool, "Algorithms for Adaptive Huffman Codes," *Inform. Process. Lett.* 18 (Mar. 1984), 159–165.

[13] P. Elias, "Interval and Recency Rank Source Coding: Two On-line Adaptive Variable Length Schemes," *IEEE Trans. Inform. Theory* IT–33 (Jan. 1987), 3–10.

[14] P. Elias, "Universal Codeword Sets and Representations of Integers," *IEEE Trans. Inform. Theory* IT–21 (Mar. 1975), 194–203.

[15] N. Faller, "An Adaptive System for Data Compression," Record of the 7th Asilomar Conference on Circuits, Systems, and Computers, 1973.

[16] Ph. Flajolet, "Approximate Counting: a Detailed Analysis," *BIT* 25 (1985), 113.

[17] Ph. Flajolet and G. N. N. Martin, "Probabilistic Counting Algorithms for Data Base Applications," INRIA, Rapport de Recherche No. 313, June 1984.

[18] R. G. Gallager, "Variations on a Theme by Huffman," *IEEE Trans. Inform. Theory* IT–24 (Nov. 1978), 668–674.

[19] M. Guazzo, "A General Minimum-Redundancy Source-Coding Algorithm," *IEEE Trans. Inform. Theory* IT–26 (Jan. 1980), 15–25.

[20] M. E. Hellman, "Joint Source and Channel Encoding," Proc. Seventh Hawaii International Conf. System Sci., 1974.

[21] R. N. Horspool, "Improving LZW," in *Proc. Data Compression Conference*, J. A. Storer and J. H. Reif, eds., Snowbird, Utah, Apr. 8–11, 1991, 332–341.

[22] P. G. Howard and J. S. Vitter, "Design and Analysis of Quasi-Arithmetic Coding," Brown University, Department of Computer Science, Technical Report, 1992.

[23] P. G. Howard and J. S. Vitter, "Analysis of Arithmetic Coding for Data Compression," in *Proc. Data Compression Conference*, J. A. Storer and J. H. Reif, eds., Snowbird, Utah, Apr. 8–11, 1991, 3–12, invited paper, also to appear as an invited paper in the special issue of *Information Processing and Management* entitled "Data Compression for Images and Texts.".

[24] P. G. Howard and J. S. Vitter, "New Methods for Lossless Image Compression Using Arithmetic Coding," in *Proc. Data Compression Conference*,

J. A. Storer and J. H. Reif, eds., Snowbird, Utah, Apr. 8–11, 1991, 257–266, also to appear as an invited paper in the special issue of *Information Processing and Management* entitled "Data Compression for Images and Texts.".

[25] D. A. Huffman, "A Method for the Construction of Minimum Redundancy Codes," *Proceedings of the Institute of Radio Engineers* 40 (1952), 1098–1101.

[26] D. E. Knuth, "Dynamic Huffman Coding," *J. Algorithms* 6 (June 1985), 163–180.

[27] G. G. Langdon, "Probabilistic and Q-Coder Algorithms for Binary Source Adaptation," in *Proc. Data Compression Conference*, J. A. Storer and J. H. Reif, eds., Snowbird, Utah, Apr. 8–11, 1991, 13–22.

[28] G. G. Langdon, "A Note on the Ziv-Lempel Model for Compressing Individual Sequences," *IEEE Trans. Inform. Theory* IT–29 (Mar. 1983), 284–287.

[29] G. G. Langdon and J. Rissanen, "Compression of Black-White Images with Arithmetic Coding," *IEEE Trans. Comm.* COM–29 (1981), 858–867.

[30] F. T. Leighton and R. L. Rivest, "Estimating a Probability Using Finite Memory," *IEEE Trans. Inform. Theory* IT–32 (Nov. 1986), 733–742.

[31] D. A. Lelewer and D. S. Hirschberg, "Streamlining Context Models for Data Compression," in *Proc. Data Compression Conference*, J. A. Storer and J. H. Reif, eds., Snowbird, Utah, Apr. 8–11, 1991, 313–322.

[32] V. S. Miller and M. N. Wegman, "Variations on a Theme by Ziv and Lempel," in *Combinatorial Algorithms on Words*, A. Apostolico and Z. Galil, eds., NATO ASI Series #F12, Springer-Verlag, Berlin, 1984, 131–140.

[33] J. L. Mitchell and W. B. Pennebaker, "Optimal Hardware and Software Arithmetic Coding Procedures for the Q-Coder," *IBM J. Res. Develop.* 32 (Nov. 1988), 727–736.

[34] A. M. Moffat, "Predictive Text Compression Based upon the Future Rather than the Past," *Australian Computer Science Communications* 9 (1987), 254–261.

[35] A. M. Moffat, "Word-Based Text Compression," *Software–Practice and Experience* 19 (Feb. 1989), 185–198.

[36] A. M. Moffat, "Linear Time Adaptive Arithmetic Coding," *IEEE Trans. Inform. Theory* IT–36 (Mar. 1990), 401–406.

[37] A. M. Moffat, "Implementing the PPM Data Compression Scheme," *IEEE Trans. Comm.* COM–38 (Nov. 1990), 1917–1921.

[38] K. Mohiuddin, J. J. Rissanen, and M. Wax, "Adaptive Model for Nonstationary Sources," *IBM Technical Disclosure Bulletin* 28 (Apr. 1986), 4798–4800.

[39] D. S. Parker, "Conditions for the Optimality of the Huffman Algorithm," *SIAM J. Comput.* 9 (Aug. 1980), 470–489.

[40] R. Pasco, "Source Coding Algorithms for Fast Data Compression," Stanford Univ., Ph.D. Thesis, 1976.

[41] W. B. Pennebaker and J. L. Mitchell, "Probability Estimation for the Q-Coder," *IBM J. Res. Develop.* 32 (Nov. 1988), 737–752.

[42] W. B. Pennebaker and J. L. Mitchell, "Software Implementations of the Q-Coder," *IBM J. Res. Develop.* 32 (Nov. 1988), 753–774.

[43] W. B. Pennebaker, J. L. Mitchell, G. G. Langdon, and R. B. Arps, "An Overview of the Basic Principles of the Q-Coder Adaptive Binary Arithmetic Coder," *IBM J. Res. Develop.* 32 (Nov. 1988), 717–726.

[44] J. Rissanen, "Modeling by Shortest Data Description," *Automatica* 14 (1978), 465–571.

[45] J. Rissanen, "A Universal Prior for Integers and Estimation by Minimum Description Length," *Ann. Statist.* 11 (1983), 416–432.

[46] J. Rissanen, "Universal Coding, Information, Prediction, and Estimation," *IEEE Trans. Inform. Theory* IT–30 (July 1984), 629–636.

[47] J. Rissanen and G. G. Langdon, "Universal Modeling and Coding," *IEEE Trans. Inform. Theory* IT–27 (Jan. 1981), 12–23.

[48] J. J. Rissanen, "Generalized Kraft Inequality and Arithmetic Coding," *IBM J. Res. Develop.* 20 (May 1976), 198–203.

[49] J. J. Rissanen and G. G. Langdon, "Arithmetic Coding," *IBM J. Res. Develop.* 23 (Mar. 1979), 146–162.

[50] J. J. Rissanen and K. M. Mohiuddin, "A Multiplication-Free Multialphabet Arithmetic Code," *IEEE Trans. Comm.* 37 (Feb. 1989), 93–98.

[51] C. Rogers and C. D. Thomborson, "Enhancements to Ziv-Lempel Data Compression," Dept. of Computer Science, Univ. of Minnesota, Technical Report TR 89–2, Duluth, Minnesota, Jan. 1989.

[52] F. Rubin, "Arithmetic Stream Coding Using Fixed Precision Registers," *IEEE Trans. Inform. Theory* IT–25 (Nov. 1979), 672–675.

[53] B. Y. Ryabko, "Data Compression by Means of a Book Stack," *Problemy Peredachi Informatsii* 16 (1980).

[54] C. E. Shannon, "A Mathematical Theory of Communication," *Bell Syst. Tech. J.* 27 (July 1948), 398–403.

[55] J. S. Vitter, "Dynamic Huffman Coding," *ACM Trans. Math. Software* 15 (June 1989), 158–167, also appears as Algorithm 673, Collected Algorithms of ACM, 1989.

[56] J. S. Vitter, "Design and Analysis of Dynamic Huffman Codes," *Journal of the ACM* 34 (Oct. 1987), 825–845.

[57] I. H. Witten and T. C. Bell, "The Zero Frequency Problem: Estimating the Probabilities of Novel Events in Adaptive Text Compression," *IEEE Trans. Inform. Theory* IT–37 (July 1991), 1085–1094.

[58] I. H. Witten, R. M. Neal, and J. G. Cleary, "Arithmetic Coding for Data Compression," *Comm. ACM* 30 (June 1987), 520–540.

[59] J. Ziv and A. Lempel, "A Universal Algorithm for Sequential Data Compression," *IEEE Trans. Inform. Theory* IT–23 (May 1977), 337–343.

[60] J. Ziv and A. Lempel, "Compression of Individual Sequences via Variable Rate Coding," *IEEE Trans. Inform. Theory* IT–24 (Sept. 1978), 530–536.

5

Context Modeling for Text Compression

Daniel S. Hirschberg† and Debra A. Lelewer‡

Abstract

Adaptive context modeling has emerged as one of the most promising new approaches to compressing text. A *finite-context model* is a probabilistic model that uses the context in which input symbols occur (generally a few preceding characters) to determine the number of bits used to code these symbols. We provide an introduction to context modeling and recent research results that incorporate the concept of context modeling into practical data compression algorithms.

1. Introduction

One of the more important developments in the study of data compression is the modern paradigm first presented by Rissanen and Langdon [RL81]. This paradigm divides the process of compression into two separate components: modeling and coding. A model is a representation of the source that generates the data being compressed. *Modeling* is the process of constructing this representation. *Coding* entails mapping the modeler's representation of the source into a compressed representation. The coding component takes information

† Department of Information and Computer Science, University of California, Irvine

‡ Computer Science Dept., California State Polytechnic University, Pomona

supplied by the modeler and translates this information into a sequence of bits. Recognizing the dual nature of compression allows us to focus our attention on just one of the two processes.

The problem of coding input symbols provided by a statistical modeler has been well studied and is essentially solved. Arithmetic coding provides optimal compression with respect to the model used to generate the statistics. That is, given a model that provides information to the coder, arithmetic coding produces a minimal-length compressed representation. Witten et al. provide a description and an implementation of arithmetic coding [WNC87]. The well-known algorithm of Huffman is another statistical coder [H52]. Huffman coding is inferior to arithmetic coding in two important respects. First, Huffman coding is constrained to represent every event (e.g., character) using an integral number of bits. While information theory tells us that an event with probability $\frac{4}{5}$ contains $\lg \frac{5}{4}$† bits of information content and should be coded in $\lg \frac{5}{4} \approx .32$ bits, Huffman coding will assign 1 bit to represent this event. The accuracy of arithmetic coding is limited only by the precision of the machine on which it is implemented. The second advantage of arithmetic coding is that it can represent changing models more effectively. Updating a Huffman tree is much more time consuming. Researchers continue to refine arithmetic coding for purposes of efficiency.

Given the existence of an optimal coding method, modeling becomes the key to effective data compression. The selection of a modeling paradigm and its implementation determine the resource requirements and compression performance of the system. Context modeling is a very promising new approach to statistical modeling for text compression. Context modeling is a special case of *finite-state modeling*, or *Markov modeling*, and in fact the term Markov modeling is frequently used loosely to refer to finite-context modeling. In

---

† lg denotes the base 2 logarithm

this section we describe the strategy of context modeling and the parameters involved in implementing context models.

1.1. Basics of Context Modeling

A *finite-context model* uses the context provided by characters already seen to determine the encoding of the current character. The idea of a context consisting of a few previous characters is very reasonable when the data being compressed is natural language. We all know that the character following *q* in an English text is all but guaranteed to be *u* and that given the context *now is the time for all good men to come to the aid of*, the phrase *their country* is bound to follow. One would expect that using knowledge of this type would result in more accurate modeling of the information source. Although the technique of context modeling was developed and is clearly appropriate for compressing natural language, context models provide very good compression over a wide range of file types.

We say that a context model predicts successive characters taking into account the context provided by characters already seen. What is meant by *predict* here is that the frequency values used in encoding the current character are determined by its context. The frequency distribution used to encode a character determines the number of bits it contributes to the compressed representation. When character x occurs in context c and the model for context c does not include a frequency for x we say that context c *fails to predict* x.

A context model may use a fixed number of previous characters in its predictions or may be a *blended* model, incorporating predictions based on contexts of several lengths. A model that always uses i previous characters to predict the current character is a *pure order-i context model*. When $i = 0$, no context is used and the text is simply coded one character at a time. When $i = 1$, the previous character is used in encoding the current character; when $i = 2$, the previous two characters are used, and so on. A *blended* model may use the previous three characters, the previous two characters when the three-character context fails to predict, and one predecessor if both the order-3 and order-2 contexts fail. A blended model is composed of two or more submodels. An order-i context model consists of a frequency distribution for each i-character sequence occurring in the input stream. In the order-1 case, this means that the frequency distribution for context q will give a very high value to u and very little weight to any other letter, while the distribution for context t will have high frequencies for a, e , i, o, u, and h among others and very little weight for letters like q, n and g.

A blended model is *fully blended* if it contains submodels for the maximum-length context and *all* lower-order contexts. That is, a fully-blended order-3 context model bases its predictions on models of orders 3, 2, 1, 0, and -1 (the model of order -1 consists of a frequency distribution that weights all characters equally). A *partially-blended* model uses some, but not all, of the lower-order contexts.

Context modeling may also be either static or dynamic. A model is *static* if the information of which it consists remains unchanged throughout the encoding process. A *dynamic*, or *adaptive*, model modifies its representation of the input as encoding proceeds and it gathers information about the characteristics of the source. Most static data compression models have adaptive equivalents and the adaptive counterparts generally provide more effective compression. In fact, Bell et al. prove that over a large range of circumstances there is an adaptive model that will be only slightly worse than *any* static model, while a static model can be arbitrarily worse than an adaptive counterpart [BCW90]. We will confine our discussion to adaptive context models.

A context model is generally combined with arithmetic coding to form a data compression system. The model provides a frequency distribution for each context (each character in the order-1 case and each pair of characters in the order-2 case). Each frequency distribution forms the basis of an arithmetic code and these are used to map events into code bits. Huffman coding is not appropriate for use with adaptive context models for the reasons given above.

1.2. Methods of Blending

Blending is desirable and essentially unavoidable in an adaptive setting where the model is built from scratch as encoding proceeds. When the first character of a file is read, the model has no history on which to base predictions. Larger contexts become more meaningful as compression proceeds. The general mechanism of blending, *weighted blending*, assigns a probability to a character by weighting probabilities (or, more accurately, frequencies) provided by the various submodels and computing the weighted sum of these probabilities. This method of blending is too slow to be practical and has the additional disadvantage that there is no theoretical basis for assigning weights to the models of various orders. In a simpler and more practical blended order-i model, the number of bits used to code character c is dictated by the preceding i characters if c has occurred in this particular context before. In this case, only the order-i frequency distribution is used. Otherwise, models of lower orders are consulted until one of them supplies a prediction. When the context of order i fails to predict the current character, the encoder emits an *escape* code, a signal to the decoder that the model of lower order is being consulted. Some lowest-order model must be guaranteed to supply a prediction for every character in the input alphabet.

The frequencies used by the arithmetic coder may be computed in a number of ways. One of the more straightforward methods is to assign to character x in context c the frequency f where f is the number of times that context c has been used to predict character x. Alternatively, f may represent the number of times that x has occurred in context c. An implementation may also require x to occur in context c some minimal number of times before it allocates frequency to the event.

1.3. Escape Strategy

In order for the encoder to transmit the escape code, each frequency distribution in the blended model must have some frequency allocated to *escape*. A simple strategy is to treat the escape event as if it were an additional symbol in the input alphabet. Like any other character, the frequency of the escape event is the number of times it occurs. Other strategies involve relating the frequency of the escape code to the total frequency of the context and the number of different characters occurring in the context. On one hand, as the number of different characters increases, the probability of prediction increases and the use of the escape code becomes less likely. On the other hand, if a context has occurred frequently and predicted the same character (or small number of characters) every time, the appearance of a new character (and the need to escape) would seem unlikely. There is no theoretical basis for selecting one of these escape strategies over another. Fortunately, empirical experiments indicate that compression performance is largely insensitive to the selection of escape strategy.

1.4. Exclusion

The blending strategy described in Section 1.3 has the effect of excluding lower-order predictions when a character occurs in a higher-order model. However, it does not exclude as much lower-order information as it might. For example, when character x occurs in context abc for the first time the order-2 context bc is consulted. If character y has occurred in context abc it can be excluded from the order-2 prediction. That is, the fact that we escape from the order-3 context abc informs the decoder that the character being encoded is not y. Thus the bc model need not assign any frequency to y in making this prediction. By excluding y from the order-2 prediction x may be predicted more accurately. Excluding characters predicted by higher-order models can double execution time. The gain in compression performance is on the order of 5%, which hardly justifies the increased execution time [BCW90]. Another type of exclusion that is much simpler and has the effect of decreasing execution time is *update exclusion*. Update exclusion means updating only those models that contribute to the current prediction. Thus if, in the above example, context bc predicts x, only the order-3 model for abc and the order-2 model for bc will be updated. The models of lower order remain unchanged.

1.5. Memory Limitations

We call an order-i context model *complete* if for every character x occurring in context c, the model includes a frequency distribution for c that contains a count for x. That is, the model retains all that it has learned. Complete context models of even order 3 are rare since the space required to store all of the context information gleaned from a large file is prohibitive. There are two obvious ways to impose a memory limit on a finite context model. The first is to monitor its size and freeze the model when the size reaches some maximum. When the model is frozen, it can no longer represent characters occurring in novel contexts, but we can continue to update the frequency values already stored in the model. The second approach is to rebuild the model rather than freeze it. The model can be rebuilt from scratch or from a buffer representing recent history. The use of the buffer may lessen the degradation in compression performance due to rebuilding. On the other hand, the memory set aside for the buffer causes rebuilding to occur earlier. A third approach, which is not strictly a solution to the problem of limited memory, is to monitor compression performance as well as the size of the data structure. Rebuilding when compression begins to degrade may be more opportune than waiting until it becomes necessary. We will say more about the data structures used to represent context models in later sections. The representation of the model clearly impacts the amount of information it can contain and the ease with which it can be consulted and updated.

1.6. Measuring Compression Performance

Compression performance is generally presented in one of two forms: *compression ratio* and *number of bits per character in the compressed representation.* Compression ratio is the ratio (size of compressed representation)/(size of original input) and may be expressed as a percentage of the original input remaining after compression. Each of these definitions has the virtue that it makes no assumptions (e.g., ergodicity) about the input stream, the model, or the coding component of the algorithm. Reporting compression in terms of number of bits per character also factors out the representation of the symbols in the original input.

The performance of a data compression technique clearly depends on the type of data to which it is applied, and further, on the characteristics of a particular data file. In order to measure the effectiveness of a data compression system, it should be applied to a large corpus of files representing a variety of types and sizes. A reliable comparison of techniques can be made only if they are applied to the same data.

2. Early Context-Modeling Algorithms

A variety of approaches to the use of context modeling have been developed. The algorithms differ in the selection of the parameters described in Section 1 and in terms of the data structures used to represent the models. We describe the most successful of the early context modeling algorithms in Sections 2.1 through 2.6. The compression performance, memory requirements, and execution speeds of the methods are compared in Section 2.7.

2.1. Algorithm LOEMA

The earliest use of context modeling for text compression is the Local Order Estimating Markov Analysis method developed by Roberts in his dissertation [R82]. Algorithm LOEMA is fully-blended and uses weights on the models to form a single prediction for each character in the text. The weights represent confidence values placed on each prediction. The data structure used to represent the model is a backward tree stored in a hash table. Roberts investigates methods of growing and storing the models so as to conserve memory. As discussed in Section 1, weighted blending is inefficient and algorithms that employ the escape strategy of blending provide a much more practical alternative to LOEMA.

2.2. Algorithm DAFC

Langdon and Rissanen's Double-Adaptive File Compression algorithm is one of the first methods that employs blending in an adaptive data compression scheme [LR83]. DAFC is a partially-blended model consisting of z order-1 contexts and an order-0 context (z is a parameter associated with the algorithm and determines its memory requirements). When encoding begins, the order-0 model is used since no characters have yet occurred in any order-1 context. In a complete order-1 model, when a character occurs for the first time it becomes an order-1 context. In algorithm DAFC, only z contexts will be constructed: corresponding to the first z characters that occur at least N times in the text being encoded (N is another parameter of the algorithm). The suggested values $z = 31$ and $N = 50$ provide approximately 50 percent compression with a very modest space requirement and very good speed [BCW90].

DAFC employs neither explicit blending nor exclusion. Explicit blending is not required because once a model is activated it can predict any character in the input. That is, every model contains a frequency value for every character in the input alphabet. If character x occurs in context c and context c has occurred N times then x is predicted by c. Otherwise x is predicted by the order-0 model.

DAFC also employs run-length encoding when it encounters a sequence of three or more repetitions of the same character. This is equivalent to employing an order-2 model for this special case. Coding is performed by decomposition using the simple binary arithmetic code [LR82].

2.3. The PPM Algorithms

The PPM, or Prediction by Partial Match, algorithms are variable-order adaptive methods that implement the escape mechanism for blending. A PPM algorithm stores its context models in a single forward tree. The maximum number of characters of context (the order of the model) is a parameter of the PPM paradigm. The optimal order varies with file type and size. For text files, three or four appears to be the best choice. PPM algorithms are fully blended, so that an order-3 model uses submodels of order 3, 2, 1, 0, and -1. Members of the PPM family differ in terms of exclusion strategy and memory management. PPMC, the best of the PPM family, employs only update exclusion and imposes a limit on model size. Other PPM algorithms use full exclusion and allow the data structure to grow without bound.

The implementation of PPMC that proves most effective is an order-3 model that permits the tree to grow to a size of 500 Kbytes. The model is rebuilt using the previous 2048 characters when it reaches this limit.

2.4. Algorithm WORD

WORD is an algorithm that employs context models based on words and non-words, where a word is a sequence of alphabetic characters and a non-word a sequence of non-alphabetic characters [Mo89]. Each of the word and non-word submodels is a partially-blended model of orders 1 and 0. Words are predicted by preceding words if possible, and if not they are predicted without context. Similarly, non-words are predicted in the context of non-words. A model of order -1 is inappropriate for words that have not been seen before since we cannot allocate frequency to all possible words. Instead, when a word (respectively, non-word) occurs for the first time, its length is encoded using an order-0 model for lengths and then its letters (non-letters) are coded using a fully-blended model of order 1. WORD employs 12 submodels in total. For each mode (word and non-word) there are: order-1 submodels for words and letters; order-0 submodels for words, letters, and lengths; and a model of order

-1 for letters. The escape mechanism is used for blending. Update exclusion is performed while prediction exclusion is rejected for reasons of efficiency. Hash tables are used to store words and non-words, and the arithmetic codes are represented by tree structures.

Moffat also experimented with including word (and non-word) models of order 2 and found that this did not provide significant improvement over the order 1 model described above. WORD provides good compression and is reasonably fast; its speed derives from the fact that the arithmetic encoder is only invoked about 20% as often as in a character-based compression system (the average length of an English word being about five characters).

2.5. Algorithm DHPC

Williams's Dynamic-History Predictive Compression technique is similar to the PPM algorithms. The two major differences explored by Williams are the use of a backward tree and the use of thresholds[W88]. Whereas PPM extends its tree every time it is presented with new context information, DHPC grows a branch only if the occurrence count of the parent node is sufficiently high. In addition, DHPC avoids rebuilding its tree by setting a limit on its size and applying no changes to the tree once the maximum size has been reached. DHPC is also similar to algorithm DAFC in that both use thresholding to limit the size of the data structure representing the context information. However, the threshold in DAFC requires that the child node occur sufficiently often to justify its inclusion as a submodel while DHPC requires the parent node to have sufficient frequency before it spawns any children. DHPC provides better compression than DAFC due to its use of higher-order context information. DHPC is faster than the PPM algorithms but yields poorer compression.

2.6. Algorithm ADSM

Abrahamson's Adaptive Dependency Source Model is a pure order-1 context model with modest memory requirements. Abrahamson describes his model as follows:

> "If, for example, in a given text, the probability that the character h follows the character t is higher than that for any other character following a t and the probability of an e following a v is higher than that for any other character following a v, then the same symbol should be used to encode an h following a t as an e following a v. It should be noted that this scheme will also increase the probability of occurrence of the encoded symbol. ... the source message *abracadabra* can be represented by the sequence of symbols *abracadaaaa*. Notice how a b following an a and an r following a b (and also an a following an r) have all been converted into an a, the most frequently occurring source character [A89, p 78]."

In simpler terms, Abrahamson's model is an order-1 context model that employs a single frequency distribution and encodes symbol y following symbol x as symbol k, where k is the position of y on x's list of successors and where successor lists are maintained in frequency count order. Thus, in the example above we think of *bra* as being encoded by 111 rather than *aaa*. The other characters in the string *abracadabra* will also be encoded as list positions, but these positions cannot be inferred from the example. While this characterization may not be obvious from the description given above, it becomes clear from the implementation details given in Abrahamson's article [A89].

Thus, Abrahamson is modifying the basic order-1 model by:

(1) employing a single frequency distribution rather than a distribution for each 1-character context and

(2) employing self-organizing lists to map characters to frequency values.

ADSM employs a *pure* order-1 context model. For any pair x, y of successive characters, y is coded using the kth frequency value where y is the kth most frequent successor of x. There is intuitive appeal in the use of the frequency count list organizing strategy in ADSM since the coding technique employed is based on frequency values. On the other hand, the frequency values used for coding are aggregate values. The frequency used for encoding character y in context x is not the frequency with which y has occurred after x, but the number of times that position k has been used to encode an event.

2.7. Comparison of the Methods

Recently, progress has been made toward standardizing methods of reporting compression performance. In the past, researchers reported performance of their approaches using only a few private data files. This provided no reliable basis for comparing algorithms. In addition, measures of compression varied widely and were occasionally either ill-defined or multiply defined (i.e., a single term was used to mean different things by different authors). Today the majority of researchers have settled on the two measures defined in Section 1.6, compression ratio and number of bits per character. In addition, a corpus of files of a variety of types and sizes is now available in the public domain to facilitate legitimate comparisons. Descriptions of the files in the corpus and methods of accessing them are given in [BCW90]. The data reported here is drawn from the books by Bell et al. and Williams[BCW90, W91A].

Algorithm PPMC represents the state of the art in context modeling. PPMC provides very good compression (approximately 30 percent on average), but has a large memory requirement (500 Kbytes) and executes slowly (encoding and decoding approximately 2000 cps on a 1MIP machine. Over the compression corpus PPMC achieves average compression of 2.48 bits per character (bpc). DAFC and ADSM fare worst among the methods described here, providing average compression of 4.02 bpc and 3.95 bpc, respectively. DAFC's performance in terms of both compression and use of memory falls between that of an order-0 and an order-1 method. ADSM also uses far less memory than a complete order-1 model, and the information lost results in a loss of compression performance. WORD provides compression close to that achieved by PPMC, 2.76 bits per character over the corpus. WORD is the fastest of the methods and achieves the 2.76 bpc performance with approximately the same memory resources as PPMC. DHPC is neither as fast nor as effective as WORD. The average number of bits per character for the corpus when compressed by DHPC is 2.98. Compression results are not available for LOEMA. Williams reports only that LOEMA provides compression comparable to that of PPMC[W91A]. LOEMA is of little practical interest, however, because it executes very slowly.

2.8. Other Competing Methods

At present, the most commonly-used data compression systems are not context-modeling methods. The UNIX utility *compress* and other Ziv-Lempel algorithms represent the current state of the art in practical data compression. Ziv-Lempel algorithms employ dictionary models and fixed codes. The books by Storer and Bell et al. provide complete descriptions of the Ziv-Lempel family of algorithms[ST88, BCW90]. We mention the Ziv-Lempel methods only as competitors that must be reckoned with. *Compress* encodes and decodes using 450 Kbytes of internal memory and at a rate of approximately 15,000 cps [BCW90]. Its speed is its primary virtue and its compression performance is less than stellar. Over the corpus *compress* reduces files to an average of 3.64 bpc. A more recent Ziv-Lempel technique, algorithm FG, provides superior compression performance (2.95 bpc over the corpus) using less memory (186 Kbytes for encoding and 130 Kbytes for decoding), but executes only about half as fast as *compress* (encoding 6,000 cps and decoding 11,000 cps)[FG89]. Research continues on the Ziv-Lempel technique. A recent article by Williams proposes techniques for combining the compression performance of algorithm FG with the throughput of *compress*[W91B].

3. Context Modeling in Limited Memory

While context-modeling algorithms provide very good compression, they suffer from the disadvantages of being relatively slow and requiring large amounts of main memory in which to execute. Algorithm PPMC achieves an average compression ratio of approximately 30 percent. However, PPMC uses 500 Kbytes to represent the model it employs. As memory becomes less expensive and more accessible, machines are increasingly likely to have 500 Kbytes of internal memory available for the task of data compression. There are applications, however, for which it is unreasonable to require this much memory. Examples of these applications include mass-produced special-purpose machines, modems, and software systems in which compressors/decompressors are embedded in larger application programs. Algorithm PPMC has the additional disadvantage of executing at only 2000 characters per second. In the next two sections we describe efforts to improve the practicality of the finite-context modeling concept by streamlining the representation of the model.

Algorithms DAFC and ADSM described in Section 2 employ simplified order-1 context models that require less run-time memory than algorithm PPMC and execute faster. These methods sacrifice a great deal of compression performance in achieving gains in memory requirement and execution speed. In Section 4 we describe an order-2 context modeling algorithm that requires far less memory and executes faster than ADSM while achieving better compression performance. In Section 5 we extend our work on context modeling in limited memory to context models of order 3. The algorithm we describe achieves much of the compression performance of PPMC using far less space and executing much faster.

The improvements provided by the algorithms described in Sections 4 and 5 are achieved through the use of self-organizing lists and a restrained approach to blending. Both ADSM and DAFC apply the strategy of limited blending. ADSM uses self-organizing lists as well, applying the frequency count list organizing strategy. Self-organizing lists have also been used as the basis of non-context-based data compression systems. A move-to-front scheme was independently invented by Ryabko, Horspool and Cormack, Bentley et al. and Elias [BCW90]. Each of these authors evaluates several variations on the basic idea. The paper by Bentley et al. provides a great deal of data structure detail for maintaining a move-to-front list of words efficiently[BSTW86]. Horspool and Cormack investigated a variety of list organizing strategies for use in a word-based model[HC87]. Finally, the common implementation of arithmetic coding uses frequency count organization for faster access.

4. Space-Limited Context Models of Order 2

The algorithm we describe in this section employs a blended order-2 context model. It can be implemented so as to provide compression performance that is better than that provided by *compress* and much better than that provided by algorithm ADSM, using far less space than either of these systems (10 percent as much memory as *compress*). In Section 4.1 we describe the method of blending we employ, and in Section 4.2 we provide more detail on our use of self-organizing lists. In Section 4.3 we describe the frequency distributions maintained by our algorithm, and Section 4.4 presents our escape strategy. We discuss the memory requirement of our algorithm and its execution speed in Section 4.5, and consider the use of dynamic memory to improve the memory requirement. In Section 4.6 we show that hashing is a much more effective means to this end. We present some experimental data on the performance of our order-2-and-0 method in Section 4.6. A more detailed comparison with competing algorithms is given in [L91], and a comparison with our order-3 algorithm appears in Section 5.

4.1. Blending Strategy

One of the ways in which we conserve on both memory and execution time is by blending only models of orders 2 and 0, rather than orders 2, 1, 0, and −1. Thus we refer to our model as an order-2-and-0 context model. We have experimented with order-2-and-1 and order-2-1-and-0 models. The order-2-and-1 model did not provide satisfactory compression performance and the order-2-1-and-0 model produces compression results that are very close to those of our order-2-and-0 algorithm. The order-2-and-0 model allows faster encoding and decoding since it consults at most two contexts per character. We provide more details on the models of orders 2 and 0 and how they are blended in Section 4.2.

4.2. Self-Organizing Lists

In our order-2-and-0 model, we maintain a self-organizing list of size s for each two-character context (s is a parameter of the algorithm). We encode z when it occurs in context xy by *event* k if z is in position k of list xy. When z does not appear on list xy we encode z itself using the order-0 model. Encoding entails mapping the event (k or z) to a frequency and employing an arithmetic coder. To complete the description of the algorithm, we need to specify a list organizing strategy and the method of maintaining frequencies. The frequency count list organizing strategy is inappropriate because of the large number of counts required. We employ the transpose strategy because it provides faster update than move-to-front.

When character z occurs in context xy and z appears on the context list for xy, the list is updated using the transpose strategy. If z does not appear on the xy list, it is added. If the size of list xy is less than s ($size < s$), the item currently in position $size$ moves into position $size + 1$ and z is stored in position $size$. If the list is full when z is to be added, z will replace the last item. An obvious disadvantage to fixing the size of the order-2 context lists is that the lists are likely to be too short for some contexts and too long for others. When an order-2 list (say, list xy) contains s items and a new character z occurs in context xy, we delete the bottom item (call it t) from the list and add z. Context xy no longer predicts t. This does not affect the correctness of our algorithm. When t occurs again in context xy it will be predicted by the order-0 model. The fact that encoder and decoder maintain identical models ensures correctness. In addition, the rationale behind the use of self-organizing lists is that we expect to have the s most common successors on the list at any point in time. As characteristics of the file change, successors that become common replace those that fall into disuse. The method of maintaining frequencies and using them to encode is described in Section 4.3.

4.3. Frequency Distributions

In order to conserve memory we do not use a frequency distribution for each context. Instead, we maintain a frequency value for each feasible event. Since there are $s + 1$ values of k (the s list positions and the escape code) and $n+1$ values for z (the n characters of the alphabet and an end-of-file character), the number of feasible events is $s+n+2$. We can maintain the frequency values either as a single distribution or as two distributions, an order-2 distribution to which list positions are mapped and an order-0 distribution to which characters are mapped. Our experiments indicate that the two-distribution model is slightly superior. When z occurs in context xy we use the two frequency distributions in the following way: if list xy exists and z occupies position k, we encode k using the order-2 distribution. If list xy exists but does not contain z, we encode an escape code (using the order-2 distribution) as a signal to the decoder that an order-0 prediction (and the order-0 frequency distribution) is to be used, and then encode the character z. When list xy has not been created yet, the decoder knows this and no escape code is necessary; we simply encode z using the order-0 distribution. Our limited use of frequency distributions is similar to that of algorithm ADSM.

4.4. Escape Strategy

We adopt the strategy of treating the escape event as if it were an additional list position. Given this decision, there are two reasonable choices for the value of *escape*. One choice is to use the value $s + 1$, as it will never represent a list position. The second choice is to use the value $size + 1$, where $size$ is the current size of list xy (and ranges from 1 to s). In the first case, the escape code is the same for every context and all of the counts for *escape* accrue to a single frequency value while in the second case, the value of *escape* depends on the context and generates counts that accrue to multiple frequency values. The two escape strategies produce similar compression results. The algorithm we describe here uses the first alternative.

We apply update exclusion in dealing with both lists and frequency distributions. That is, a list or frequency distribution is updated when it is used. Thus, when list xy exists, both the list and the frequency distribution are updated after being used to encode either a list position or an escape. The order-0 distribution is used and updated each time context xy fails to predict.

4.5. Memory Requirement and Execution Speed

The data stored for our method is a function of the list size s and the alphabet size n. When the self-organizing lists are implemented as arrays, the total memory requirement of our method is $O(n^2s)$. With an s value as low as 2, our method is faster than ADSM and provides better compression with less storage required. Based on empirical data, $s = 7$ provides the best average compression over a suite of test files. With $s = 7$ we use approximately three times as much memory as ADSM but achieve compression that is 20 percent better on average (3.16 bits per character as opposed to 3.94) and execute faster. Our method also provides better compression than *compress* (approximately 15 percent better with $s = 7$) using essentially the same memory requirement for $n = 256$ and far less for $n = 128$.

Using dynamic memory allocation to implement the self-organizing lists results in a much more efficient use of space. We allocate an array of n^2 pointers to potential lists, and allocate space for list xy only if xy occurs in the text being compressed. The memory requirement becomes $O(n^2 + us)$, where u represents the number of distinct character pairs occurring in the text. In our suite of test files, the maximum value of u was 4721. This value was encountered in a 0.69 megabyte file of messages extracted from the bulletin board *comp.windows.x*. Even in this worst case, the dynamic-memory version of the order-2-and-0 algorithm results in a 95 Kbyte space savings over ADSM (when both methods use $k = 256$ and with $s = 7$, our space requirement is approximately half that of ADSM). The compression performance is, of course, the same as that provided by an array-based implementation.

The dynamic-memory implementation is slightly slower than the static version due to overhead incurred by dynamic allocation, but this algorithm is still faster than Abrahamson's implementation of ADSM. C-language versions of ADSM and our dynamic-memory implementation compress approximately 1900 and 3000 characters per second, respectively. We estimate that if our implementation were optimized, its speed would be competitive with that of algorithm FG. The execution time of our algorithm is determined by the size of the input file, the size of the output file, and the lengths of the self-organizing lists. For each input character we consult and update one or both models, and use and update the corresponding frequency distribution(s). The time contribution due to the order-2 model consists of the time required to search for the current character on an order-2 list and the time required to update the order-2 list and corresponding frequency distribution. The time to search and update the model is limited by the current size of the self-organizing list (which is in turn bounded by the maximum list size, s). We expect the frequently-occurring characters to be near the front of the context lists, so that the average time spent in manipulating the order-2 model should be much less than the maximum list length, s. When the order-2 model does not supply a prediction, the order-0 model must be consulted. Consulting the order-0 model requires very little time since the order-0 model is stored in frequency-count order. Updating the order-0 model involves maintaining frequency-count order, so that a single update could require n operations. However, the frequency-count strategy maintains popular characters near the front of the list so that the average cost is again much less than the maximum.

When an order-2 list contains fewer than s items, we are subject to the criticism that we are not putting our memory resources where we need them. In fact, fixing the number of successors represents a tradeoff of the ability to predict any character against the ability to predict quickly while using a reasonable amount of space. Fixing the number of successors suggests the use of an array data structure rather than a linked structure; thus we avoid the space required for links and the time involved in creating and updating linked nodes. The links in a linked structure may also be viewed as consuming memory without directly representing information needed for prediction. Another disadvantage of the linked structure is that it is more difficult to control its growth. In algorithm PPMC, the tree is simply allowed to grow until it reaches a limit and then is discarded and rebuilt. Rebuilding can result in loss of prediction accuracy. When rebuilding takes place, all of the information constructed from the prefix of the file is lost. By contrast, our model loses only the ability to predict certain successors in certain contexts, and only when they have ceased to occur frequently. Finally, we must keep in mind that a dynamic data compression system attempts to "hit a moving target". When characteristics of the file being compressed change, it may be advantageous to lose some of the data collected in compressing the early part of the file. Unfortunately, we can only make an intelligent guess at what information to collect and when to discard it.

4.6. Using Hashing to Improve Memory Requirement

We have described an algorithm that allocates n^2 self-organizing lists of size s and another that uses dynamic memory to allocate lists of size s only when they are needed. The second algorithm, however, statically allocates n^2 pointers, one for each of the n^2 possible contexts. In this section we describe an order-2-and-0 strategy that uses hashing rather than dynamic memory. This algorithm employs a hash table into which all n^2 contexts are hashed. Each hash table entry is a self-organizing list of size s. An implementation of this strategy provides better average compression than the earlier methods and requires much less memory.

Encoding and decoding proceed as in the earlier algorithms. When z occurs in context xy and no xy list exists we encode z using the order-0 frequency distribution. When an xy list exists but does not contain z, we emit an escape code and then code z using the order-0 distribution. When z is contained on the list for xy we code its position. An obvious disadvantage of the use of hashing is the possibility of collision. If two or more contexts (say xy and ab) hash to the same table position, the lists for these contexts are coalesced into a single self-organizing list used to represent both contexts. We can view this as xy's successors vying with those of ab for position on the list. Intuitively, it would seem that our predictions are more accurate when xy and ab are represented by separate lists. However, we repeat our admonitions on the unreliability of intuition in compressing text. It is possible that when we expect it least the characteristics of our file change and our good statistics become bad. Thus, we can be optimistic and hope that coalescing two lists will a) happen infrequently, b) not degrade performance, or c) will actually improve performance.

Hash conflicts have no impact on the correctness of the approach; they may, however, impact compression performance. We mitigate the negative effects of hashing in three ways. First, we select the hash function so as to minimize the occurrence of collisions. Second, we use double hashing to resolve collisions. In order to resolve collisions, we must be able to detect them. We detect collisions by storing with the self-organizing list an indication of the context to which it corresponds. When context xy hashes to position h but the check value at position h does not correspond to context xy, we know that we have collision. In order to maintain reasonable running time we perform only a small number of probes. If the short probe sequence does not resolve the hash conflict, we allow the two lists to coalesce.

The third way in which we minimize the negative effects of hashing is to use some of the space gained by eliminating n^2 pointers to provide $m > 1$ order-2 frequency distributions. The value of m is is significantly smaller than the size of the hash table (H) so that we are coalescing H/m lists into each frequency distribution. Thus the cost is less than that of providing a frequency distribution for each context while compression results are better than those achieved when we use a single frequency distribution for all lists.

The disadvantages of coalescing frequency distributions are the same as those of coalescing lists except that coalescing lists is likely to cause loss of the ability to predict some characters (since two contexts now have s list positions between them instead of s each), and limiting the number of frequency distributions does not cause this problem. Because the number of frequency distributions is very small relative to the number of contexts of order 2 we consider hash collisions to be inevitable and do not attempt to resolve them.

An implementation of the hash-based algorithm with $H = 4800, m = 70$, $s = 7$, and $n = 256$ provides approximately 6 percent more compression than the order-2-and-0 algorithm described above and uses only 45 Kbytes of memory (less than half of the requirement of the dynamic-memory method). The use of hashing provides improved compression performance overall.

5. Space-Limited Context Models of Order 3

In this section we extend our work on context modeling in limited memory to context models of order 3†. The use of hashing to store context information permits the extension of the strategy developed in Section 3 to blended models of arbitrary order. The primary problem in designing an order-3 algorithm with modest memory requirements is that of deciding which lower-order models to blend with the order-3 model. We concentrate our discussion on the blended order-3 context model that gives the best overall results. Our algorithm has a much more modest memory requirement than competing algorithms FG and PPMC and provides compression performance that is superior on average to that provided by FG. In addition, it runs much faster than PPMC. When tuned, we expect encode speed comparable to that of the faster algorithm FG. In Section 5.1 we discuss the method of blending we employ, and in Section 5.2 the data structures used. Section 5.3 details the way in which the predictions

† A preliminary version of these results was presented at the 1991 Data Compression Conference[LH91]

supplied by our model are coded, and in Section 5.4 we discuss the memory requirements and execution speed of our order-3 algorithm. Section 5.5 contains experimental results comparing our order-3 method with PPMC and FG.

5.1. Blending Strategy

The best algorithm in our family is based on an order-3-1-and-0 context model. That is, we construct a prediction for the character being encoded by blending predictions based on the previous three characters, the previous character, and unconditioned character counts. We considered order-3-and-0 models and order-3-2-and-0 models as well as the order-3-1-and-0 approach that we describe here. The addition of order-2 context information to the order-3-and-0 model generally did not improve compression performance, while the addition of contexts of order 1 does provide significantly better results. Eliminating some of the models of lower order contributes to both the decreased memory requirement and increased speed of our methods. Thus we limited the total number of contexts to be blended to three, and did not consider models that blended orders 3, 2, 1, and 0, for example. In Section 5.2 we describe the way in which we store context information.

5.2. Data Structures

We use self-organizing lists to maintain the order-3 and order-1 context information. As in the order-2-and-0 model, we employ the transpose list organizing strategy. The order-3 context information is stored in two hash tables, table $H3$ of size $h3$ whose elements are self-organizing lists of size $s3$, and table $F3$ containing $f3$ frequency distributions. Thus, each trigram (i.e., context of order 3) appearing in the file being compressed is mapped to a position in table $H3$, where a list of $s3$ successor characters is stored. A second hash function maps the trigram to a position in table $F3$ that stores the frequency distribution corresponding to the $s3$ successor characters. Since there are only n order-1 lists (where n is the size of the alphabet), hashing is not used to store the self-organizing lists of order 1. We maintain a list of $s1$

successors for each single character (context of order 1). However, we maintain just $f1$ (where $f1 < n$) frequency distributions for the collection of order-1 lists. Thus while the order-3 model is a essentially a two-level hashing scheme, where a context hashes first to a position in the table of self-organizing lists and then to a smaller table of frequency distributions, the order-1 model employs just one level of hashing, mapping the n contexts to $f1$ frequency distributions. The order-0 data consists of frequency data for the n symbols of our alphabet.

5.3. Coding the Model

The models of order 3, 1, and 0 are used to form a prediction of the current character in much the same way as we used them in the order-2-and-0 algorithm. We encode character z occurring in context wxy by event k if z occurs in position k of the list for context wxy. If z does not appear on wxy's list, we code an escape and consult the list for the order-1 context y. An order-3 frequency distribution is used to code either k or *escape*. When the order-1 model is consulted, an order-1 frequency distribution is used to code either j (if z occurs in position j of list y) or *escape*. When neither context wxy nor context y predicts z we follow the two escape codes with an order-0 prediction (i.e., we code the character itself). If the list for context wxy (likewise context y) is empty, the corresponding escape code is not necessary.

The escape codes are represented as list positions $s3 + 1$ and $s1 + 1$, respectively. As in our order-2-and-0 algorithm we apply update exclusion so that lists and frequency distributions are updated only when they contribute to the prediction of the current character, z. If list wxy exists, we update it using the transpose heuristic. If no wxy list exists one will be created. If context wxy does not predict z, then the y list is updated using the transpose method. If list y is not used in the prediction, it is not updated. When list wxy exists, the wxy frequency distribution is updated after it is used to encode either a list position or an escape. When context wxy does not predict and list y exists, the y frequency distribution is updated. The order-0 frequency distribution is updated whenever the character itself is coded.

5.4. Memory Requirement and Execution Speed

Our order-3-1-and-0 algorithm is in fact a family of algorithms where each algorithm in the family corresponds to a different set of values for the parameters $s3$, $h3$, $f3$, $s1$, and $f1$. The space requirements, speed, and compression performance of a particular algorithm depend on the values of these parameters. We report results in Section 5.5 for an algorithm that executes in 100 Kbytes of memory and encodes and decodes approximately 2800 cps. Bell et al. report compression speeds for competing algorithms running on a 1-MIP VAX 11/780 [BCW90]. In order to provide a meaningful comparison of running times, we execute on our research machine the order-3-1-and-0 algorithm and the version of *compress* used by Bell et al. Using the execution time of *compress* as a baseline, we adjust the running time of our algorithm to reflect the difference in machines. While this approach is obviously imperfect, it provides a reasonable basis for comparison. Our programs are part of a research testbed and have not been optimized for speed. We believe that with some attention to optimization they can be tuned to compress at approximately the same rate as algorithm FG.

5.5. Experimental Results

We compare the performance of our order-3-1-and-0 model to that of *compress* and the 45-Kbyte method of Section 3 on the corpus used by Bell et al. to measure the performance of a collection of data compression methods [BCW90]. The files represent a variety of sizes and types: **obj1** and **obj2** are executable files for two different machines, **geo** is a file of 32-bit numbers representing seismic data, **pic** is a bit map of a black and white facsimile picture. The remaining files are ASCII files of various types including program source files (the **prog** files). In Table 1 we display compression ratios for the order-2-and-0 model, the order-3-1-and-0 models, and *compress*. The order-3-1-and-0 model used here has parameter settings: $s3 = 3$, $h3 = 12000$, $f3 = 900$,

File	Order 2-0	Order 3-1-0	Unix Compress
bib	3.27	2.49	3.35
book1	3.39	3.03	3.46
book2	3.23	2.69	3.28
geo	5.18	5.14	6.08
news	3.68	3.22	3.86
obj1	4.28	4.10	5.23
obj2	3.55	3.33	4.17
paper1	3.34	2.87	3.77
paper2	3.27	2.85	3.52
pic	0.91	0.90	0.97
progc	3.24	2.87	3.87
progl	2.58	2.06	3.03
progp	2.63	2.12	3.11
trans	2.71	2.02	3.27
Averages	3.23	2.84	3.64
Memory (Kbytes)	45	186	500

Table 1

Comparison of order-3-1-0, order-2-and-0, and *compress*

$s1 = 20$, $f1 = 256$ and uses less than 100 Kbytes of internal memory. The performance of the order-3-1-and-0 model is significantly better than that of the order-2-and-0 method which, in turn, provides significant gains over the state-of-the-art *compress*. The order-3-1-and-0 algorithm reduces a file to an average of 35 percent of its original size, while the order-2-and-0 method reduces

File	Original Size	Order 3-1-0	FG	PPMC
bib	111261	2.49	2.90	2.11
book1	768771	3.03	3.62	2.48
book2	610856	2.69	3.05	2.26
geo	102400	5.14	5.70	4.78
news	377109	3.22	3.44	2.65
obj1	21504	4.10	4.03	3.76
obj2	246814	3.33	2.96	2.69
paper1	53161	2.87	3.03	2.48
paper2	82199	2.85	3.16	2.45
pic	513216	0.90	0.87	1.09
progc	39611	2.87	2.89	2.49
progl	71646	2.06	1.97	1.90
progp	49379	2.12	1.90	1.84
trans	93965	2.02	1.76	1.77
Averages		2.84	2.95	2.48
Memory (Kbytes)		100	186	500

Table 2

Comparison of order-3-1-0, FG, and PPMC

files to 39 percent of original size on average, and *compress* leaves 45 percent of the original size.

We compare the performance of our order-3-1-and-0 model to the performance of algorithms FG and PPMC in Table 2. The data for FG and PPMC is taken from [BCW90]. The compression performance of our method is superior on average to that provided by algorithm FG. The order-3-1-and-0 model requires only 20 percent as much internal memory as algorithm PPMC and half as much as FG. Without tuning, the speed of the order-3-1-and-0 algorithm is superior to that of PPMC, and we expect to achieve speed comparable to that of FG when we take advantage of optimization techniques such as the use of registers and incorporating assembly-language code.

6. The Future of Context Modeling

Context modeling is a relatively new and very promising method for data compression. Early context modeling algorithms require large amounts of runtime memory and execute slowly. Algorithm PPMC, for example, provides excellent compression when 500 Kbytes of memory are available and speeds of 2000 cps are adequate. Our work provides an alternative to PPMC for applications in which 500 Kbytes of internal memory is not a reasonable requirement. Our order-3-1-and-0 method achieves much of the compression performance of PPMC without the large memory requirement (in fact it requires only one-fifth as much run-time memory). Our method has the additional advantage of executing much faster. We are able to achieve compression factors of less than 2.4 bits per character for source code files and less than 2.8 bits per character for a large variety of file types using less than 100 Kbytes of internal memory. The order-2-and-0 algorithm of Section 3 achieves respectable compression using only 48 Kbytes of internal memory. The compression performance of this method is superior to that of *compress* and uses less than 10 percent as much memory.

The work we describe in Sections 3 and 4 is applicable to context models of any order. The use of self-organizing lists and hashing provides a means of representing context models of any order in any available amount of memory. The restriction on internal memory and the values of the parameters (list sizes, hash table sizes, blending method, etc.) must be carefully balanced so as to achieve satisfactory performance in terms of compression ratio and execution speed. We have conducted limited experiments in order-4 context modeling. We have not yet identified a combination of parameters that provides performance that is consistently superior to that of our order-3-1-and-0 algorithm, given approximately the same restrictions on the use of internal memory. It is possible that with a different selection of parameter values an order-4 model may provide improved performance.

Finite-state modeling is an extension of finite-context modeling that permits exploitation of characteristics of the input that cannot be represented in finite-context models. For instance, finite-state models can represent information such as "every fourth character in the file is a zero" or "every sequence of a's has even length". Horspool and Cormack describe an adaptive finite-state model DMC (for dynamic Markov compression) [HC86, CH87]. Due to the way in which states are added to the model, however, DMC does not attain the potential power of finite-state models. Bell and Moffat show that the DMC model is equivalent to a finite-context model [BM89].

Thus, the increased power of the finite-state model is attractive but, to date, successful use of this increased potential has eluded researchers. Developing methods of constructing finite-state models dynamically is an open problem the solution to which has considerable value. It is likely that straightforward methods of representing finite-state models, when they are developed, will consume large amounts of memory like the early context models. Representing finite-state models in limited memory is another challenging open problem.

References

[A89] Abrahamson, D. M. An adaptive dependency source model for data compression. *Commun. ACM 32*, 1 (Jan., 1989), 77–83.

[BCW90] Bell, T., Cleary, J. G., and Witten, I. H. *Text Compression*, Prentice-Hall, Englewood Cliffs, N.J., 1990.

[BM89] Bell, T. and Moffat, A. A note on the DMC data compression scheme. *Comput. J. 32*, 1 (Feb., 1989), 16–20.

[BSTW86] Bentley, J. L., Sleator, D. D., Tarjan, R. E., and Wei, V. K. A locally adaptive data compression scheme. *Commun. ACM 29*, 4 (Apr., 1986), 320–330.

[CH87] Cormack, G. V. and Horspool, R. N. S. Data compression using dynamic Markov modeling. *Comput. J. 30*, 6 (Dec., 1987), 541–550.

[FG89] Fiala, E. R. and Greene, D. H. Data compression with finite windows. *Commun. ACM 32*, 4 (Apr., 1989), 490–505.

[HC86] Horspool, R. N. and Cormack, G. V. Dynamic Markov modelling — a prediction technique. *Proc. International Conference on the System Sciences*, Honolulu, HA. (Jan., 1986).

[HC87] Horspool, R. N. and Cormack, G. V. A locally adaptive data compression scheme. *Commun. ACM 16*, 2 (Sept., 1987), 792–794.

[H52] Huffman, D. A. A method for the construction of minimum-redundancy codes. *Proc. IRE 40*, 9 (Sept., 1952), 1098–1101.

[LR82] Langdon, G. G. and Rissanen, J. J. A simple general binary source code. *IEEE Trans. Inf. Theory 28*, 5 (Sept., 1982), 800–803.

[LR83] Langdon, G. G. and Rissanen, J. J. A double-adaptive file compression algorithm. *IEEE Trans. Comm. 31*, 11 (Nov., 1983), 1253–1255.

[L91] Lelewer, D. A. *Data Compression on Machines with Limited Memory*, Ph. D. dissertation, Department of Information and Computer Science, University of California, Irvine, 1991.

[LH91] LELEWER, D. A. AND HIRSCHBERG, D. S. Streamlining Context Models for Data Compression. *Proc. Data Compression Conference*, Snowbird, Utah (Apr., 1991), 313–322.

[Mo89] MOFFAT, A. Word-based text compression. *Software - Practice and Experience 19*, 2 (Feb., 1989), 185–198.

[RL81] RISSANEN, J. J. AND LANGDON, G. G. Universal modeling and coding. *IEEE Trans. Inf. Theory 27*, 1 (Jan., 1981), 12–23.

[R82] ROBERTS, M. G. *Local Order Estimating Markovian Analysis for Noiseless Source Coding and Authorship Identification*, Ph. D. dissertation, Computer Science Dept., Stanford Univ.,, Stanford, 1982.

[St88] STORER, J. A. *Data Compression: Methods and Theory*, Computer Science Press, Rockville, Md., 1988.

[W88] WILLIAMS, R. N. Dynamic-history predictive compression. *Information Systems 13*, 1 (Jan., 1988), 129–140.

[W91A] WILLIAMS, R. N. *Adaptive Data Compression*, Kluwer Academic Publishers, Boston, 1991.

[W91B] WILLIAMS, R. N. An extremely fast Ziv-Lempel data compression algorithm. *Proc. Data Compression Conference*, Snowbird, Utah (Apr., 1991), 362–371.

[WNC87] WITTEN, I. H., NEAL, R. M., AND CLEARY, J. G. Arithmetic coding for data compression. *Commun. ACM 30*, 6 (June, 1987), 520–540.

6

Ziv-Lempel Compressors with Deferred-Innovation

Martin Cohn
Computer Science Department
Brandeis University, Waltham, MA 02254
marty@cs.brandeis.edu

I. Introduction

The noiseless data-compression algorithms introduced by Ziv and Lempel [ZL77,78] parse an input data string into successive substrings, each consisting of two parts: The *citation*, namely the longest prefix that has appeared earlier in the input, and the *innovation*, the symbol immediately following the citation. Thus the citation has appeared earlier, but was not then followed by the innovation symbol. In "extremal" versions of the LZ algorithm the citation may have begun anywhere in the input; in "incremental" versions it must have begun a previously parsed substring. Originally the citation and the innovation were encoded, individually or jointly, into an output word to be transmitted or stored. Subsequently, several authors [MW85, ZL78, SS82, W84] speculated that the cost of this encoding might be excessive because the coded innovation contributes roughly $\lg(\alpha)$ bits, where α is the size of the input alphabet, regardless of the compressibility of the source. To remedy the possible excess, these authors suggested storing the parsed substring as usual, but encoding for output only the citation, deferring the encoding of the innovation as the first symbol of the next parsed substring. Thus the innovation might participate in whatever compression that substring enjoyed. We call this strategy *deferred innovation.* It is exemplified in the algorithm described by Welch [W84] and implemented in UNIX `compress` and its progeny.

This work was supported by the Unisys Corporate Technology Center, Salt Lake City, Utah, and by NSF Grant NCR-8917489

While `compress` and other deferred-innovation compressors achieve respectable performance on highly compressible data (say two-to-one or better), their compression is disappointing on relatively incompressible data. In the extreme of total incompressibility, such as previously compressed or encrypted data, `compress` frequently expands the input by about 45% when the output word size is 12 bits and by about 90% when the output word size is 16, to mention two common options. These figures stand in contrast to LZ realizations without deferred innovation, where random data are expanded by about 5% for output words of 12 or more bits. The purpose of this paper is to explain the expansion caused by deferred innovation.

II. Compression with Deferred Innovation

A. Novel Pairs

Suppose a deferred-innovation LZ algorithm operates on a string of b-bit input characters producing B-bit output words,* and assume that, as in most implementations, the dictionary of citations is initiallized with all individual symbols of the input alphabet. For an input string "x y x z . . ." such an algorithm will output B bits for the first "x", and store "xy"; then output B bits for "y", and store "yx"; then output B bits for "x", and store "xz"; and so on. In general, B bits will be output for every single symbol that initiates a *novel pair*, an ordered pair not encountered earlier. We will show that if the input length is much less than the square of the alphabet size, the typical string of length N has almost N novel pairs; therefore the output length must be almost NB, and the compression ratio almost B/b. Now, when the input is a string over the alphabet of 256 bytes, the input length would have to be comparable to 2^{16} to avoid this condition; otherwise the compression ratio will likely be close to $12/8 = 1.5$ or $16/8 = 2.0$ for common choices of B. This is just the behavior mentioned above

* To begin with, we make the simplifying assumption of fixed-length outputs, relaxing it later to include progressive output lengths.

for the program compress. The "typical" string is generated by a uniform, independent source over the alphabet, or just selected uniformly from among the α^N possible strings of length N.

B. Counting Arguments

To compute the average number of novel pairs in typical sequences of length N over an alphabet of size α, we can consider any sequence to be a sequence of $N-1$ overlapping symbols from a higher-order alphabet of size α^2. Over the higher-order alphabet, the symbols are still drawn uniformly, but no longer independently. A novel pair in the original alphabet corresponds to a novel singleton in the higher-order alphabet. The number of novel singletons in a sequence is just the number of distinct symbols, because each symbol is novel at (and only at) its first occurrence. We compute the *expected value* of the number of novel pairs by a technique well known in the context of the "birthday paradox."

Lemma: The expected number of novel pairs in a sequence of length N over an alphabet of size α is

$$\alpha^2\left(1-\left(\frac{\alpha^2-1}{\alpha^2}\right)^{N-1}\right) \doteq \alpha^2\left(1-e^{-(N-1)/\alpha^2}\right)$$

Proof: Define the random binary variable b_i to be 1 if the ith pair appears in the sequence, and 0 if not. The probability that a given pair never appears in $N-1$ trials is $(\frac{\alpha^2-1}{\alpha^2})^{N-1}$, so the expected value of b_i is $\left(1-(\frac{\alpha^2-1}{\alpha^2})^{N-1}\right)$. The expected number of novel pairs is $\Sigma_i b_i$, proving the lemma. For the reasonably large α's that interest us, the exponential approximation is valid. An analogous lemma, with square replaced by $k-$th power and $N-1$ replaced by $N-k+1$, describes the expected number of $k-$tuples for any k.

As a consequence, if a deferred-innovation compressor outputs a fixed-length, B-bit token for each novel pair, the compression ratio will remain close to B/b so long as the input length is less than the *square* of the alphabet size. For $\alpha = 256$, this means files on the order of 64Kbytes. Alternatively, if the output-token size grows progressively from $\lg \alpha$ bits to B bits, and remains fixed thereafter, we need to compute the sum

$$\sum_{i=1}^{N} \lceil \lg(N+\alpha) \rceil \approx N \lg(N+\alpha) + \frac{\alpha}{N} \lg(\frac{N+\alpha}{\alpha}), \quad N < 2^B - \alpha.$$

So the average compression ratio is approximately:

$$\frac{1}{b}\left(\lg(N+\alpha) + \frac{\alpha}{N} \lg(\frac{N+\alpha}{\alpha})\right), \quad N < 2^B - \alpha, N < \alpha^2.$$

The correction for $N \geq 2^B$ is obvious.

III. Distribution of Memory Contents

We next consider the distribution of citation lengths, that is the proportions of pairs, triples, and higher-order tuples in the Ziv-Lempel compressor memory during two regimes: when the memory has just filled; and while the memory is in equilibrium after filling. Our assumption remains that input symbols are selected uniformly and independently over a finite alphabet of size α. Another assumption must now be made, regarding possible deletions from the memory once it has filled. In practice a variety of deletion strategies have been used, notably *l.r.u.* whereby the least-recently used entry is deleted to make room for the newest insertion, and *l.f.u.* whereby the least-frequently used entry is deleted. A third assumption, to simplify the analysis, might be that the entry to be deleted is chosen uniformly and independently from among the non-singletons; in other words, that deletion is random except for alphabet symbols, which are never deleted.

Initially the compressor memory (or dictionary) contains α singletons, namely the alphabet symbols themselves. Each time a match to a singleton is found, a pair is inserted; when a pair is matched, its extension to a triple is inserted; and so on. Under the uniform, independent input assumption, it is clear that the likelihood of matching a given pair is only $1/\alpha$ times the likelihood of matching a singleton, and the likelihood of matching a triple is $1/\alpha$ that of matching a pair. Since we are interested mainly in alphabet sizes like α =32, 64, 128, 256, we will ignore the possibility of creating quadruples and higher-order tuples, or simply lump them together with the triples. Thus the memory at any time contains α singletons, β pairs, and γ triples. Let $\mu + \alpha$ be the memory size (measured in strings), so that μ locations are available for pairs, and triples (the latter type including all higher orders.) Then at the time of the tth insertion we have $\beta + \gamma = t$ for $t < \mu$, $\beta + \gamma = \mu$ for $t \geq \mu$.

B. Transient Period under L.R.U. Deletion

We are particularly interested in memory sizes that are roughly integer powers of the alphabet size. Jim Storer and I conjectured that when the memory has just filled with pairs and triples there could be interaction between insertion and deletion rule that could cause temporary instability and a loss of compression performance. In particular, if the memory size is close to α^2 then most of the earliest insertions will have been pairs, while most of the recent arrivals will be triples, the results of pairs being matched and extended. This suggests a "gradient" from least recent to most recent, shading from pairs to triples. Under such a scenario, using l.r.u. deletion, disproportionately more pairs would be deleted, increasing the proportion of triples until the inability to match pairs led to pairs being re-created and re-inserted. Although the alternation in proportions of pairs versus triples would damp out, pair re-creation phases would reduce compression temporarily owing to the mechanism discussed earlier. Likewise, for *any* memory size close to an integral power of α, oscillation might be expected. We have not, however, confirmed this transient behavior. The graphs shown in Appendix II depict simulations of deferred innovation compression, plotting the percentages of pairs found in memory during filling and just thereafter, under uniform i.i.d. inputs. The three cases depicted correspond to alphabet sizes of 2^5, 2^6, and 2^7; in each case the memory size is the square of the alphabet size, namely 2^{10}, 2^{12}, and 2^{14} respectively. When the memories have just filled, the percentages of pairs are 0.626,

0.638, and 0.633, all close approximations to $1 - e^{-1} = 0.632$. It's possible (with a bit of wishful thinking) to detect some transient periodicity just after filling, but in any case the effect is not great enough to make a significant difference in overall compression. The distinction between the transient-behavior hypothesis and the equilibrium analysis below stems from the l.r.u. deletion policy, which makes critical not just the *distribution* of pairs and non-pairs, but also their times of insertion, hence their locations in the l.r.u. queue. In the analysis to follow, deletions are chosen randomly.

C. Equilibrium State and Distribution

Next we consider the distribution of memory contents and the compression ratio at equilibrium. this time we invoke the assumption of random deletion (in contrast to the l.r.u. rule discussed in the previous section.) First we solve for an equilibrium state, that is, a stable ratio of pairs to non-pairs; then we generalize to an equilibrium distribution of ratios between pairs and triples.

C1. Equilibrium State, Random Deletion

Suppose that the memory is full, that it contains β pairs (and thus $\mu - \beta$ triples) and that a randomly chosen input pair is read. The probability that the input matches some pair in the memory is β/α^2. The probability that a pair, rather than a triple, is randomly chosen for deletion is β/μ. Since these are independent events, the four joint probabilities for the change in β are:

$$\Delta\beta = \begin{cases} (+1) \cdot \frac{\alpha^2-\beta}{\alpha^2} \cdot \frac{\mu-\beta}{\mu} & \text{gain a pair, lose a triple} \\ 0 \cdot \frac{\alpha^2-\beta}{\alpha^2} \cdot \frac{\beta}{\mu} & \text{gain a pair, lose a pair} \\ 0 \cdot \frac{\beta}{\alpha^2} \cdot \frac{\mu-\beta}{\mu} & \text{gain a triple, lose a triple} \\ (-1) \cdot \frac{\beta}{\alpha^2} \cdot \frac{\beta}{\mu} & \text{gain a triple, lose a pair} \end{cases}$$

At equilibrium the first and last probabilities must be equal, and we can solve for β: (Recall that we are ignoring the creation of quadruples or high-orders.)

$$\beta = \frac{\alpha^2\mu}{\alpha^2+\mu}; \qquad \gamma = \mu - \alpha = \frac{\mu^2}{\alpha^2+\mu}.$$

Using these values we can estimate the compression ration achieved in this equilibrium state by `compress`, which will output B bits for each $2b$ bits when a pair can be matched in memory, and will output B bits for each b bits when the pair cannot be matched. The compression ratio at equilibrium is thus

$$\rho_{eq} = \frac{(\alpha^2 - \beta)B + \beta B}{(\alpha^2 - \beta)b + 2\beta b} = \frac{\alpha^2 B}{(\alpha^2 + \beta)b} = \frac{(\alpha^2 + \mu)lg\mu}{(\alpha^2 + 2\mu)lg\alpha} + O(1/\alpha).$$

From this expression we would expect `compress` in default mode $(b = 8, B = 12)$ to yield $\rho_{eq} \doteq 1.42$, which is quite close to experience.

For $\mu = \alpha^2$ the equilibrium conditions predict that pairs and non-pairs will be equinumerous. Simulations of l.r.u. deletion, such as those graphed in Appendix II, show proportions more like 0.55 pairs to 0.45 non-pairs. This apparent discrepancy is explained by recalling that during every match some pair is examined and promoted to most-recently-used, no matter whether a pair or a non-pair gets inserted, so the least-recently-used candidate remains biased toward non-pairs.

C2. Equilibrium Distribution

Having found the equilibrium state, we next consider the equilibrium distribution governing the number of pairs present in the compressor memory. Again we assume that the probabilities of creating quadruples, quintuples, and so on are negligible, so that they can be lumped together with the triples. As before, the singletons are permanent residents.

With memory size μ consider the random variable β describing the number of pairs present. β ranges from 0, when the memory has no pairs, to $\min(\mu, \alpha^2)$ when either the memory is full of pairs, or all pairs are present. At each parse of the uniform, independent input, β may increase by 1 or decrease by 1 (except at the extremes) or stay the same, with the respective probabilities given in Section C1 above. This gives us a Markov process with transition matrix

$$T_{i,j} = \begin{cases} \frac{i^2}{\alpha^2 \mu} & \text{j = i-1,} \\ \frac{i}{\alpha^2} + \frac{i}{\mu} + -2\frac{i^2}{\alpha^2 \mu} & \text{j = i,} \\ (1 - \frac{i}{\alpha^2})(1 - \frac{i}{\mu}) & \text{j = i+1,} \\ 0 & \text{otherwise.} \end{cases}$$

Because this is a connected Markov process, it has an equilibrium distribution p which satisfies $pT = p$, or $(pT)_i = p_i$. It is actually the componentwise product of two Ehrenfest models.[F50] We show in Appendix I that

$$p_i = \binom{\alpha^2}{i}\binom{\mu}{i} \Big/ \sum_{i=0}^{min(\alpha^2,\mu)} \binom{\alpha^2}{i}\binom{\mu}{i}.$$

For a very simple example, let $\alpha^2 = 4, \quad \mu = 5$. Then

$$T = \frac{1}{20}\begin{pmatrix} 0 & 20 & 0 & 0 & 0 \\ 1 & 8 & 12 & 0 & 0 \\ 0 & 4 & 10 & 6 & 0 \\ 0 & 0 & 9 & 9 & 2 \\ 0 & 0 & 0 & 16 & 4 \end{pmatrix}$$

$$p(i) = \frac{1}{16}(1,4,6,4,1) \otimes \frac{1}{31}(1,5,10,10,5) = \frac{1}{126}(1,20,60,40,5).$$

This means that asymptotically the distribution is the product of two Gaussian distributions, whose means are relatively displaced unless $\alpha^2 = \mu$.

Acknowledgement
I am grateful for conversations during this work with Ira Gessel and Jim Storer.

References

F50: Feller, William, *An Introduction to Probability Theory and its Applications, Vol. 1* New York, John Wiley & Sons Inc. 1950

MW85: Miller, V.S, and Wegman, M.N., Variations on a theme by Lempel and Ziv. *Combinatorial Algorithms on Words,* Springer-Verlag (A. Apostolico and Z. Galil, editors) (1985) 131-140.

R58: Riordan, John *An Introduction to Combinatorial Analysis* New York, John Wiley & Sons, Inc. (1958).

SS82: Storer, J.A., Szymanski, T.G., Data compression via textual substitution, *J.ACM 29, 4,* (1982) 928-951.

W84: Welch, T.A., A technique for high-performance data compression, *IEEE Computer 17, 6* (1984) 8-19.

ZL77: Ziv, J. and Lempel, A., A universal algorithm for sequential data compression. *IEEE Trans. Inf. Theory IT-23, 3* (1977) 337-343.

ZL78: Ziv, J. and Lempel, A., Compression of individual sequences via variable-rate coding. *IEEE Trans. Inf. Theory IT-24, 5* (1978) 530-536.

AppendixI

Let $q_i = \binom{\alpha^2}{i}\binom{\mu}{i}$, so that $p_i = q_i / \sum q_i$. It suffices to show that

$$(qT)_i = q_{i-1}T_{i-1,i} + q_i T_{i,i} + q_{i+1}T_{i+1,i} = \binom{\alpha^2}{i}\binom{\mu}{i} = q_i.$$

$$\begin{aligned}
(qT)_i &= \binom{\alpha^2}{i-1}\binom{\mu}{i-1}\left(1-\frac{i-1}{\alpha^2}\right)\left(1-\frac{i-1}{\mu}\right) \\
&+ \binom{\alpha^2}{i}\binom{\mu}{i}\left[\frac{i}{\alpha^2}+\frac{i}{\mu}-\frac{2i^2}{\alpha^2\mu}\right] + \binom{\alpha^2}{i+1}\binom{\mu}{i+1}\frac{(i+1)^2}{\alpha^2\mu} \\
&= \frac{1}{\alpha^2\mu}\left[\binom{\alpha^2}{i-1}\binom{\mu}{i-1}(\alpha^2-i+1)(\mu-i+1)\right. \\
&+ \left.\binom{\alpha^2}{i}\binom{\mu}{i}(\alpha^2 i+\mu i-2i^2) + \binom{\alpha^2}{i+1}\binom{\mu}{i+1}(i+1)^2\right] \\
&= \frac{1}{\alpha^2\mu}\binom{\alpha^2}{i}\binom{\mu}{i}\left[i^2+\alpha^2 i+\mu i-2i^2+(\alpha^2-i)(\mu-i)\right] \\
&= \binom{\alpha^2}{i}\binom{\mu}{i}.
\end{aligned}$$

Q.E.D.

AppendixII Statistics at Filling for Three Alphabet/Memory Sizes:

$A = 2^5$, $M = 2^{10}$: 641 pairs, 380 triples, 3 quadruples.
$A = 2^6$, $M = 2^{12}$: 2614 pairs, 1480 triples, 2 quadruples
$A = 2^7$, $M = 2^{14}$: 10378 pairs, 5992 triples, 14 quadruples

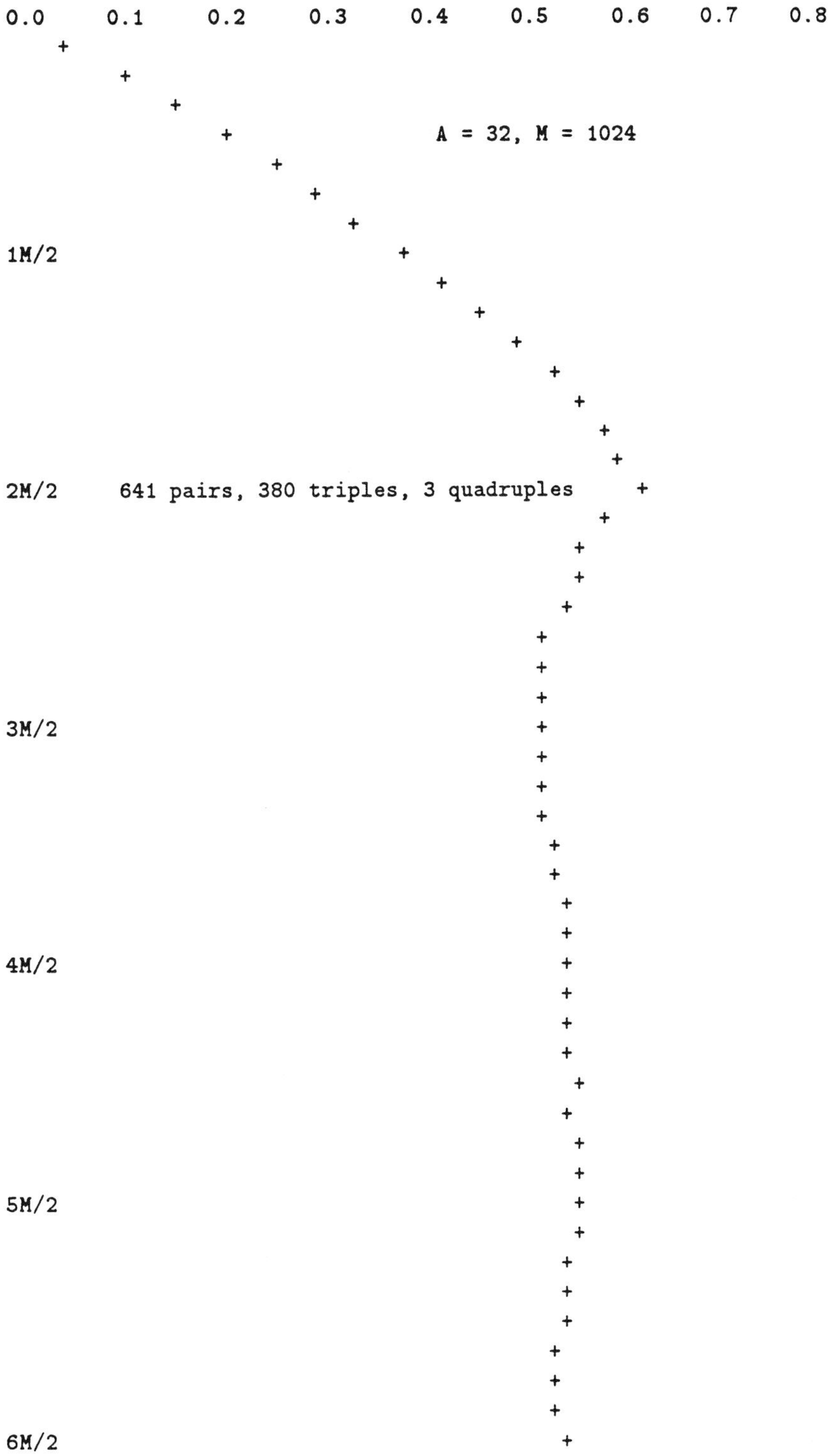

```
0.0     0.1     0.2     0.3     0.4     0.5     0.6    0.7    0.8
   +
        +
            +
                +                   A = 32, M = 1024
                    +
                       +
                          +
1M/2                          +
                                 +
                                    +
                                       +
                                          +
                                            +
                                              +
                                               +
2M/2     641 pairs, 380 triples, 3 quadruples    +
                                              +
                                            +
                                            +
                                           +
                                         +
                                         +
                                         +
3M/2                                     +
                                         +
                                         +
                                         +
                                          +
                                          +
                                           +
                                           +
4M/2                                       +
                                           +
                                           +
                                           +
                                            +
                                           +
                                            +
                                            +
5M/2                                        +
                                            +
                                           +
                                           +
                                           +
                                          +
                                          +
                                          +
6M/2                                       +
```

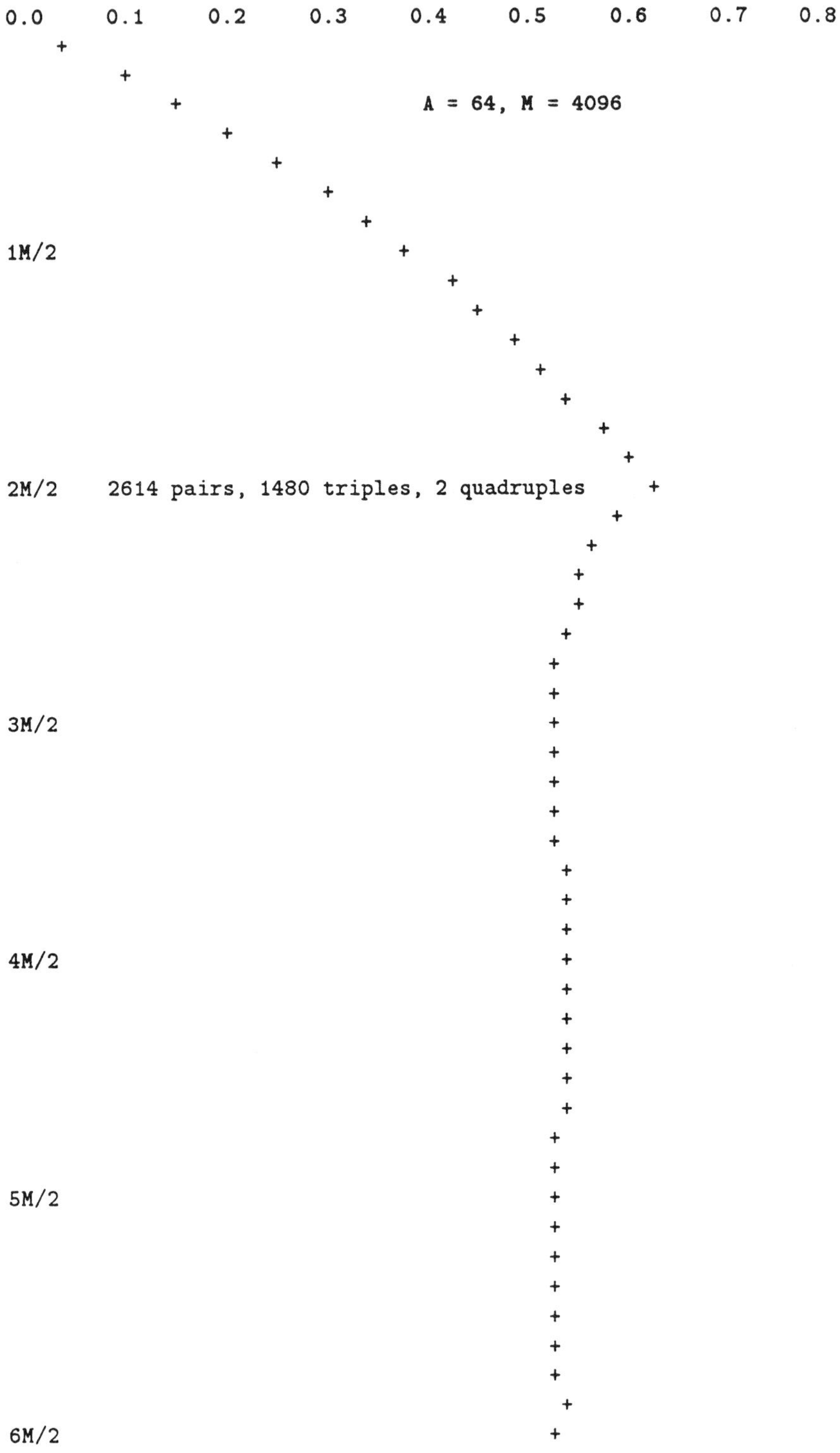

0.0 0.1 0.2 0.3 0.4 0.5 0.6 0.7 0.8

A = 64, M = 4096

1M/2

2M/2 2614 pairs, 1480 triples, 2 quadruples

3M/2

4M/2

5M/2

6M/2

```
0.0     0.1     0.2     0.3     0.4     0.5     0.6     0.7    0.8
   +
        +
            +
                +                   A = 128, M = 16384
                    +
                        +
                           +
1M/2                          +
                                 +
                                    +
                                      +
                                         +
                                           +
                                             +
                                               +
2M/2   10378 pairs, 5992 triples, 14 quadruples  +
                                               +
                                             +
                                            +
                                            +
                                           +
                                           +
                                           +
3M/2                                      +
                                          +
                                          +
                                           +
                                           +
                                           +
                                           +
                                           +
4M/2                                       +
                                           +
                                           +
                                           +
                                           +
                                           +
                                           +
                                           +
5M/2                                       +
                                           +
                                           +
                                           +
                                           +
                                           +
                                           +
                                           +
6M/2                                       +
```

7

Massively Parallel Systolic Algorithms for Real-Time Dictionary-Based Text Compression

James A. Storer
Computer Science Department
Brandeis University
Waltham, MA 02254

Abstract: *Textual substitution* is a powerful and practical method of lossless data compression, where repeated substrings are replaced by pointers into a dynamically changing dictionary of strings. They are often called *dictionary methods* or "LZ" methods after the important work of Lempel and Ziv. With many applications, high speed hardware that can perform compression or decompression in real time is essential. We present massively parallel approaches for real-time textual substitution.

1. Introduction

Lossless data compression is the process of encoding ("compressing") a body of data into a smaller body of data which can be uniquely decoded ("decompressed") back to the original data. *Textual Substitution* methods, often called *Dictionary Methods* or "LZ" methods after the important work of Lempel and Ziv [1976] and many of their subsequent papers (e.g., Ziv and Lempel 1977,1978]), maintain a constantly changing dictionary of strings to adaptively compress a stream of characters by replacing common substrings by indices into the dictionary. Here we present massively parallel *systolic pipe* architectures for textual substitution that can be the basis of fast real-time compression and decompression hardware for high bandwidth channels where data travels at millions or billions of bits per second.

In the next section we review the basic serial model of textual substitution on which our parallel implementations will be based. Section 2 reviews the *systolic pipe* model. Section 3 considers the *Static Dictionary* methods, where the dictionary is fixed in advance; this material will motivate the methods to be considered in later sections. Section 4 considers *Sliding Dictionary* methods, where the dictionary is a shifting window of the last n input characters, and overviews the work of Gonzalez and Storer [1985], and ZitoWolf [1990, 1990b]. Section 5 considers *Dynamic Dictionary* methods, where the encoder and decoder dictionaries are continually modified by update and deletion heuristics, and overviews an approach that is proposed in Storer [1988]. The discussion of the Dynamic Dictionary Method will be more detailed than for the Static and Sliding dictionary methods, and Section 7 will overview a recently completed data compression board that is introduced in Reif and Storer [1991]; it is based on a custom VLSI chip that currently runs at 160 million bits per second and appears to be extendible to over 600 million bits per second with current CMOS technology.

2. Textual Substitution Methods

We assume all *data* to be a sequence of *characters* and we refer to the set of all possible characters as the *input alphabet*, Σ. For example, in a typical application, data is a sequence of bytes and Σ is the set of integers 0 through 255.

We assume that all handling of the data while it is in compressed form is *noiseless* (no bits are added, lost, or changed). We make this assumption for convenience; a host of techniques are available in the literature for error detection and correction; see Reif and Storer [1991b] for a discussion of how to make textual substitution methods "resilient" to errors on the communication line or storage medium.

We employ an *on-line* model for data compression where the *encoder* and *decoder,* each have a fixed (finite) amount of local memory, which we refer to as the *local dictionary*, D. We assume that the two local dictionaries may be initialized at the beginning of time to be empty. What distinguishes this model from an *off-line* model is that neither the sender or the receiver can see all of the data at once; data must be constantly passing from the sender through the encoder, through the decoder, and onto the receiver. In fact, the algorithms that we shall present satisfy the requirements of the following definition:

Definition: An on-line compression /decompression algorithm is *real-time* if there exists a constant k (which does not depend on the data being processed or the size of the dictionary) such that for every k units of time, exactly one new character is read by the encoder and exactly one character is written by the decoder. The only exception to this rule is that we may allow a "lag" between the encoder and decoder. That is, there is another constant l (that is independent of the data being processed) such that during the first l units of time the decoder produces no characters and, in the case that the input is finite in length, l time units may pass between the time that the encoder has finished reading characters and the decoder is finished outputting characters.

2.1. Generic Encoding and Decoding Algorithms

With *textual substitution methods*, a local dictionary D is used to store a constantly changing set of strings. Data is compressed by replacing substrings of the input stream that also occur in D by the corresponding index into D; we refer to such indices as *pointers*. Thus, the input to an on-line data compression algorithm employing textual substitution is a sequence of characters and the output is a sequence of pointers (where typically, most of the pointers specify strings of length greater than 1 but a few are essentially character codes).

The following page presents generic encoding and decoding algorithms for on-line textual substitution. Most on-line textual substitution algorithms can be viewed as instances of one general approach. The generic encoding algorithm reads a stream of characters and writes a stream of bits, and the generic decoding algorithm receives a stream of bits and outputs a stream of characters. We use the notation D_{max} to denote the maximum number of entries that D may contain.

(1) Initialize the *local dictionary* D to have one entry for each character of the input alphabet (these characters must always be in D and can never be removed).

(2) repeat forever

(a) {Get the current match string s:}

Use a *match heuristic* MH to read s from the input.

Transmit $\lceil log_2|D|\rceil$ bits for the index of s.

(b) {Update D:}

Add each of the strings specified by an *update heuristic* UH to D (if D is full, then first employ a *deletion heuristic* DH to make space).

Generic Encoding Algorithm

(1) Initialize the local dictionary D by performing Step 1 of the encoding algorithm.

(2) repeat forever

(a) {Get the current match string s:}

Receive $\lceil log_2|D|\rceil$ bits for the index of s.

Retrieve s from D and output the characters of s.

(b) {Update D:}

Perform Step 2b of the encoding algorithm.

Generic Decoding Algorithm

We have intentionally talked about the maximum number of entries that D may contain without any reference to how long individual entries may be. This is because D can be stored in a data structure that represents each entry with a constant amount of space (that is independent of the length of the string corresponding to the entry). Practical values for D_{max} range from 2^{10} to 2^{16}.

The Generic Encoding and Decoding Algorithms work in lockstep to maintain identical copies of D (which is constantly changing). The encoder repeatedly finds a match between the incoming characters of the input stream and the dictionary by applying a *match heuristic* (MH), deletes these characters from the input stream, transmits the index of the corresponding dictionary entry, and updates the dictionary with an *update heuristic* (UH) that depends on the current contents of the dictionary and the match that was just found; if there is not enough room left in the dictionary, a *deletion heuristic* (DH) is used to delete an existing entry. Similarly, the decoder repeatedly receives an index, retrieves the corresponding dictionary entry as the current match, and then performs the same algorithm as the encoder to update its dictionary. Note that because of this "lock-step" relationship between the encoder and decoder, it is never necessary to transmit any explicit information about the dictionary from the encoder to the decoder.

Let us now consider more closely the heuristics left out of the Generic Encoding and Decoding Algorithms:

The match heuristic, MH: A function that removes from the input stream a string s that is in D. Since the characters of the input alphabet are always in D and can never be deleted, there is always at least one such s.

The update heuristic, UH: A function that takes the local dictionary D and returns a set of strings that should be added to the dictionary if space can be found for them.

The deletion heuristic, DH: A function that takes the local dictionary D and returns a string of D that may be deleted (this string may not be one of the characters of the input alphabet).

Typically, the match heuristic is simply the *greedy* heuristic; that is, always take the longest possible match between the input stream and a string in D. For some update and deletion heuristics there are optimum on-line parsing strategies that sometimes take shorter matches so that longer ones can be taken later (e.g., Wagner [1973] considers the Static Dictionary Method, Szymanski and Storer [1978,1982] consider the Sliding Dictionary Method, and Hartman and Rodeh [1985], Storer [1988], and De Agostino and Storer [1992] consider the Dynamic Dictionary Method).

For a general treatment of textual substitution techniques as they apply to both on-line and off-line algorithms, as well as further references on the subject, see the book of Storer [1988]. We now present three basic categories of update and deletion heuristics.

2.2. The static Dictionary Method

With the *Static Dictionary Method*, we augment Step 1 of the Generic Encoding and Decoding Algorithms to initialize D to contain not only the characters of the input alphabet, but also a fixed set of strings; the update and deletion heuristics do nothing. In practice, the methods discussed later in this section can out-perform the Static Dictionary Method, even on data for which the static dictionary has been specially constructed. However, we shall briefly present a systolic architecture for the Static Dictionary Method as a way of motivating other methods.

2.3. The Sliding Dictionary Method

With the *Sliding Dictionary method*, the dictionary is a window of the last n characters and pointers are $(displacement, length)$ pairs that indicate a substring of the window. The value of $displacement$ can be any integer between 2 and the length of the window. The value of $length$ can be and integer between 2 and a an agreed upon maximum; this maximum can be any fixed length up to the length of the window or an amount that is adjusted dynamically, The update and deletion heuristics are to "slide" the window (i.e., delete the entries that include the last character of the window and insert new entries that include the new character). To insure that a match of at least one character can always be found, a pointer value is reserved for each character of the alphabet.

In a simple practical implementation might use a window in the range $1,024$ to $4,096$ with a maximum displacement in the range 16 to 64, and encode pointers with a fixed number of bits for each field. However, it can pay to use "fancier" methods of encoding pointers because the distribution of pointer values (particularly the length field) tends not to be uniform. Better methods of also avoid the inefficiency of having to divide the pointer into two fields and allow long but infrequently occurring lengths to be encoded using more bits. See Fiala and Greene [1989] for further discussion of this issue.

If the window extends back to the beginning of the input string, a linear time implementation is to build a position tree as you go (McCreight [1976]) can be used to compute longest matches. When the window does not extend all the way to the beginning of the input string, three overlapping copies of the position tree can be employed (Rodeh, Pratt, and Even [1980]) or the McCreight Algorithm can be modified to delete strings (Fiala and Greene [1989]).

The Sliding Dictionary method can be viewed as a practical interpretation of Lempel and Ziv's first method (Ziv and Lempel [1978]), and was proposed in Storer and Szymanski [1978,1982].

2.4. Dynamic Dictionary Method

Choices for the update heuristic include: The *first character* (*FC*) heuristic; add the last match concatenated with the current match. The *next character* (*NC*) heuristic; add the last match concatenated with the next character of the input stream (decoding is slightly more complicated - see Miller and Wegman [1985]). The *identity* (*ID*) heuristic; add the previous match concatenated with the current match. The *All Prefixes* (*AP*) heuristic; add the set of strings consisting of the previous match concatenated with all prefixes of the current match (note that this set includes the strings added by the FC and ID heuristics).

Choices for the deletion heuristic include: The freeze (*FREEZE*) heuristic; once the dictionary is full, "freeze" it and do not allow any further entries to be added. The restart (*RESTART*) heuristic; once the dictionary is full, monitor the amount of compression being achieved and if it drops below an agreed upon percentage of what it was when the dictionary first filled up, empty the dictionary and start learning new strings. The least recently used (*LRU*) heuristic; delete the entry that has been least recently used. The *swap* (*SWAP*) heuristic works as follows: When the *primary* dictionary first becomes full, an *auxiliary* dictionary is started, but compression based on the primary dictionary is continued. From this point on, each time the auxiliary dictionary becomes full, the roles of the primary and auxiliary dictionaries are reversed, and the auxiliary dictionary is reset to be empty. Although this heuristic does not fit directly into the Generic Encoding and Decoding Algorithms, they can be modified to accommodate it.

Different combinations of choices for the above three heuristics give rise to a variety of methods, many of which have been studied extensively in the literature. The book of Storer [1988] considers trade-offs between different heuristics (in terms of both computing resources and amount of compression) and the relationship between various heuristics and work by past authors. Most authors in the past have used for MH the *greedy heuristic*. Using the traditional serial RAM model of computation, the FC and AP heuristics can be easily implemented in real time using constant space per entry by representing the dictionary with a standard trie data structure (dictionary entries correspond to nodes of the trie). The ID heuristic requires a much more complicated data structure that has some undesirable worst-case properties. However, the situation changes with a massively parallel model of computation (with some appropriate modifications to the ID heuristic). SWAP can be viewed as a discrete version of LRU. The FREEZE heuristic can be "unstable" in practice when the data characteristics change after the dictionary is full. The LRU and SWAP heuristics typically perform equivalently in practice. The serial nature of a doubly-linked list use by LRU is "natural" for a serial implementation whereas SWAP appears to be more natural for parallel implementations. The FC-FREEZE heuristic can be viewed as a practical interpretation of the second model proposed by Lempel and Ziv (Ziv and Lempel [1978]), the FC-RESTART heuristic models the "LZW" method discussed by Welch [1984], and the FC-LRU is a natural generalization FC-FREEZE and FC-RESTART. The ID heuristic was considered by Seery and Ziv [1977,1978] and the ID-LRU heuristic models the work of Miller and Wegman [1985]. The AP-LRU heuristic is discussed in Storer [1988].

3. Systolic Pipes

The model of parallel computation employed by our algorithm is a *systolic pipe*; that is, a linear array of processors, each connected only to its left and right neighbors.

A real-life example of a systolic pipe is an automobile assembly line that may produce a new car every few minutes even though each car is in the assembly line for a day. Although each station in the automobile assembly line performs a different task, the stations are at least conceptually identical, if we view them all as taking as input a partially built car, performing an elementary operation (such as welding), and then outputting a partially completed car to the next station.

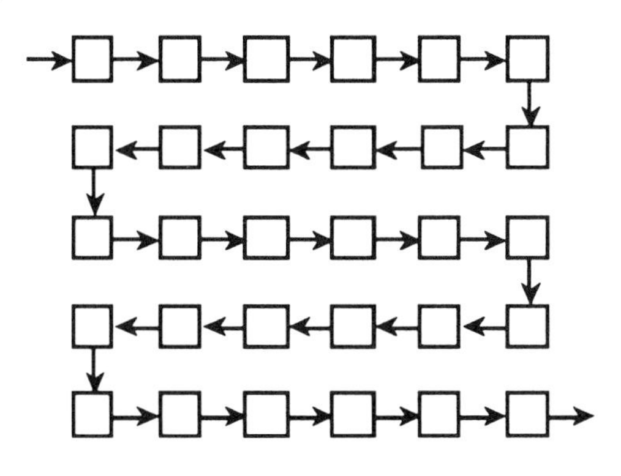

Figure 1
Systolic Pipe

From the point of view of VLSI, a systolic pipe has the following desirable properties:

- All processors are identical and the length of connections between adjacent processors can be bounded by a constant.
- The structure can be laid out in linear area and power and ground can be routed without crossing wires.
- The layout strategy can be independent of the number of chips used. A larger pipe can be obtained by placing as many processors as possible on a chip and then, using the same layout strategy, placing as many chips as possible on a board.

We have intentionally left out of our definition of a pipe the specification of what constitutes a "processor". In principle, any computational device, including a mainframe computer, could be used. However, it is typical for processors to be extremely simple. Here a processor consists of only a few registers, a comparator, and some finite-state logic.

4. Static Dictionary Method

In this section we overview a simple systolic architecture that can be used to compress or decompress with a static dictionary.

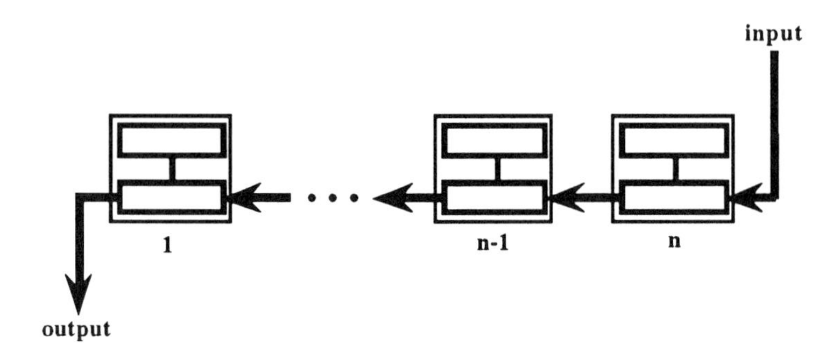

Figure 2
Architecture for the Static Dictionary Method

As depicted in Figure 2, we employ a systolic array of n processors, where each processor has an input buffer and a storage buffer, each capable of holding a maximum of k pointers. Each processor holds a dictionary entry in its storage buffer; different static dictionaries can be loaded in at start-up or a fixed static dictionary can be "hard-wired". Data enters from the right, travels along the bottom row of input buffers, and leaves to the left. The characters of the input alphabet are not explicitly represented by a processor of the pipe; rather, as a character enters the encoder pipe, it is padded with high order zeros to form an index that represents that character; similarly, the decoder strips high order zeros from indices as they leave the pipe. In compression mode, substrings of the input stream (which is a stream of indices) that match a processor's storage buffer are replaced by the processor index. In decompression mode, indices are replaced by storage register strings. In order to keep hardware costs down, the maximum length of dictionary entries must be kept relatively small (e.g., $k \leq 6$); longer entries can be represented by entries that contain indices of processors to their right.

The key observation about this architecture is that a new character can enter the encoder on every clock cycle (an a new character can be produced by the decoder on every clock cycle), and it is clearly possible to build a system where the clock is quite short, since the computation performed by a processor on a given clock cycle amounts to little more than a parallel comparison of a number of indices.

5. Sliding Dictionary Method

In this section we overview three systolic architectures for the Sliding Dictionary Method. Our approach will be to "cheat" slightly on our basic model of a systolic pipe and add extra connections. However, all of the architectures to be presented will still have linear area layouts.

Each of the three methods has its advantages depending on the application; we shall compare them at the end of this section. Three measures of performance will be of interest to us for a system that implements a window of length n with a maximum match length of m:

$TIME(n, m)$ The worst case time for the encoder to process an input character or the decoder to produce an output character.

$LAG(n, m)$ The worst case time between when a character enters the pipe and when it leaves the pipe.

$RESTART(n, m)$ The worst case time to initialize the dictionary to be empty and start processing a new data stream.

We shall also be interested in issues such as the amount of hardware required per cell, how well the implementation lends itself to packaging on more than one chip (which is likely to be necessary for large pipes), and how well the architecture lends itself to implementation on existing massively parallel computers.

5.1. The Match Tree Architecture

As depicted in Figure 3, the *Match Tree* architecture for the Sliding Dictionary Method employs a pipe of $2n$ processors with a binary *match tree* attached to the pipe, where each processor corresponds to a distinct leaf of the tree (for simplicity, we assume n to be a power of 2).

Each processor has a *data register* and a *storage register*, each capable of holding a single character; data flows along the bottom through the data registers. On each "phase" of operation, the data registers are copied into the storage registers, and then n shifts of the input stream are performed; this has the effect of passing each of the characters in data registers $n+1$ through $2n$ past each of the n characters that preceded that character in the input stream that are stored in the storage registers.

Each character traveling through the data registers "carries with it" a pair of integers $(position, length)$ indicating the starting position and length of the longest match found thus far starting with this character. After each shift of the input stream, for each data register that matches its storage register, the match tree is used to determine how far back to the right the data registers continue to match their storage registers; if this length is more that the current value of the $(position, length)$ associated with the character in this data register, then the current position and this new length become the new $(position, length)$ associated with the character in this data register.

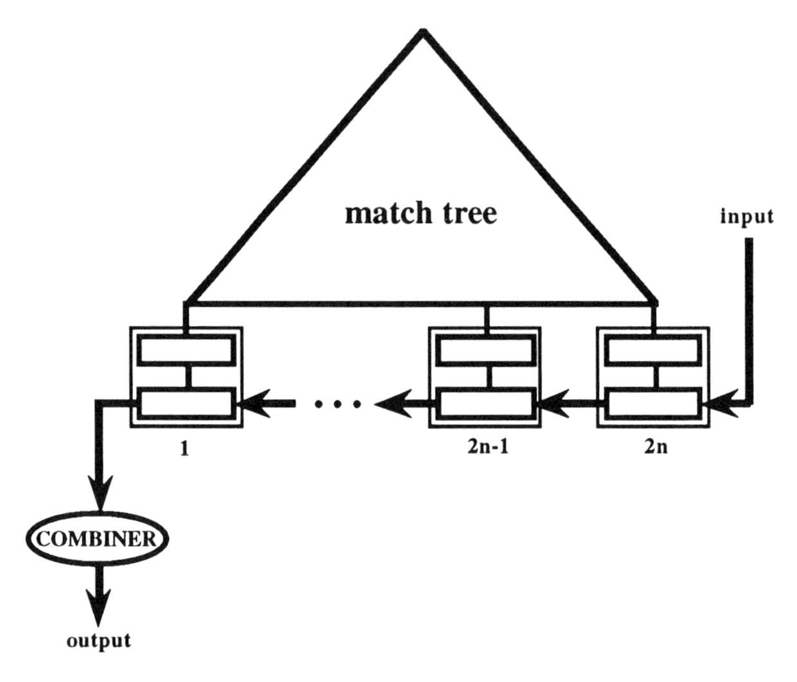

Figure 3
Match Tree Architecture
for the Sliding Dictionary Method

The tree works by placing in a leaf a left [if it matched but the processor to the left did not, a right] if it matched but the processor to its right did not, and nothing otherwise; "attached" to each bracket is the index of the processor. The brackets are passed up and then back down the tree according to the following rules:

nothing	Take no action.
[	Send [to parent.
]	Send] to parent.
][	Send][to parent.
[]	Compute length and send it to the processor that sent [.
[][	Send [to parent and handle [] as in case [].
][]	Send] to parent and handle [] as in case [].
][][	Send][to parent and handle [] as in case [].

As characters leave the encoding pipe, contiguous groups of characters that belong to the same match (the group boundaries can be determined from the (*position*, *length*) pairs attached to the characters) are replaced by the appropriate index by the special *combiner* processor.

The match tree encoding architecture has $TIME(n,m) = O(log(n))$; in fact, it is possible to modify the architecture presented above to have a collection of trees of height $log(m)$ and thus reduce this time to $TIME(n,m) = O(log(m)$. The decoder, which we do not describe here (Gonzalez and Storer [1985] or Storer [1988]), does not need a tree and has $TIME(n,m) = O(1)$. Both the encoder and decoder have $LAG(n,m) = O(n)$. A restart can be implemented by sending a special restart character ahead of the new incoming data to "flush" out the storage registers; new data can follow after $log(m)$ steps (the time needed to flush out the tree). Hence for the decoder $RESTART(n,m) = O(log(n))$ and for the decoder $RESTART(n,m) = O(1)$.

5.2. The Broadcast-Reduce Tree Architecture

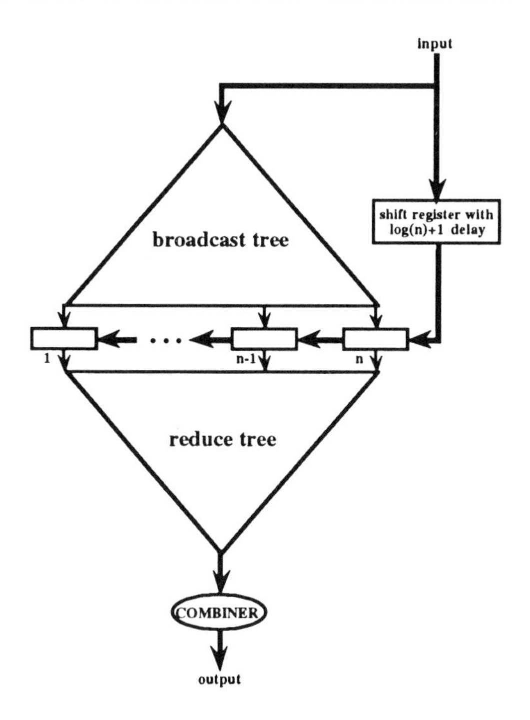

Figure 4
Broadcast-Reduce Architecture
for the Sliding Dictionary Method

As depicted in Figure 4, the *Broadcast-Reduce* architecture for the Sliding Dictionary Method employs a pipe of n processors with a binary *broadcast tree* attached to the top of the pipe and a binary *reduce tree* attached to the bottom of the pipe; each processor corresponds to a distinct leaf of these trees (for simplicity, we assume n to be a power of 2). On each step, data in the broadcast tree shifts down a level and a new input character enters the top of the broadcast tree. The input character also enters a shift register that will cause it to enter the right of the pipe one step after the copy entered into the broadcast tree makes it to all of the processors. Each character comes out of the bottom of the broadcast tree at the moment that the pipe contains the n characters preceding it in the input stream, and so it can be compared with all n preceding characters simultaneously. By a one-directional version of the bracket passing mechanism used by the match tree method, the reduce tree can send the character with a $(length, position)$ pair attached to it on to the *combiner*, which works similarly to the combiner for the match tree method. Note that unlike the match-tree method, the $(length, position)$ pair refers to a match *ending* at that character.

Both the broadcast-reduce tree encoding and decoding architecture (which we do not discuss here - see Zito-Wolf [1990]) has $TIME(n, m) = O(1)$. Both the encoder and decoder have $LAG(n, m) = O(log(n))$ and $RESTART(n, m) = O(log(n))$. Zito-Wolf[1990] discusses how to reduce the number of bits needed to pass information up the reduce tree.

5.3. The Wrap Architecture

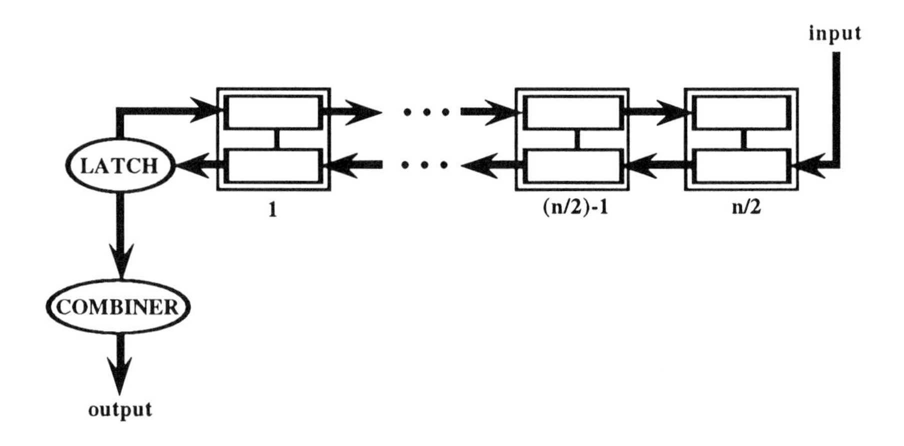

Figure 5
Wrap Architecture
for the Sliding Dictionary Method

As depicted in Figure 5, the *Wrap* architecture for the Sliding Dictionary Method employs a pipe of $n/2$ processors with bi-directional data paths (for simplicity, we assume n to be a power of 2).

Each processor has a *data register* and a *storage register*, each capable of holding a single character; data flows from right to left along the bottom through the data registers and then wraps back and flows from left to right along the top through the storage registers.

On even numbered steps, the characters in the data registers shift left, where the character in the data register of processor 1 shifts into the latch register. On odd numbered steps, the characters in the storage registers shift right, where the character in the latch register is sent both to the storage register of processor 1 and to the *combiner*. When an input character travels from processor $n/2$ to processor 1 through the data registers, the n characters that preceded it in the input stream pass by it in the storage registers. A $(position, length)$ pair can be attached to each character that is reset when the character does not match and incremented if it does. Thus, when characters enter the combiner, they are labeled by the $(position, length)$ of the longest match that *ends* with this character. Similar to the broadcast-reduce tree method, the combiner can replace characters that belong to the same match by a pointer.

The way we have described it, the Wrap encoding architecture reads an input character on every other step. By having each data register compare both with its own storage register and with the storage register of the processor to its left (or with the latch register for processor 1), the architecture can read a character at every step. However, this must be weighed against the increase complexity of a processor and the increased data path between processors.

Both the Wrap encoding and decoding architecture (which we do not discuss here - see Zito-Wolf [1990b]) has $TIME(n, m) = O(1)$. Both the encoder and decoder have $LAG(n, m) = O(n)$ and $RESTART(n, m) = O(1)$.

5.4. Comparison of Sliding Dictionary Architectures

From a theoretical point of view, the Wrap architecture is a clear winner, unless LAG is a critical issue, in which case broadcast-reduce wins. Although all three architectures can be fabricated with a chip area that grows linearly with n (see Storer [1988] for a discussion of the layout of tree-pipe structures), the Wrap Architecture is clearly simpler to lay out (in addition, trees must have long edges - see Storer [1988] for a discussion of this issue). Match Tree architecture is the loser since it uses a non-constant amount of time per character for the encoder.

In practice, however, there are other factors to consider. If the architectures are going to be simulated on an existing parallel machine, the trees may come for "free". In addition, the tree of the Match Tree architecture is not really needed for small values of m that may be encountered in practice. The match tree may also have an advantage when the array must be placed on multiple chips because delays due to pads can be more easily "latched out" when m is small and the data stream is essentially one-way.

6. Dynamic Dictionary Method

Here we describe an architecture for the Dynamic Dictionary Method based on a variant of the *ID* update heuristic, the *SWAP* deletion heuristic, and a bottom up parallel version of the greedy match heuristic. Since this architecture appears to be more practical than those discussed for the Sliding Dictionary Method and has been successfully built (as will be discussed in the last section), we will describe it in more detail.

We employ a simple systolic array where all data travels from left to right except for two bits, called *leader* and *stop*, which travel from right to left. Figure 6 depicts an individual processor, which contains 5 registers that are capable of holding a pointer and some finite state control.

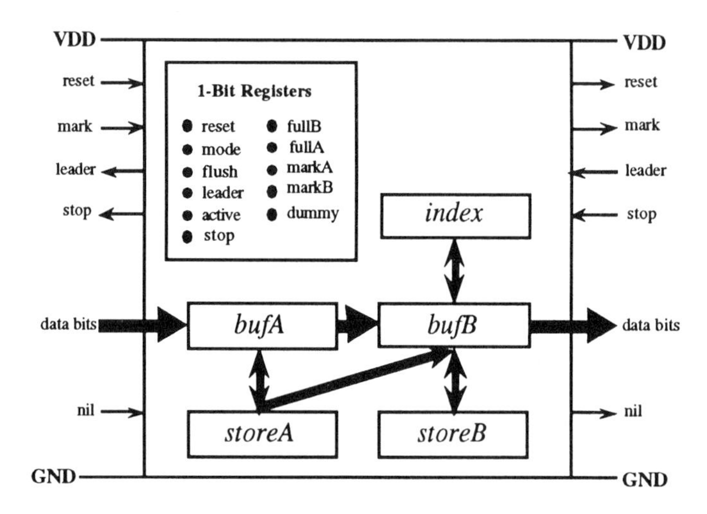

Figure 6
An Individual Processor

6.1. Encoding

We start by describing a systolic pipe implementation only as it pertains to encoding; unlike the serial case, here, decoding is the more complicated operation. All pointers are represented by the same number of bits; the bits for each input character are padded to the left with zeros to form the corresponding pointer and sent into the left end of the pipe (for example, with 8-bit input characters and a dictionary of 4,096 entries, each input character is padded with 4 leading 0's to form a 12-bit pointer). The pipe consists of $D_{max} - |\Sigma| - 1$ processors numbered $|\Sigma|$ through $D_{max} - 2$ going from left to right; there are no processors for indices 0 through $|\Sigma| - 1$ since the characters of Σ are always implicitly in the dictionary and there is no processor numbered $D_{max} - 1$ (this pointer value is used for the nilpointer, to be discussed later). The processors are numbered so that if $i < j$, then processor i is to the left of processor j (and occurs earlier in the pipe than processor j).

The pipe stores a "pair forest" representation of the dictionary (each dictionary entry is represented by a pair of pointers to two other entries or to single characters) and implements a modified version of the *ID* heuristic. Each processor is capable of holding a pair of pointers (corresponding to the left and right pointers into the pair forest), but is initialized to be empty. The *leader* bit is initially 1 for the leftmost processor and 0 for all others. It is always the case that at most one processor is the leader, that the processors to its left contain dictionary entries, and that processors to its right are empty.

As data passes through the processors to the left of the leader, it is encoded. That is, whenever a pair of pointers enter a processor and matches the pair of pointers stored at that processor, this pair of pointers is removed from the data stream and replaced by a single pointer (the index of that processor). Data passes unchanged through the processors to the right of the leader. When a pair of pointers enters the leader, they represent adjacent substrings of the original data and can be viewed as the "previous" and "current" match. The leader can simply adopt this pair of pointers as its entry, set its *leader* bit to 0, and send a signal to the processor to its right to set its *leader* bit to 1. Note that unlike the serial algorithm, this is a bottom up approach to finding a longest match, since longer matches are built from smaller ones. We do not want the new leader to adopt the same pair of pointers that was just adopted by the processor to its left. Hence, when a processor first becomes the leader, it must allow one pointer to pass through before proceeding to "learn".

When the rightmost processor becomes the leader, it adopts a pair of pointers as its entry and then sends a signal to the right to pass on the *leader* bit; this signal indicates that the dictionary is full. At this point, the *SWAP* deletion heuristic can be implemented with a switch between two copies of the pipe that routes input/output lines appropriately as the dictionaries turn over.

6.2. Decoding

Consider now the operation of the decoding pipe. Although the same pipe is used for both encoding and decoding and when decoding data still enters from the left and leaves to the right, the numbering of processors is reversed; the rightmost processor is numbered $|\Sigma|$ (processor indices are sent down the pipe by the controller into the *index* registers upon power up or upon a restart that changes between encode and decode mode). Operation of the decoding pipe is essentially the reverse of that of the encoding pipe. Processor $|\Sigma|$ is initially the leader, and the remaining processors are empty. Compressed data enters from the left, passes through the empty processors unchanged, and is decoded in the non empty portion of the pipe; that is, when a pointer arrives to a processor that is equal to the processor index, it is replaced by the two pointers stored by that processor. The leader bit propagates from right to left (the opposite direction from the encoder).

It is an inherent aspect of data compression there may be no reasonable bound on how many more decompressed bits there are than compressed bits. We address this issue with a *stop* bit that is be sent from a processor to the processor to its left to signal that no more data should be sent until the bit is turned off. A typical sequence of events for a decoding processor would be the following (each processor has a buffer capable of holding at most two pointers in its *BuffA* and *BuffB* registers):

1. The input buffer currently contains one pointer that is not the same as the processor index and the stop bit from the processor to the right is not set.
2. A clock tick occurs and the current pointer in the buffer is sent to the right and a new pointer is read into the buffer.
3. The new pointer in the buffer is the same as the processor index. This pointer is replaced by the first of the two pointers stored in the processor memory and the stop bit is set.
4. A clock tick occurs. The pointer in the buffer is transmitted to the right. Although a stop bit is now being transmitted, the processor to the left has not seen it yet and another pointer arrives. The second pointer in the processor memory is placed in the buffer before the pointer that just arrived. The buffer now contains two pointers.
5. A clock tick occurs and the first of the two pointers in the buffer is transmitted to the right (no new pointer arrives from the left since the stop bit was set on the last clock tick). The stop bit is unset.
6. A clock tick occurs, the pointer in the buffer is sent to the right, and a new pointer arrives from the left.

If the stop bit is sent from the leftmost processor to the communication line, then this must handled with an xon / xoff protocol to the sender. Note that the use of the stop bit does not require any signal propagation; on each clock tick, the *only* communication is between a processor and the processors to its immediate left and right.

6.3. A Modified Pointer Adoption Strategy

We have not yet discussed in detail how the decoder "learns" entries. The naive method is to do essentially the same as the encoder. Suppose, for example, that four pointers are coming down the pipe: pointer x, followed by pointer w, followed by pointer v, followed by pointer u. Now consider what happens when pointers w and x have arrived at the leader processor. The leader processor adopts the entry wx and then passes leadership to the left. The processor to the left waits until pointer u and v have arrived and then adopts the entry uv. The encoder would have also learned the entries uv and wx; but in addition, it would have learned the entry vw. The entry vw has been "lost" by the above naive implementation of the decoder. The situation can be remedied by interspersing a nilpointer between every pointer in the input stream to the decoder. The effect is to slow down the operation of the decoder to allow time for the "overlapping" entries such as vw to be learned. The interspersion of nilpointers is necessary only during the learning period when not all processors have adopted an entry. Note also that the output data rate from the decoder is unaffected so long as the data has been compressed by at least a factor of 2.

We do not consider the details of such an implementation further because empirical tests indicate that learning every other entry achieves equivalent compression in practice. That is, our strategy is to use the naive decoder implementation described above and modify the encoder to also learn only every other entry. The old rule for a leader in the encoder was to adopt a pointer pair only if at least one of them was not adopted by the preceding leader. The modified rule is:

> *Adopt a pointer pair only if neither pointer was adopted by the preceding processor.*

We implement this modified version of the ID heuristic by marking a pointer (passing an extra bit along with each pointer) once it has been adopted by a leader. This could also be done by waiting for two pointers to pass by before trying to adopt; however, the mark bit proves useful for other purposes when the details of the finite state control of a processor are considered.

6.4. The Flush Operation

As mentioned earlier, the processor numbered $D_{max} - 1$ has been left out of the pipe. The corresponding pointer value (which is all 1's when D_{max} is a power of 2) is called the *nilpointer*. Nilpointers are sent into the encoding pipe on every clock tick for which no input is present. Another role of the nilpointer is to implement *flush* operations and record them in compressed data. A flush operation is a signal sent down the encoding pipe to "push" all data out. Flushes can be used to start a new input stream (that will learn a new dictionary). Note that it is not necessary to wait for a flush to make it through the pipe before starting a new input stream; after one clock tick, new data can "follow" the flush that is pushing out old data. Flush operations can also be used to signal that there is a pause in the input stream. For example, consider a satellite that transmits data to earth continuously for a day but then must pause for a short period of time while a sensor is adjusted. Rather than make the receiver on earth wait for the data that is left in the pipe, a flush signal can be sent to push it out (but the dictionary that has been learned remains in the pipe). Each time such a flush occurs, a single nilpointer is stored in the compressed data stream so that when decoding, flushes can be performed at the same points that they were during encoding (thus ensuring proper decoding of the data).

7. Prototype Massively Parallel Hardware

Here we briefly overview a custom VLSI chip and a host compression board for these chips that have been built for the Dynamic Dictionary architecture of the last section. The current prototype chip has the following specifications:

- 1.0 micron double metal CMOS technology.
- 68-pin LCC package (48 pins for signals, 20 pins for power and ground).
- Die size = 9,650 by 10,200 microns (.38 by .40 inches).
- 350,000 transistors.
- Power consumption = 750mW (the consumption is very evenly distributed; the chip has no "hot" spots).

- 128 processors per chip (30 chips form a complete systolic pipe of 3,839 processors).
- 20 mhz clock, 8 bits per cycle, 160 million bits per second.

To speed completion time and reduce costs, the design process has been kept simple (standard cell construction for the most part) and the size of the chip has been kept relatively modest. Less area per processing element, more processing elements per chip and higher processing speeds are clearly possible, even with current technology. In fact, simulations show that the maximum delay time of a single cell is only 13 nano seconds, so a clock rate of up to 75 MHz may be possible with current technology, yielding a data rate of 600 million bits per second. A new design, that employs techniques of Storer [1991] to isolate pad delay times, that runs at 320 million bits per second and has 256 processors per chip (so that 15 chips are needed for a complete system) has already been completed, but yet to be fabricated. In addition, Storer, Reif and Markas [1990] describe a variant of this architecture that can greatly reduce the size of a chip while maintaining the speed by storing several entries (e.g. 16) entries per processor and employing a parallel match of these entries on a single clock cycle. It is also possible to modify the design to have a degree of fault-tolerance where faulty cells are detected and bypassed at startup.

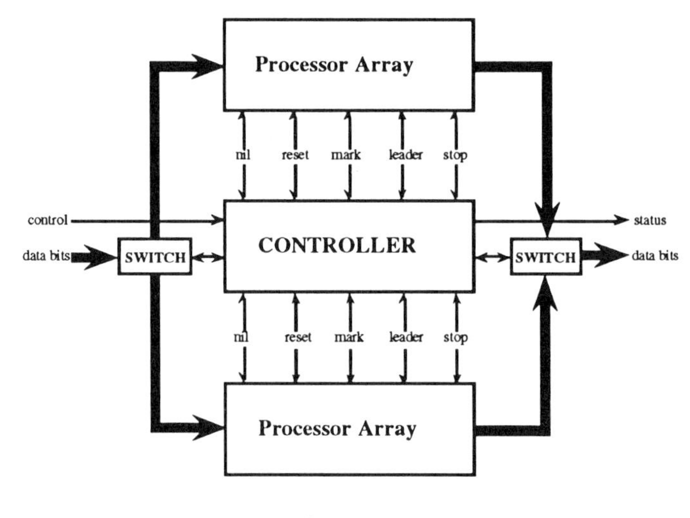

Figure 7
Controller

The packaged chips are less than 1 inch square. Due to the extremely simple pattern of interconnections (communication is only between a chip and the chips to its immediate left and right), a board area of approximately 5 by 6 inches suffices to accommodate a basic set of 30 chips that forms a complete systolic array of 3,839 processing elements (smaller areas are possible with multiple chip packaging). This chip set along with some control logic can easily be placed on a small board, as depicted in Figure 7; note that this figure shows two copies of the array for implementation of the SWAP deletion heuristic.

8. References

S. De Agostino and J. A. Storer [1992]. "Parallel Algorithms for Optimal Compression using Dictionaries with the Prefix Property", Proceedings IEEE Data Compression Conference, Snowbird, Utah.

E. R. Fiala and D. H. Greene [1989]. "Data Compression with Finite Windows", *Communications of the ACM 32:4*, 490-505.

M. Gonzalez and J. A. Storer [1985]. "Parallel Algorithms for Data Compression", *Journal of the ACM* 32:2, 344-373.

A. Hartman and M. Rodeh [1985], Optimal Parsing of Strings, *Combinatorial Algorithms on Words*, Springer - Verlag (A. Apostolico and Z. Galil, editors), 155 - 167.

A. Lempel and J. Ziv [1976]. "On the Complexity of Finite Sequences", *IEEE Transactions on Information Theory* 22:1, 75-81.

E. M. McCreight [1976]. "A Space-Economical Suffix Tree Construction Algorithm", *Journal of the ACM* 23:2, 262-272.

V. S. Miller and M. N. Wegman [1985]. "Variations on a Theme by Lempel and Ziv", *Combinatorial Algorithms on Words,* Springer-Verlag (A. Apostolico and Z. Galil, editors), 131-140.

J. H. Reif and J. A. Storer [1991]. "A Parallel Architecture for High Speed Data Compression", *Journal of Parallel and Distributed Computing* 13, 222-227.

J. H. Reif and J. A. Storer [1991b]. "Adaptive Lossless Data Compression over a Noisy Channel", *Proceedings Communication, Security, and Sequences Conference*, Positano, Italy.

M. Rodeh, V. R. Pratt, and S. Even [1980]. "Linear Algorithms for Compression Via String Matching", *Journal of the ACM* 28:1, 16-24.

J. B. Seery and J. Ziv [1977]. "A Universal Data Compression Algorithm: Description and Preliminary Results", Technical Memorandum 77-1212-6, Bell Laboratories, Murray Hill, N.J.

J. B. Seery and J. Ziv [1978]. "Further Results on Universal Data Compression", Technical Memorandum 78-1212-8, Bell Laboratories, Murray Hill, N.J.

J. A. Storer [1988]. *Data Compression: Methods and Theory,* Computer Science Press, Rockville, MD.

J. Storer [1991]. "Massively Parallel System for High Speed Data Compression", patent pending.

J. A. Storer, J. H. Reif, and T. Markas [1990]. "A Massively Parallel VLSI Design for Data Compression using a Compact Dynamic Dictionary", *Proceedings IEEE VLSI Signal Processing Conference*, San Diego, CA.

J. A. Storer and T. G. Szymanski [1978]. "The Macro Model for Data Compression", *Proceedings Tenth Annual ACM Symposium on Theory of Computing*, San Diego, CA, 928-951.

J. A. Storer and T. G. Szymanski [1982]. "Data Compression Via Textual Substitution", *Journal of the ACM* 29:4, 928-951.

R.A. Wagner [1973], Common Phrases and Minimum Text Storage, *Communications of the ACM* **16**, 148 - 152.

T. A. Welch [1984]. "A Technique for High-Performance Data Compression", *IEEE Computer* 17:6, 8-19.

R. Zito-Wolf [1990]. "Broadcast / Reduce Architecture for High Speed Data Compression", Proceedings Second IEEE Symposium on Parallel and Distributed Processing", Dallas, TX, 1990, 174-181.

R. Zito-Wolf [1990b]. "A Systolic Architecture for Sliding Window Data Compression", *Proceedings IEEE VLSI Signal Processing Conference*, San Diego, CA, 1990, 339-351.

J. Ziv and A. Lempel [1977]. "A Universal Algorithm for Sequential Data Compression", *IEEE Transactions on Information Theory* 23:3, 337-343.

J. Ziv and A. Lempel [1978]. "Compression of Individual Sequences Via Variable-Rate Coding", *IEEE Transactions on Information Theory* 24:5, 530-536.

Part 3: CODING THEORY

8

Variations on a Theme by Gallager*

Renato M. Capocelli
Department of Computer Science
Università di Roma "La Sapienza"
00185 Roma, Italy

Alfredo De Santis
Department of Computer Science
Università di Salerno
84081 Baronissi (Salerno), Italy

Abstract

Upper and lower bounds on the redundancy of Huffman codes, in terms of the most likely and of the least likely letter probability are discussed. Whenever it is possible the general situation of codes with arbitrary alphabet size is considered.

1 Introduction

Let A be a discrete *source* with N, $2 \leq N < \infty$, letters and let p_k denote the probability of the letter a_k, $1 \leq k \leq N$. Let $x_1, x_2, \ldots, x_N$ be the *codewords* used for encoding the letters of the source A. Let D be the size of the code *alphabet.* Let $n_1, n_2, \ldots, n_N$ be the codewords lengths. The *Huffman* encoding algorithm [16] provides an *optimal prefix code C* for the source A. The encoding is optimal in the sense that codewords lengths minimize the *redundancy* r of the code, defined as the difference between the *average codeword length* E of the code and the *entropy* $H(p_1, p_2, \ldots, p_N)$ of the source:

$$r_D = E - H(p_1, p_2, \ldots, p_N) = \sum_{i=1}^{N} p_i n_i + \sum_{i=1}^{N} p_i \log_D p_i$$

where $\log_D$ denotes the logarithm to base D (in the binary case we omit the subscript). Indeed, as the entropy is constant for a given probability distribution, minimizing the average codeword length minimizes the redundancy of the code.

The redundancy is a basic concept in Information Theory. It plays a relevant rôle in Data Compression where it may be used to measure the achievable compression since the goal is to reduce the redundancy leaving only the informational content. The Huffman algorithm provided the first solution to the

*This work was partially supported by the National Council of Research (C.N.R.) under grants 90.00829.CT07 and 90.011552.CT12.

problem of constructing minimum redundancy codes. It attains the best possible compression under the constraints that each code source message is mapped to a unique codeword and that the compressed text is the concatenation of the codewords for the source messages. The algorithm runs in $O(N \log N)$. If letters probabilities are already sorted it can be implemented to run in $O(N)$.
It has been also proposed an adaptive version of the algorithm. Independently conceived by Faller [11] and Gallager [12], later improved by Knuth [17] and refined subsequently by Vitter [23], it exhibits the interesting feature of requiring only one pass over the source messages.
According to Shannon's first theorem, the redundancy of any Huffman code is always nonnegative and less than or equal to one. These limits can be reached only by special distributions. Although it is generally assumed that all statistical parameters of a source message are known with perfect accuracy, there are situations in which only a partial (but exact) knowledge of the source is available or the required measurements for determining all statistical parameters accurately become increasingly difficult. In such conditions the best code could be less efficient than simpler codes.
A typical example in which only partial information is available is the situation, often of practical interest, where only some letters probabilities are known. For example, the most likely probability, the least likely probability, the first j or the last j probabilities, the extreme probabilities p_1 and p_N, and so on. In such cases it is possible to determine a sharper estimate of the redundancy that permits to make a good guess of the efficiency of the best procedure and to appraise whether or not a less-than-optimal code is acceptable. The guess allows a simple evaluation of the performance of a suboptimal code without going to the trouble of constructing the optimal code.
In a remarkable paper [12], Gallager considered for the first time the problem of determining upper bounds on the redundancy of the Huffman code when it is known the probability of the most likely source letter. He proved that for the binary case the following upper bounds hold

$$r \leq \beta + p_1 \qquad \text{if } p_1 < 1/2$$

$$r \leq 2 - p_1 - \mathcal{H}(p_1) \qquad \text{if } p_1 \geq 1/2,$$

where p_1 denotes the probability of the most likely source letter, $\beta = 1 - \log e + \log(\log e)$, and $\mathcal{H}(p_1) = -p_1 \log p_1 - (1 - p_1)\log(1 - p_1)$. Gallager considered also the case of arbitrary code alphabet size D and proved that

$$r_D \leq \sigma_D + p_1 D / \ln D$$

where $\sigma_D = \log_D(D-1) + \log_D(\log_D e) - \log_D e + (D-1)^{-1}$.
After the nice paper of Gallager many scientists have been working along the same lines and considered in detail and extensively all different possible various situations of incomplete or inaccurate probability knowledge [2]–[9], [14], [18]–[21], [24].
It has been also considered the problem of bounding the redundancy of a code with limited length [10], [13] or with other special features [1], [7], [8] in order to estimate the efficiency of the optimal code with the desired feature. Indeed,

a simple solution to the problem of incomplete knowledge of the source characteristics could be to avoid long codewords, thereby minimizing the error of badly underestimating the probability of a message.

In this paper we shall limit ourselves to bound the redundancy of sources for which we only know either the probability of the most likely letter or the probability of the least likely letter. Whenever it is possible we will consider the case of arbitrary code alphabet size D.

The organization of the paper is the following. In Section 2 we state all mathematical machinery necessary to prove the results. In Section 3 we provide upper bounds on the redundancy of binary Huffman codes as function of p_1. In Section 4 we consider the D-ary case and present upper bounds and lower bounds as function of p_1. Finally in Section 5 we provide upper bounds and lower bounds of binary Huffman codes as function of p_N.

2 Mathematical preliminaries

Let A be a discrete *source* with N, $2 \leq N < \infty$, letters and let p_k denote the probability of the letter a_k, $1 \leq k \leq N$. Let $\{x_1, x_2, \ldots, x_N\}$ be the set of *codewords* used for encoding the letters of the source A and let $n_1, n_2, \ldots, n_N$ be the codeword lengths. $C = \{x_1, x_2, x_3, \ldots, x_N\}$ is said to be a *prefix* (condition) *code* if no codeword is a prefix of any other codeword. Let D be the size of the code alphabet. It is well-known that a D-ary prefix code can be represented as a conveniently labeled rooted D-ary code tree in which each source letter corresponds to a *leaf* on the tree and where the associated codeword is the sequence of labels on the path from the root to the leaf. The *level* of a node is the number of links from the root to it. A node a is called *parent* of node b, which is the *child* of the node a, if there is a link between them and the level of a increased by 1 is equal to the level of b. Two nodes with a common parent are called *siblings*.

Each of the leaves of the code tree has a probability assigned to it, namely, the probability of the corresponding source letter. We also assign a probability to each intermediate node, defined as the sum of the probabilities of all leaves for which the path from root to leaf passes through the given node.

We allow at most $D-1$ of the source messages to have zero probability. Thus we can assume without loss of generality that $N = c(D-1)+1$ for some integer c. The *Huffman* encoding algorithm provides an *optimal* prefix code C for the source A. Such an encoding rule can be regarded as the following bottom-up tree construction algorithm.

1. Assign probabilities $p_1, p_2, p_3, \ldots, p_N$ to nodes $t_1, t_2, t_3, \ldots, t_N$, respectively.

2. Let $t_1, t_2, t_3, \ldots, t_M$ be the roots of subtrees constructed until now and let $w_1, w_2, w_3, \ldots, w_M$ be the probabilities assigned to such roots. Take the D least likely roots $t_{i_1}, t_{i_2}, \ldots, t_{i_D}$ and make them siblings by generating a new node t as their parent node. Assign probability $w_{i_1} + w_{i_2} + \cdots + w_{i_D}$ to t. Assign label 0 to the edge (t, t_{i_1}), label 1 to the edge (t, t_{i_2}), $\ldots$, label $(D-1)$ to the edge (t, t_{i_D}).

3. If t is the only root left, then stop. Otherwise repeat step 2.

Property 1. In any D-ary Huffman code tree, for each $l \geq 1$, the probability of each node at level l is less than or equal to the probability of each node at level $l-1$.

Property 2. Let a and b be two nodes on the same level in a D-ary Huffman code tree and p_a and p_b their probabilities. If b is not a leaf then $p_b \leq Dp_a$.

Definition. [12] A D-ary code tree with $N = c(D-1)+1$ nodes has the *sibling property* if the nodes (except the root) can be listed in order of nonincreasing probability such that for each i, $1 \leq i \leq c$, nodes iD, $iD-1$, ..., $iD-D+1$ are all siblings of each other.
The following theorem, due to Gallager [12], holds

Theorem 1 *A D-ary prefix code is an Huffman code iff the code tree has the sibling property.*

The sibling property allows numbering all the nodes (except the root) in order of decreasing probability and increasing level so that for each k, $1 \leq k \leq c$, nodes kD, $kD-1$, ..., $kD-D+1$ are siblings. Let q_k be the probability of the kth node on the list, $1 \leq k \leq N+c-1$. The expected codeword length and the entropy can be written in terms of q_k as

$$E = \sum_{k=1}^{N+c-1} q_k$$

and

$$H(p_1, p_2, ..., p_N) = \sum_{k=1}^{c} t_k \, H\left(\frac{q_{kD}}{t_k}, \frac{q_{kD-1}}{t_k}, \ldots, \frac{q_{kD-D+1}}{t_k}\right),$$

where $t_k = q_{kD} + q_{kD-1} + \cdots + q_{kD-D+1}$. Combining above formulas one has

$$r_D = \sum_{k=1}^{c} t_k \left[1 - H\left(\frac{q_{kD}}{t_k}, \frac{q_{kD-1}}{t_k}, \ldots, \frac{q_{kD-D+1}}{t_k}\right)\right].$$

Let $l \geq 1$ be the maximum level at which the tree has $L = D^l$ nodes, but for which there are also nodes at level $l+1$. If $N \geq D$ then such an l must exist. Let m be the smallest integer for which node $mD-D+1$ is at level $l+1$, and let $q'_1, q'_2, \ldots, q'_L$ be the probabilities of nodes at level l. One has

$$r_D = l - H(q'_1, q'_2, \ldots, q'_L) + \sum_{k=m}^{c} t_k \left[1 - H\left(\frac{q_{kD}}{t_k}, \frac{q_{kD-1}}{t_k}, \ldots, \frac{q_{kD-D+1}}{t_k}\right)\right].$$

Let $y_1 \geq y_2 \geq \cdots \geq y_D$ be probabilities. Since $1 - H(y_1, y_2, \ldots, y_D) \leq (y_1 - y_D)D/\ln D$ [12], one has

$$r_D \leq l - H(q'_1, q'_2, \ldots, q'_L) + q_{mD-D+1} D/\ln D.$$

In the binary case, i.e. $D = 2$, above formula simplifies into

$$r \leq l - H(q'_1, q'_2, ..., q'_L) + q_{2m-1}. \tag{1}$$

The following lemma holds.

Lemma 1 *If a D-ary Huffman code for the source A has a minimum codeword length of l, $l \geq 1$, then the probability p_1 of the most likely letter of A satisfies*

$$\frac{1}{D^2 - D + 1} \leq p_1 \leq 1 \qquad \text{if} \quad l = 1 \tag{2}$$

$$\frac{1}{D^{l+1} - D + 1} \leq p_1 \leq \frac{D}{D^l + D - 1} \qquad \text{if} \quad l > 1. \tag{3}$$

Proof. Let $C = \{x_1, x_2, ..., x_N\}$ be a D-ary Huffman code for A and let x_1 be the codeword with minimum length l that encodes the source symbol with probability p_1. Let $q_1 \geq q_2 \geq \ldots \geq q_L$, $L = D^l$, be the probabilities of nodes at level l in the Huffman code tree. Let g be the node with smallest probability, namely $q_g = q_{L-D+1} + ... + q_L$, at level $l-1$. Then all codewords that correspond to the leaves of the subtree rooted at g have the common prefix σ of length $l-1$. We may assume without loss of generality that g is not the father of the node corresponding to x_1.
From Property 2 we get $q_i \leq Dp_1$, $1 \leq i \leq L$. Thus obtaining $(1-p_1)/(D^l-1) \leq \max q_i \leq Dp_1$, where the maximization is over all $i \in [1, L]$, $q_i \neq p_1$. This proves (2) and the left inequality of (3).
Now assume $l > 1$. Consider the encoding of A by means of the code C' with codewords $x'_1 = \sigma$, $x'_i = x_1\tau$ if $x_i = \sigma\tau$ and $x'_i = x_i$ otherwise. In words C' is obtained by C interchanging the subtree rooted at g with the node corresponding to x_1. C' is a prefix code. The difference between the expected codeword lengths of C and C' is $E(C) - E(C') = p_1 - q_g \geq p_1 - (1-p_1)D/(D^l - 1)$. From the optimality of C we get $E(C) - E(C') \leq 0$ that, in turn, implies the right inequality of (3). □

A different but equivalent statement of the relation between the probability p_1 of the source A and the minimum codeword length in the associated D-ary Huffman code is stated in the following lemma that reduces for $D = 2$ to a lemma by Montgomery and Kumar [20] which, in turn, is an extension of Johnsen's Theorem 1 [18] and Capocelli et al.'s Lemma 1 [2].

Lemma 2 *If for some $l \geq 1$,*

$$\frac{D}{D^{l+1} + D - 1} < p_1 < \frac{1}{D^l - D + 1}, \tag{4}$$

then the minimum codeword length of any D-ary Huffman code for a source A, where the most likely letter has probability p_1, is l. Furthermore, if

$$\frac{1}{D^{l+1} - D + 1} \leq p_1 \leq \frac{D}{D^{l+1} + D - 1}, \tag{5}$$

for some $l \geq 1$, then any D-ary Huffman code for A has a minimum codeword length of either l or $l + 1$.

Proof. The lemma can be easily proved by distinguishing the two cases (4) and (5), and by making use of Lemma 1. □

Consider the source B with alphabet $\{a_2, a_3, \ldots, a_N\}$, probability distribution $\{p_2/(1-p_1), \ldots, p_N/(1-p_1)\}$ and entropy

$$H_B = H\left(\frac{p_2}{1-p_1}, \ldots, \frac{p_N}{1-p_1}\right).$$

The following lemmas provide an useful link between the expected codeword length L and L_B of the Huffman codes C and C_B for sources A and B, respectively.

Lemma 3 *The expected codeword lengths L and L_B of the D-ary Huffman codes C and C_B corresponding respectively to source $A = \{a_1, a_2, ..., a_N\}$ and to subsource $B = \{a_2, a_3, .., a_N\}$ satisfy*

$$L \leq 1 + (1-p_1)(L_B - 1 + 2/D). \tag{6}$$

Moreover, for every $p_1 \geq 1/(D+1)$, a source A exists that satisfies (6) with equality.

Proof. If $N \leq D$ then $L = L_B = 1$ and the lemma is true. Suppose $N > D$. Let $x_{B,i}$, $i = 2, \ldots, N$, be the codewords for the D-ary Huffman code of the source B. Its expected codeword length is

$$L_B = \sum_{i=2}^{N} \frac{p_i}{1-p_1} |x_{B,i}|.$$

Let $q_1 \geq q_2 \geq \ldots \geq q_D$ be the probabilities of nodes $g_1, g_2, \ldots, g_D$ at level 1 in the Huffman code tree of the source B.
Let $C[i]$, $1 \leq i \leq D$, be the set of codewords that correspond to the leaves of the subtree rooted at g_i. All codewords in $C[i]$ have as prefix the same letter σ_i. Consider the code R for the source A with codewords $x'_1 = \sigma_D$, $x'_i = \sigma_{D-1}\, x_{B,i}$ if the first letter of $x_{B,i}$ is either σ_{D-1} or σ_D and with codeword $x'_i = x_{B,i}$ otherwise. R is a prefix code and its expected codeword length is

$$\begin{aligned} L_R &= \sum_{i=1}^{N} p_i\, |x'_i| \\ &= p_1 + \sum_{x_{B,i} \notin C[D] \cup C[D-1]} p_i\, |x_{B,i}| + \sum_{x_{B,i} \in C[D] \cup C[D-1]} p_i\big(|x_{B,i}| + 1\big) \\ &= p_1 + (1-p_1)L_B + (1-p_1)(q_{D-1} + q_D) \end{aligned}$$

From $\sum_{i=1}^{D} q_i = 1$ and $q_1 \geq q_2 \geq \ldots \geq q_D$ it follows $q_D + q_{D-1} \leq 2/D$, and thus

$$L_R \leq 1 + (1-p_1)(L_B - 1 + 2/D).$$

Since the expected codeword length of the Huffman code is minimum one has $L \leq L_R$, and (6) follows.
Bound (6) is tight for every $p_1 \geq 1/(D+1)$. Indeed the source consisting of $N = D+1$ letters having probabilities $p_1 = p$ and $p_i = (1-p)/D$, $i = 2, 3, ..., D+1$, satisfies (6) with equality. $\square$

The entropy $H = H(p_1, p_2, \ldots, p_N)$ of A is related to the entropy H_B of B by

$$H = \mathcal{H}_D(p_1) + (1 - p_1)H_B \tag{7}$$

where $\mathcal{H}_D(x)$ is the Shannon function $-x \log_D x - (1-x)\log_D(1-x)$. From Lemma 3 and (7) one has

$$r_D \leq 1 - \mathcal{H}_D(p_1) + (1 - p_1)(L_B - H_B - 1 + 2/D). \tag{8}$$

If the maximum probability $p_2/(1-p_1)$ of B is less than or equal to p and an upper bound $L_B - H_B \leq r_{max}(p)$ is known, then (8) implies for $p_1 \leq 1/2$

$$r_D \leq 1 - \mathcal{H}_D(p_1) + (1 - p_1)\big(r_{max}(p_1/(1-p_1)) - 1 + 2/D\big). \tag{9}$$

Lemma 4 *[4] If the probability p_1 of the most likely source letter of a discrete source A consisting of letters $a_1, a_2, \ldots, a_N$, satisfies*

$$\frac{1}{1 + 2^{l+1}} < p_1 \leq \frac{1}{1 + 2^l}$$

for some $l \geq 1$, then the expected codeword lengths L and L_B of the binary Huffman codes C and C_B corresponding respectively to source A and to subsource $B = \{a_2, a_3, \ldots, a_N\}$ satisfy

$$L \leq (1 - p_1)L_B + 2^{-l} + (l + 1 - 2^{-l})p_1. \tag{10}$$

The bound is tight, for every $p_1 \leq 1/3$ a source A exists that satisfies (10) with equality.

Define X_j, $j \geq 2$, as the set of nonnegative real numbers $x_1, x_2, \ldots, x_j$ that satisfy $x_1 \geq x_2 \geq \ldots \geq x_j \geq x_1 - x_{j-1}$ along with $\sum x_i = 1$. Moreover, define W_j, $j \geq 2$, as the set of positive real numbers $w_1, w_2, \ldots, w_j$ that satisfy $w_1 \geq w_2 \geq \ldots \geq w_j \geq w_1/2$ along with $\sum w_i = 1$.

Lemma 5 *[4], [9] The minimum*

$$M_j = \min_{\{x_i \in X_j\}} \Big\{ H(x_1, x_2, \ldots, x_j) - x_j \Big\}$$

is equal to $\log(j-1)$ for $2 \leq j \leq 9$. Moreover, for $j \geq 10$ it holds $M_j = A_j$, where A_j is the minimum

$$A_j = \min_{\{w_i \in W_j\}} \Big\{ H(w_1, w_2, \ldots, w_j) - w_j \Big\}.$$

The quantity A_j satisfies

$$0 < A_j - \log(2j - 1) + 1 + \beta < \frac{(\log e)^3}{(2j-1)(2j - 1 + \log e)} \tag{11}$$

where $\beta = 1 - \log e + \log(\log e) \approx 0.0860$.

For illustrative purposes, values of M_j and of $\log(2j-1)-1-\beta$ up to $j=511$, together with their difference ϵ_j are presented in Table 1.

j	$M_j = A_j$	$\log(2j-1)-1-\beta$	ϵ_j
10	$\log 13 - 7/13$	3.161856181387651286	0.000121998291902411
11	$\log 15 - 9/15$	3.306246090722826082	0.000644504885692447
12	$55/16$	3.437490624001078665	0.000009375998921334
13	$\log 17 - 9/17$	3.557784857718790488	0.000266218825666566
14	$\log 19 - 11/19$	3.668816170107534337	0.000163974914998524
15	$\log 5 + 29/20$	3.771909663071637913	0.000018431815724434
31	$\log 21 + 19/42$	4.844666005506952069	0.000032369652760600
63	$\log 87 - 49/87$	5.879712952606152836	0.000012152437977764
127	$\log 175 - 97/175$	6.896922242638375921	0.000003154908238595
255	$\log 353 - 197/353$	7.905450514019761088	0.000000204860484357
511	$\log 177 + 1021/708$	8.909695818821867280	0.000000126741356148

Table 1.

For future reference we report the values of $A_2 = \log 3 - 1$, $A_3 = 5/4$ and $A_7 = 2\log 3 - 5/9$.

3 Upper bounds as function of p_1. The binary case

Assume $p_1 \geq 1/2$. From Lemma 2 one has that the minimum codeword length n_1 is 1. Taking $l = 1$ in (1), we obtain

$$r \leq 1 - \mathcal{H}(p_1) + q_{2m-1}.$$

Since $q_{2m-1} \leq 1 - p_1$, above inequality implies

$$r \leq 2 - \mathcal{H}(p_1) - p_1.$$

Assume $p_1 < 1/2$. If all codewords are of the same length $n_1 = l$, choose l in (1) to be $n_1 - 1$; otherwise choose $l = n_1$. In both cases $q_{2m-1} \leq p_1$, and in both cases we can order $q'_1, ..., q'_L$ to satisfy $q'_1 \geq ... \geq q'_L \geq q'_1$. Let Q be the set of choices for $q'_1, ..., q'_L$ that satisfy $q'_1 \geq ... \geq q'_L \geq q'_1$ along with $\sum_{i=1}^{L} q'_i = 1$. Then

$$r \leq l - \min_{\{q'_i \epsilon Q\}} H(q'_1, ..., q'_L) + p_1.$$

Since the entropy function is convex $\cap$, the minimum must occur at an extreme point of Q. The extreme points are those for which for some n, $q'_i = q'_1$ for $i \leq n$, and $q'_i = q'_1/2$ for $n < i \leq L$. Hence, for a given n, all q'_i are equal to either $2/(L+n)$ or $1/(L+n)$, and

$$\min_{\{q'_i \epsilon Q\}} H(q'_1, ..., q'_L) = \min_{1 \leq n \leq L} \Big\{ -\log \frac{2}{L+n} + \frac{L-n}{L+n} \Big\}.$$

Allowing n to take nonintegral values one finds, by elementary calculus, that the minimum is not greater than $l - \beta$. The following theorem holds

Theorem 2 *[12] Let p_1 be the probability of the most likely source letter of a discrete source A. The redundancy of the corresponding Huffman code is upper bounded by*

$$r \leq \begin{cases} 2 - \mathcal{H}(p_1) - p_1, & \text{if } 0.5 \leq p_1 < 1 \\ \beta + p_1, & \text{if } p_1 < 0.5. \end{cases} \tag{12}$$

Above bound is tight if $p_1 \geq 0.5$. Indeed the Huffman code for the source with probabilities p_1, $1 - p_1 - \epsilon$, and ϵ, has a redundancy that approaches $2 - \mathcal{H}(p_1) - p_1$ when $\epsilon \to 0$. It is possible to give a bound on r stronger than $r \leq \beta + p_1$ for several intervals of p_1, $p_1 < 1/2$.
For $1/3 \leq p_1 < 0.5$ the bound (12) has been improved in [18] and subsequently refined in [2]. The result is summarized in the following theorem

Theorem 3 *Let p_1 be the probability of the most likely source letter of a discrete source A. The redundancy of the corresponding binary Huffman code is upper bounded by*

$$r \leq \begin{cases} 3 - 5p_1 - \mathcal{H}(2p_1), & \text{if } \delta \leq p_1 < 0.5 \\ 1 + 0.5(1 - p_1) - \mathcal{H}(p_1), & \text{if } 1/3 \leq p_1 < \delta. \end{cases} \tag{13}$$

where $\delta \approx 0.4505$. The bounds are tight.

Proof. Using (12) one obtains the following upper bound on $r_{max}(p)$

$$r_{max}(p) \leq \begin{cases} 2 - \mathcal{H}(p) - p, & \text{if } 0.8535 \leq p < 1 \\ 0.5454, & \text{if } 0.5 \leq p \leq 0.8535 \\ \beta + p, & \text{if } p_1 < 0.5. \end{cases} \tag{14}$$

Substituting (14) into (9) one has

$$r \leq \begin{cases} 3 - 5p_1 - \mathcal{H}(2p_1), & \text{if } 0.4605 \leq p_1 < 0.5 \\ 1 - \mathcal{H}(p_1) + (1 - p_1)0.5454, & \text{if } 1/3 \leq p_1 \leq 0.4605. \end{cases} \tag{15}$$

Using (15) one obtains the following upper bound on $r_{max}(p)$

$$r_{max}(p) \leq \begin{cases} 2 - \mathcal{H}(p) - p, & \text{if } 0.821 \leq p < 1 \\ 0.5, & \text{if } p \leq 0.821. \end{cases} \tag{16}$$

Substituting (16) into (9) one finally gets the bound (13).
The upper bounds are tight. Indeed, for $\delta \leq p_1 < 0.5$ the bound is reached by a source of 4 letters having probabilities p_1, p_1, $1 - 2p_1 - \epsilon$, and ϵ, where ϵ approaches zero. For $1/3 \leq p_1 < \delta$ the bound is reached by a source of 4 letters having probabilities p_1, $(1-p_1-\epsilon)/2$, $(1-p_1-\epsilon)/2$, and ϵ, where ϵ approaches zero. $\square$

The following lemma gives a bound on r when the Huffman code has all codewords of the same length.

Lemma 6 *Let p_1 be the probability of the most likely source letter of a discrete source A. If the corresponding binary Huffman code has all codewords of the same length $l \geq 1$ and $p_1 < 1/(2^l - 1)$, then $p_1 \geq 2^{-l}$ and the redundancy of the code is upper bounded by*

$$r \leq l - p_1(2^l - 1)\log(2^l - 1) - \mathcal{H}\big((2^l - 1)p_1\big). \tag{17}$$

Proof. Suppose all codewords have the same length $l \geq 1$. Then the Huffman code consists of $N = 2^l$ codewords and

$$r = l - H(p_1, p_2, \ldots, p_N).$$

From $1 - p_1 = \sum_{i=2}^{N} p_i \leq (N-1)p_1$ one has $p_1 \geq 1/N$.
Let Q be the set of positive real numbers $p_2, p_3, \ldots, p_N$ that satisfy $p_1 \geq p_2 \geq \ldots \geq p_N$ along with $\sum_{i=1}^{N} p_i = 1$. Then

$$r \leq l - \min_{\{p_i\} \in Q} H(p_1, p_2, \ldots, p_N).$$

Since H is convex $\cap$, the minimum must occur at an extreme point of Q. The extreme points of Q are those for which for some integer k, $1 \leq k \leq N-1$, $p_i = p_1$ if $i \leq k$ and $p_i = p_N$ if $N \geq i > k$.

Define the probability distribution $P^k = (p_1^k, \ldots, p_N^k)$, $1 \leq k \leq N-1$, as $p_i^k = p_1$ if $i \leq k$ and $p_i^k = (1 - kp_1)/(N-k)$ if $i > k$.
Then $H(p_1^k, p_2^k, \ldots, p_N^k)$

$$\begin{aligned}
&= H(p_1^k, \ldots, p_k^k, 1 - kp_1) + (1 - kp_1) H\Big(\frac{p_{k+1}^k}{1 - kp_1}, \ldots, \frac{p_N^k}{1 - kp_1}\Big) \\
&= H(p_1^k, \ldots, p_k^k, 1 - kp_1) + (1 - kp_1) H\Big(\frac{1}{N-k}, \ldots, \frac{1}{N-k}\Big) \\
&\geq H(p_1^k, \ldots, p_k^k, 1 - kp_1) + (1 - kp_1) H\Big(\frac{p_{k+1}^{N-1}}{1 - kp_1}, \ldots, \frac{p_N^{N-1}}{1 - kp_1}\Big) \\
&= H(p_1^{N-1}, \ldots, p_1^{N-1}, 1 - kp_1^{N-1}) + (1 - kp_1) H\Big(\frac{p_{k+1}^{N-1}}{1 - kp_1}, \ldots, \frac{p_N^{N-1}}{1 - kp_1}\Big) \\
&= H(p_1^{N-1}, p_2^{N-1}, \ldots, p_N^{N-1}).
\end{aligned}$$

Therefore the minimum is achieved by the probability distribution P^{N-1} and

$$\begin{aligned}
r &\leq l - H(p_1, p_1, \ldots, p_1, 1 - (N-1)p_1) \\
&= l - p_1(2^l - 1)\log(2^l - 1) - \mathcal{H}\big((2^l - 1)p_1\big).
\end{aligned}$$

□

The following theorem gives a tight upper bound on the redundancy of Huffman codes for $2/9 < p_1 < 1/3$.

Theorem 4 *Let p_1 be the probability of the most likely source letter of a discrete source A. The redundancy of the corresponding binary Huffman code is upper bounded by*

$$r \leq \begin{cases} 3 - 3(1+\log 3)p_1 - \mathcal{H}(3p_1), & \text{if } \theta \leq p_1 < 1/3 \\ 3/4 + 5p_1/4 - \mathcal{H}(p_1), & \text{if } 2/9 < p_1 < \theta \end{cases}$$

where $\theta \approx 0.3138$. The bounds are tight.

Proof. If $2/9 < p_1 < 1/3$ then the minimum codeword length of the Huffman code is 2, by Lemma 2.
Let t be the number of codewords of length 2. Recalling (1) it follows that if $t \leq 3$

$$r \leq 2 - H(p_1, q'_2, q'_3, q'_4) + q_{2m-1} \tag{18}$$

where we have set, without loss of generality, $q'_1 = p_1$.
We distinguish four cases: $t = 1$, $t = 2$, $t = 3$ and $t = 4$.
Suppose $t = 1$. Making use of Property 2 it is possible to order q'_i to satisfy $q'_2 \geq q'_3 \geq q'_4 \geq q'_2/2$. Let Q_1 be the set of positive real numbers q'_2, q'_3 and q'_4 that satisfy above linear inequality constraints along with $p_1 + q'_2 + q'_3 + q'_4 = 1$. Define $s_i = q'_i/(1-p_1)$, i=2,3,4. Recalling Property 1 one gets $q_{2m-1} \leq q'_4$ and hence

$$\begin{aligned} r &\leq 2 - \min_{\{q'_i \in Q_1\}} \Big\{ H(p_1, q'_2, q'_3, q'_4) - q'_4 \Big\} \qquad (19) \\ &= 2 - \mathcal{H}(p_1) - (1-p_1) \min_{\{s_i \in W_3\}} \Big\{ H(s_2, s_3, s_4) - s_4 \Big\}. \end{aligned}$$

Recalling that $A_3 = 5/4$, (19) thus yields

$$\begin{aligned} r &\leq 2 - \mathcal{H}(p_1) - (1-p_1)5/4 \\ &= 3/4 + 5p_1/4 - \mathcal{H}(p_1). \qquad (20) \end{aligned}$$

Suppose $t = 2$. From Property 2 it is possible to order q'_i to satisfy $2p_1 \geq 2p_2 \geq q'_3 \geq q'_4 \geq q'_3/2$. Let Q_2 be the set of positive real numbers p_2, q'_3 and q'_4 that satisfy above linear inequality constraints along with $p_1 + p_2 + q'_3 + q'_4 = 1$. Recalling (18) one then finds

$$r \leq 2 - \min_{\{p_2, q'_i \in Q_2\}} \Big\{ H(p_1, p_2, q'_3, q'_4) - q'_4 \Big\}. \tag{21}$$

Since the function to minimize is convex $\cap$, the minimum occurs at an extreme point of Q_2. Any extreme point must satisfy at least one of the two equalities: $2p_2 = q'_3$; $p_1 = p_2$.
If $2p_2 = q'_3$ then the extreme point is in Q_1 and therefore (20) holds.
Suppose $p_1 = p_2$ and define $v_i = q'_i/(1-2p_1)$, $i = 3, 4$. From (21) it follows that

$$\begin{aligned} r &\leq 2 - H(p_1, p_1, q'_3, q'_4) + q'_4 \\ &= 2 - \mathcal{H}(2p_1) - 2p_1 - (1-2p_1)\Big(H(v_3, v_4) - v_4 \Big) \\ &\leq 2 - \mathcal{H}(2p_1) - 2p_1 - (1-2p_1) \min_{\{v_i \in W_2\}} \Big\{ H(v_3, v_4) - v_4 \Big\} \\ &= 2 - \mathcal{H}(2p_1) - 2p_1 - (1-2p_1)A_2, \end{aligned}$$

that since $A_2 = \log 3 - 1$ gives

$$\begin{aligned} r &\leq 2 - \mathcal{H}(2p_1) - 2p_1 - (1-2p_1)(\log 3 - 1) \\ &= 2 - H\left(p_1, p_1, \frac{1-2p_1}{3}, \frac{2}{3}(1-2p_1)\right) + \frac{1-2p_1}{3}. \end{aligned} \tag{22}$$

Let $\Delta_1(p_1)$ be the difference between (22) and (20). $\Delta_1(p_1)$ is a strictly convex $\cup$ function of p_1. Since $\Delta_1(0.2) = 0$ and $\Delta_1(1/3) \approx -0.028$, it follows $\Delta_1(p_1) \leq 0$ for $0.2 \leq p_1 \leq 1/3$, which implies that (20) holds also for $t = 2$.

Suppose now $t = 3$. It is possible to write (18) as

$$r \leq 2 - H(p_1, p_2, p_3, q) + q_{2m-1} \tag{23}$$

with $p_1 \geq p_2 \geq p_3$. Making use of Properties 1 and 2 one finds $q_{2m-1} \leq \min\{q, p_3\}$ and $p_3 \geq q/2$; hence

$$r \leq 2 - \min_{\{p_i, q, u\} \in Q_3} \left\{ H(p_1, p_2, p_3, q) - u \right\} \tag{24}$$

where Q_3 is the set of positive real numbers p_2, p_3, q, u that satisfy $p_1 \geq p_2 \geq p_3 \geq q/2$, $u \leq p_3$, $u \leq q$, along with $p_1 + p_2 + p_3 + q = 1$. Since the function $H(s_1, s_2, s_3, s_4) - s_5$ is convex $\cap$, the minimum above must occur at an extreme point of Q_3. The extreme points of Q_3 are three, all have $u = \min\{q, p_3\}$, and are given by: $p_1 = p_2 = p_3$; $p_1 = p_2$, $p_3 = q/2$; and $p_2 = p_3 = q/2$.
If $p_1 = p_2 = p_3$ then

$$H(p_1, p_2, p_3, q) - u = \begin{cases} H(p_1, p_1, p_1, 1-3p_1) - p_1, & \text{if } p_1 \leq 1/4 \\ H(p_1, p_1, p_1, 1-3p_1) - 1 + 3p_1, & \text{if } p_1 > 1/4. \end{cases} \tag{25}$$

If $p_1 = p_2$ and $p_3 = q/2$ then

$$H(p_1, p_2, p_3, q) - u = H\left(p_1, p_1, \frac{1-2p_1}{3}, \frac{2}{3}(1-2p_1)\right) - \frac{1-2p_1}{3}. \tag{26}$$

If $p_2 = p_3 = q/2$ then

$$H(p_1, p_2, p_3, q) - u = H\left(p_1, \frac{1-p_1}{4}, \frac{1-p_1}{4}, \frac{1-p_1}{2}\right) - \frac{1-p_1}{4}. \tag{27}$$

The difference between (27) and (26) is $-\Delta_1(p_1)$, therefore the minimum of (24) is determined by (25) and (27). Let $\Delta_2(p_1)$ be the difference between (27) and (25). $\Delta_2(p_1)$ is a strictly convex $\cup$ function of p_1 in each of the intervals $[0.2,\ 1/4]$ and $[1/4,\ 1/3]$. Since $\Delta_2(0.2) = 0$, $\Delta_2(1/4) \approx -0.0012$ and $\Delta_2(1/3) \approx 0.16633$, it follows that $\Delta_2(p_1)$ has exactly one zero $\theta \approx 0.3138$ in the interval $1/4 \leq p_1 < 1/3$ and is negative for $0.2 \leq p_1 \leq 1/4$. Hence (27) is less than (25) iff $p_1 < \theta$. Therefore if $t = 3$, from (24) one obtains

$$r \leq \begin{cases} 3/4 + 5p_1/4 - \mathcal{H}(p_1), & \text{if } 2/9 < p_1 < \theta \\ 3 - (3 + 3\log 3)p_1 - \mathcal{H}(3p_1), & \text{if } \theta \leq p_1 < 1/3. \end{cases} \tag{28}$$

Finally suppose $t = 4$. From Lemma 6 one has $r \leq 2 - (3\log 3)p_1 - \mathcal{H}(3p_1)$ which can be written as

$$r \leq 2 - H(p_1, p_1, p_1, 1 - 3p_1),$$

that is easily seen to be not greater than (28).

In conclusion the redundancy r is upper bounded by (28); i.e., it is not greater than the maximum bound for the four distinct cases.

We now show that the bounds are tight. Indeed, for $2/9 < p_1 < \theta$ the bound is reached by a source consisting of 6 letters with probabilities p_1, $(1-p_1)/4$, $(1-p_1)/4$, $(1-p_1)/4$, $(1-p_1)/4-\epsilon$ and ϵ as ϵ approaches zero. For $\theta \leq p_1 < 1/3$ the bound is reached by a source consisting of 5 letters with probabilities p_1, p_1, p_1, $1 - 3p_1 - \epsilon$ and ϵ as ϵ approaches 0. □

The technique used in Theorem 4 could also be used to prove tight upper bounds in each of the intervals $2/(2^{l+1}+1) < p_1 < 1/(2^l - 1)$, for $l = 3, 4, 5, 6, \ldots$. However, this can be done with a reasonable effort (proofs are *case-by-case proofs*) only for the first few values of l. For $l \geq 7$, calculations are very long and laborious. To provide upper bounds for each of the above intervals we resort to a general technique. Lemma 5 will result very useful for achieving this general result.

Theorem 5 *Let p_1 be the probability of the most likely source letter of a discrete source A. If for some l, $l \geq 3$,*

$$\frac{2}{2^{l+1}+1} < p_1 < \frac{1}{2^l - 1}, \tag{29}$$

then the redundancy of the corresponding binary Huffman code is upper bounded by

$$r \leq 3 - \mathcal{H}(p_1) - (1-p_1)(2\log 3 - 5/9), \qquad \text{if } 2/17 < p_1 < \gamma \tag{30}$$

$$r \leq 4 - 7(1 + \log 7)p_1 - \mathcal{H}(7p_1), \qquad \text{if } \gamma \leq p_1 < 1/7 \tag{31}$$

$$r \leq l - \mathcal{H}(p_1) - (1-p_1)A_{2^l-1} \qquad \text{if } l \geq 4 \tag{32}$$

where $\gamma \approx 0.1422$. The bounds are tight.

Proof. Let $l \geq 3$ and suppose $2/(2^{l+1}+1) < p_1 < 1/(2^l - 1)$. Then the minimum codeword length of the Huffman code is l. Let $t \geq 1$ be the number of codewords of length l in the Huffman code. From (1) it follows that if $t \leq 2^l - 1$ then

$$r \leq l - H(p_1, q'_2, \ldots, q'_{2^l}) + q_{2m-1} \tag{33}$$

where we have set, without loss of generality, $q'_1 = p_1$.
We distinguish three cases: $t \leq 2^l - 2$, $t = 2^l - 1$ and $t = 2^l$.

Suppose $t \leq 2^l - 2$. We can set q'_i, $i \leq t$, as the probabilities of the t codewords of length l, i.e., $q'_i = p_i$. By Properties 1 and 2 it is possible to order p_i and q'_j to satisfy $p_1 \geq p_2 \geq \ldots \geq p_t$, $q'_{t+1} \geq q'_{t+2} \geq \ldots \geq q'_{2^l} \geq q'_{t+1}/2$ and $2p_t \geq q'_{t+1}$.

Moreover, since $\sum p_i + \sum q'_j = 1$ the point $p_1, ..., p_t, q_{t+1}, ..., q_{2^l}$ belongs to R^l_{t+1}. Property 1 also provides $q_{2m-1} \leq q'_{2^l}$. Making use of Lemma 5 one then finds

$$\begin{aligned} r &\leq l - H(p_1, ..., p_t, q'_{t+1}, ..., q'_{2^l}) + q'_{2^l} \\ &\leq l - \min_{\{p_i, q'_j \in R^l_{t+1}\}} \left\{ H(p_1, ..., p_t, q'_{t+1}, ..., q'_{2^l}) - q'_{2^l} \right\} \\ &\leq l - \mathcal{H}(p_1) - (1 - p_1) A_{2^l - 1}. \end{aligned}$$

Suppose now $t = 2^l - 1$. As in the previous case, we set $q'_i = p_i$, $i \leq t$; we can thus rewrite relation (33) as

$$r \leq l - H(p_1, p_2, ... p_{2^l-1}, q'_{2^l}) + q_{2m-1} \tag{34}$$

with $p_1 \geq p_2 \geq ... \geq p_{2^l-1}$. By Properties 1 and 2 one obtains $q_{2m-1} \leq \min\{q'_{2^l}, p_{2^l-1}\}$ and $p_{2^l-1} \geq q'_{2^l}/2$, and hence

$$r \leq l - \min_{\{p_i, q'_{2^l}, u \in R'\}} \left\{ H(p_1, p_2, ... p_{2^l-1}, q'_{2^l}) - u \right\} \tag{35}$$

where R' is the set of positive real numbers $p_2, p_3, ..., p_{2^l-1}, q_{2^l}, u$ that satisfy $p_1 \geq p_2 \geq ... \geq p_{2^l-1} \geq q'_{2^l}/2$ and $u \leq \min\{q'_{2^l}, p_{2^l-1}\}$, along with $\sum p_i + q'_{2^l} = 1$. Since the function to minimize is convex $\cap$, the minimum above must occur at an extreme point of R'. The extreme points of R' are $2^l - 1$, all have $u = \min\{q'_{2^l}, p_{2^l-1}\}$, and are given by $p_1 = p_i$, $i \leq k$ and $p_i = q_{2^l}/2$, $i > k$, for $k = 1, 2, ..., 2^l - 1$; i.e.,

$$p_i = \begin{cases} p_1, & \text{if } i \leq k \\ (1 - kp_1)/(t - k + 2), & \text{if } i > k \end{cases} \tag{36}$$

and $q_{2^l} = 2(1 - kp_1)/(t - k + 2)$.
If $p_1 = p_2 = ... = p_{2^l-1}$ then the quantity $H(p_1, p_2, ... p_{2^l-1}, q'_{2^l}) - u$ is greater than or equal to

$$\begin{cases} H\big(p_1, ..., p_1, 1 - (2^l - 1)p_1\big) - p_1, & \text{if } p_1 \leq 1/2^l \\ H\big(p_1, ..., p_1, 1 - (2^l - 1)p_1\big) - 1 + (2^l - 1)p_1, & \text{if } p_1 > 1/2^l. \end{cases} \tag{37}$$

Let us consider now the remaining $2^l - 2$ extreme points. They all belong to $R^l = \cup_{i=2}^{2^l-1} R^l_i$. Indeed, $p_1 > 2/(2^{l+1} + 1)$ and $t = 2^l - 1$ imply $p_1 > 1/(t + 2)$, and thus, recalling (36), $2p_1 > q_{2^l}$. Since extreme points (36), for $k \leq 2^l - 2$, satisfy $2p_1 = 2p_2 = ... = 2p_k > q_{2^l} > q_{k+1} = p_{k+2} = ... = p_{2^l-1} = q_{2^l}/2$ one obtains that they really belong to R^l_{k+1}. Applying Lemma 5 leads thus to

$$H(p_1, p_2, ... p_{2^l-1}, q'_{2^l}) - u \geq \mathcal{H}(p_1) + (1 - p_1) A_{2^l-1}. \tag{38}$$

Consequently the minimum of (35) is determined by (37) and (38). The difference, $\Delta_l(p_1)$, $l \geq 3$, between (37) and (38) can be written as $\Delta_l(p_1) = \mathcal{H}((2^l - 1)p_1) - \mathcal{H}(p_1) - (1 - p_1)A_{2^l-1} + (2^l - 1)p_1 \log(2^l - 1) - u$ where $u = p_1$ if $p_1 \leq 1/2^l$ and $u = 1 - p_1$ if $p_1 > 1/2^l$. $\Delta_l(p_1)$ is a convex $\cap$ function of p_1, in each of the intervals $[2/(2^{l+1} + 1), 1/2^l]$ and $[1/2^l, 1/(2^l - 1)]$. The next three claims prove that, for $l \geq 4$, $\Delta_l(p_1)$ is positive at the extreme points of the two intervals. Hence from the convex $\cap$ property, $\Delta_l(p_1)$ is positive in both the intervals, proving (38) true for $t = 2^l - 1$ and $l \geq 4$, and therefore (32) for the case $t = 2^l - l$. In the case $l = 3$ the function $\Delta_3(p_1)$ is positive in $p_1 = 2/(2^{l+1}+1) = 2/17$ and $p_1 = 1/2^l = 1/8$, negative in $p_1 = 1/(2^l - 1) = 1/7$. Then by making use of the convexity $\cap$ property, $\Delta_3(p_1)$ is positive in the intervals $[2/17, 1/8]$ and $[1/8, \gamma]$ and negative in the interval $]\gamma, 1/7]$, where $\gamma \approx 0.1422$ is the unique zero of $\Delta_3(p_1)$ in the interval $1/8 \leq p_1 \leq 1/7$. Therefore the minimum of (35) is not smaller than

$$\begin{cases} \mathcal{H}(p_1) + (1-p_1)A_7, & \text{if } p_1 < \gamma \\ \mathcal{H}(7p_1) + (7\log 7)p_1 - 1 + 7p_1, & \text{if } p_1 \geq \gamma; \end{cases} \tag{39}$$

that because of (35), recalling $A_7 = 2\log 3 - 5/9$, proves (30) and (31). Define $\Delta'_l(p_1)$, $l \geq 3$, as

$$\Delta'_l(p_1) = \Delta_l(p_1) + (1-p_1)[A_{2^l-1} - (\log(2^{l+1}-3) - 1 - \beta) - \bar{\epsilon}_{2^l-1}], \tag{40}$$

where $\bar{\epsilon}_{2^l-1} = (\log e)^3/((2^{l+1}-3)(2^{l+1}-3+\log e))$. Lemma 5 implies

$$\Delta'_l(p_1) \leq \Delta_l(p_1). \tag{41}$$

Claim 1. $\Delta_l(2/(2^{l+1}+1)) > 0$, *for* $l \geq 3$.

If we allow l to take nonintegral values, then $\Delta'_l(2/(2^{l+1}+1))$ is a continuous function of l. An elementary calculus argument shows that it is increasing with l, $l \geq 3$. Since its evaluation in $l = 3$ gives $\Delta'_3(2/17) \approx 0.023$, it results increasing and positive for any $l \geq 3$. Recalling (41) the claim follows.

Claim 2. $\Delta_l(1/2^l) > 0$, *for* $l \geq 3$.

As in the proof of the previous claim, we allow l to take nonintegral values. Then one has that $\Delta'_l(1/2^l)$ is increasing with l, $l \geq 3$. Since its evaluation at $l = 3$ gives $\Delta_3(1/8) \approx 0.0298$, it results increasing and positive for any $l \geq 3$. The claim then follows from (41).

Claim 3. $\Delta_l(1/(2^l-1)) > 0$ *for* $l \geq 4$, *and* $\Delta_3(1/7) < 0$.

A direct calculation shows that $\Delta_3(1/7) \approx -0.0252$. Suppose now $l \geq 4$. The function $\Delta'_l(1/(2^l-1))$ can be written as $(1-(2^l-1)^{-1})(1+\beta-\log(2+(2^l-2)^{-1})) - (\log e)^3/((2^{l+1}-3)(2^{l+1}-3+\log e)))$ and hence it is easily seen increasing with l, $l \geq 4$. Moreover, evaluated at $l = 4$ it gives $\Delta'_4(1/15) \approx 0.0298$. From (41) it follows that $\Delta_l(1/(2^l-1))$ is positive if $l \geq 4$, proving the claim.

We are left with the case $t = 2^l$. Suppose $t = 2^l$. From Lemma 6 we have

$$r \leq l - H(p_1, ..., p_1, 1-(2^l-1)p_1); \tag{42}$$

that, by (37) and by Claims 1, 2, and 3 leads to (30), (31) and (32).

We now show that bounds (30), (31) and (32) are tight.

Bound (30) is reached by a source consisting of 11 letters with probabilities p, $(1-p)/9$, $(1-p)/9$, $(1-p)/9$, $(1-p)/9$, $(1-p)/9$, $(1-p)/9$, $(1-p)/9$, $(1-p)/9$, $(1-p)/9-\epsilon$ and ϵ as ϵ approaches 0.
Bound (31) is reached by a source consisting of 9 letters with probabilities p, p, p, p, p, p, p, $1-7p-\epsilon$ and ϵ as ϵ approaches 0.
Let k'_j be the integer value which minimizes A_j, i.e. $A_j = \log(j+k'_j)-1+(j-1-k'_j)/(j+k'_j)$ [4]. Then consider the source consisting of $1+2^l+k'_l$ letters, three of which with probabilities p, $(1-p)/(2^l+k'_l-1)-\epsilon$, ϵ, and each of the remaining with probability $(1-p)/(2^l+k'_l-1)$, where p satisfies (29) with $l \geq 4$. If ϵ approaches zero then the redundancy approaches (32). □

Remark 1. As l becomes large then p_1 approaches zero and the difference between upper bound (32) and upper bound

$$l-\mathcal{H}(p_1)-(1-p_1)\big(\log(2^{l+1}-3)-1-\beta\big) \tag{43}$$

becomes negligible because of Lemma 5. Therefore, for practical purposes, we can consider the bound (43) equivalent to (32), for small values of p_1. Moreover as p_1 approaches zero, bound (43), and consequently bound (32), approaches the bound of Gallager $r \leq \beta+p_1$. The difference between the two bounds tends to zero as l increases. Therefore, Theorem 5 provides also a proof of the fact that, as claimed by Gallager [12], the bound (12) is tight in the limit $p_1 \to 0$.

Remark 2. The derivation of all previous bounds does not depend on the finiteness of the source. Hence these upper bounds hold also for optimal infinite encodings. They are the tightest possible bounds as well. Infinite sources that achieve above upper bounds can be obtained by replacing, in the sources considered in the corresponding finite cases, the symbol with probability ϵ with a countably number of symbols with probabilities $\epsilon/2, \epsilon/4, \epsilon/8, \epsilon/16, ..., \epsilon/2^i, \ldots$

The following theorem gives a tight upper bound on r as a function of the minimum codeword length l of the Huffman code [9].

Theorem 6 *Let l be the minimum length of the Huffman code C. Then the redundancy of C is upper bounded by*

$$r \leq l - M_{2^l}. \tag{44}$$

The bound is the best possible bound expressed only in terms of l.

Proof. Let $q_1 \geq q_2 \geq ... \geq q_L$ be the $L = 2^l$ probabilities at level l in the Huffman code. Recalling (1) and minimizing over all possible choices of q_i's we obtain

$$r \leq l - \min_{q_i \in X_L}\left\{H(q_1, q_2, ..., q_L) - q_L\right\}$$

that, by Lemma 5, yields $r \leq l - M_{2^l}$.

We now prove that the bound is the best possible bound expressed only in terms of l. First, suppose $l \leq 3$. Consider the source consisting of 2^l letters, one of them with probability ϵ, one of them with probability $(2^l - 1)^{-1} - \epsilon$, and each of the remaining letters with probability $1/(2^l - 1)$. The Huffman code has 2^l codewords of length l. Its redundancy approaches $l - \log(2^l - 1)$ as $\epsilon \to 0$. Finally, suppose $l \geq 4$. Let k'_j be the value for which A_j reaches its minimum, $A_j = \log(j + k'_j) - (1 + 2k'_j)/(j + k'_j)$ [9]. Consider the source consisting of $2^l + k'_{2^l} + 1$ letters, two of them with probabilities ϵ and $(2^l + k'_{2^l})^{-1} - \epsilon$, and each of the remaining $2^l + k'_{2^l} - 1$ letters with probability $(2^l + k'_{2^l})^{-1}$. The Huffman code has $2^l - k'_{2^l} - 1$ codewords of length l and $2k'_{2^l} + 2$ of length $l+1$. Its redundancy approaches $l - M_{2^l}$ as $\epsilon \to 0$. □

The first nine values of $l - M_{2^l}$ are presented in Table 2.

l	$l - M_{2^l}$
1	1
2	$2 - \log 3$
3	$3 - \log 7$
4	$95/21 - \log 21$
5	$157/44 - \log 11$
6	$313/88 - \log 11$
7	$1338/177 - \log 177$
8	$2675/354 - \log 177$
9	$6776/709 - \log 709$

Table 2.

A property of the bound stated in Theorem 6 is that it is decreasing with l.

Theorem 7 *The function $l - M_{2^l}$ is strictly decreasing with l.*

Proof. From Table 2, the theorem is true for the first few values. We prove that $l-1-M_{2^{l-1}} > l - M_{2^l}$, for $l \geq 5$. From (11) we have $M_{2^l} \geq \log(2^{l+1}-1)-1-\beta$ and $M_{2^{l-1}} \leq \log(2^l - 1) - 1 - \beta + (\log e)^3/((2^l - 1)(2^l - 1 + \log e))$. Thus, for proving the theorem it is enough to prove that, for $l \geq 5$,

$$\log(2^{l+1} - 1) - 1 - \beta > 1 + \log(2^l - 1) - 1 - \beta + \frac{(\log e)^3}{(2^l - 1)(2^l - 1 + \log e)}.$$

To prove above inequality, denote by $\omega(l)$ the function $\log(2^{l+1} - 1) - 1 - \log(2^l - 1)$. The function $\omega(l)$ can be rewritten as $\log((2^{l+1} - 1)/(2^{l+1} - 2))$ which is $\geq (\log e)/(2^{l+1} - 1)$, because of the inequality $\ln x \geq 1 - 1/x$. In turn, $(\log e)/(2^{l+1} - 1)$ is $\geq 0.2/(2^l - 1)$, for $l \geq 5$. Since $(\log e)^3/(2^l - 1 + \log e)$ is a decreasing function of $l \geq 4$, it reaches its maximum at $l = 4$ where it is equal to $0.18262 < 0.2$. Hence, $\omega(l) \geq (\log e)^3/((2^l - 1)(2^l - 1 + \log e))$, which concludes the proof of the theorem. □

Remark 3. Since the difference between M_j and $\log(2j - 1) + 1 + \beta$ becomes negligible as $j \to 0$, the upper bound (44) approaches $l - \log(2^{l+1} - 1) + 1 + \beta$, for large l. Thus, as $l \to \infty$ the upper bound approaches β.

Making use of Theorem 6 and Lemma 2, we can prove the following theorem.

Theorem 8 *Let p_1 be the probability of the most likely source letter of a discrete source A. The redundancy of the corresponding binary Huffman code is upper bounded by*

$$r \leq \beta + p_1, \quad \text{if } 1/(2^l - 1) \leq p_1 < l - 1 - M_{2^{l-1}} - \beta,\ l \geq 4, \tag{45}$$

$$r \leq l - 1 - M_{2^{l-1}}, \quad \text{if } l - 1 - M_{2^{l-1}} - \beta \leq p_1 \leq 2/(2^l + 1),\ l \geq 4. \tag{46}$$

Proof. From Theorem 2 one has that $r \leq p_1 + \beta$ for $1/(2^l - 1) \leq p_1 \leq 2/(2^l + 1)$, $l \geq 4$. From Lemma 2 we know that for $1/(2^l - 1) \leq p_1 \leq 2/(2^l + 1)$, $l \geq 4$, the Huffman code has a minimum codeword length of either l or $l - 1$. Because of Theorem 6, and since the function $l - M_{2^l}$ is a decreasing function of $l \geq 1$, it follows that for $1/(2^l - 1) \leq p_1 \leq 2/(2^l + 1)$, $l \geq 4$, one has $r \leq l - 1 - M_{2^{l-1}}$. Therefore, $r \leq \min\{\beta + p_1, l - 1 - M_{2^{l-1}}\}$, for $1/(2^l - 1) \leq p_1 \leq 2/(2^l + 1)$, $l \geq 4$. Let us now compute this minimum.
First, notice that from (11) we get $l - 1 - M_{2^{l-1}} < l - \log(2^l - 1) + \beta$. Making use of the inequality $\ln x \leq x - 1$, it follows $l - \log(2^l - 1) + \beta = \beta + \log(2^l/(2^l - 1)) \leq \beta + (\log e)/(2^l - 1)$ which in turn results $< \beta + 2/(2^l + 1)$. Thus, $l - 1 - M_{2^{l-1}} < \beta + 2/(2^l + 1)$.
Next, notice that from (11) we also have $l - 1 - M_{2^{l-1}} > l - \log(2^l - 1) + \beta - (\log e)^3/((2^l - 1)(2^l - 1 + \log e))$. Because of the inequality $\ln x \geq 1 - 1/x$, it follows that $l - \log(2^l - 1) = \log(2^l/(2^l - 1)) \geq (\log e)/2^l$. Since $(\log e)^3/(2^l - 1 + \log e)$ is a decreasing function of $l \geq 4$ it reaches its maximum at $l = 4$ where it equals $0.18262 < 0.2$. Hence, $l - 1 - M_{2^{l-1}} > (\log e)/2^l + \beta - 0.2/(2^l - 1)$. Finally, since $(\log e)/2^l \geq 1.2/(2^l - 1)$, we have $l - 1 - M_{2^{l-1}} > \beta + 1/(2^l - 1)$. Since $\beta + 1/(2^l - 1) < l - 1 - M_{2^{l-1}} < \beta + 2/(2^l + 1)$, then the two bounds $p_1 + \beta$ and $l - 1 - M_{2^{l-1}}$ intersect exactly at a single point in the interval $[1/(2^l - 1), 2/(2^l + 1)]$, $l \geq 4$. Thus, $p_1 + \beta < l - 1 - M_{2^{l-1}}$ if $1/(2^l - 1) \leq p_1 < l - 1 - M_{2^{l-1}} - \beta$, $l \geq 4$, and $p_1 + \beta > l - 1 - M_{2^{l-1}}$ if $l - 1 - M_{2^{l-1}} - \beta \leq p_1 \leq 2/(2^l + 1)$, $l \geq 4$. Bound $r \leq l - 1 - M_{2^{l-1}}$ improves Gallager's bound for $l \geq 4$ when $p_1 \in [l - 1 - M_{2^{l-1}} - \beta, 2/(2^l + 1)]$. □

Recently Manstetten [19] refining the Gallager's bound $r \leq l - H(q'_1, q'_2, ..., q'_L) + q_{2m-1}$ proved that the redundancy of binary Huffman codes can be estimated as

$$r \leq s + \sum_{k=v}^{m} q'_k - H(q'_1, q'_2, ..., q'_m) + q'_m,$$

where $1/(j + 1) \leq p_1 < 1/j$, s is the integer such that $2^s - 1 \leq j < 2^{s+1} - 1$, $m = j + 1$, $v = 2^{s+1} - j$, and the sum is set to zero if $v > m$.
Above expression permits for each j (with *a case-by-case* computation) to compute optimal bounds on the redundancy of binary Huffman codes for $1/(j+1) \leq p_1 < 1/j$. Manstetten performed the computation down to $p_1 \geq 1/127$. We refer the reader to [19] for the explicit values of such bounds.
Beyond $p_1 = 1/127$ and outside the intervals (29) the best upper bounds for the redundancy of binary Huffman codes to date are provided by (45) and (46).

Figure 1 shows the upper bound on the redundancy of binary Huffman codes versus the probability p_1 of the most likely letter.

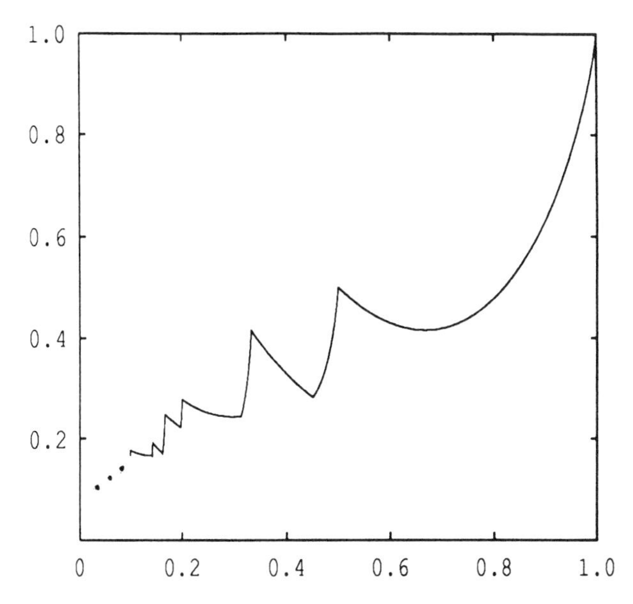

Figure 1. Upper bound of binary Huffman codes as function of p_1.

4 Bounds as function of p_1. The D-ary case

4.1 An upper bound as function of p_1

Throughout this section we assume that only the probability of the most likely source letter p_1 is known and that the size of the code alphabet is D. We consider first the problem of determining upper bounds on the redundancy. Gallager [12] proved that

$$r_D \leq \sigma_D + p_1 D / \ln D \tag{47}$$

where $\sigma_D = \log_D(D-1) + \log_D(\log_D e) - \log_D e + (D-1)^{-1}$. As a numerical example for σ_D, $\sigma_3 = 0.135$, $\sigma_5 = 0.194$, $\sigma_{10} = 0.269$, $\sigma_{20} = 0.335$ have been calculated in [12]. The bound has been improved subsequently by Manstetten [19] who showed that

$$r_D \leq \sigma_D + p_1 D / (e \ln D). \tag{48}$$

Bound (48) improves the Shannon limit, $r_D \leq 1$, only when $p_1 < \gamma_D = (1 - \sigma_D)(e \ln D)/D$.

Manstetten [19] also proved that for $0.5 \leq p_1 < 1$ and $D \geq 3$ the following tight bound holds

$$r_D \leq 1 - \mathcal{H}_D(p_1). \tag{49}$$

Recalling inequality (8) and using $L_B \leq H_B + 1$, one has [5]

$$r_D \leq 1 - \mathcal{H}_D(p_1) + (1 - p_1)2/D. \tag{50}$$

The bound (50) is better than (48) for $\theta_D < p_1 < 1/2$, where θ_D is the unique solution of the equation $\sigma_D + xD/(e \ln D) = 1 - \mathcal{H}_D(x) + (1-x)2/D$ in the interval $x \in]0, 0.5[$. As D gets large, bounds (49) and (50) approach 1. The same is true for Gallager's (47) and Manstetten's bounds (48) (we only consider the interval where they improve the Shannon limit, since as stressed in [12], $\sigma_D \to 1$ as D increases). In all cases the convergence is not rapid. This is not just the case of (47)–(50), but it is rather inherent to the definition of the redundancy, as the following theorem shows

Theorem 9 *Let $\phi^+(D, p_1) \leq 1$ be an upper bound on the redundancy of any D-ary Huffman code for the source A in terms of the probability p_1 of the most likely letter. Then for any p_1, $1 > p_1 > 0$, it results* $\lim_{D\to\infty} \phi^+(D, p_1) = 1$.

Proof. Let A be a source consisting of N letters. For every $D > N$, the average codeword length of the corresponding D-ary Huffman code is 1. For a fixed N, the entropy of the source A approaches 0 as D gets large, as it is computed with base D logarithms. Therefore any upper bound on r_D must approach 1 as D increases. □

Lemma 7 *For any $D \geq 2$, the function*

$$f_D(x) = 1 - \mathcal{H}_D(x) + (1-x)2/D$$

is a continuous and convex ∪ function of x that reaches its minimum at point $x_{D,min} = D^{2/D}/(1 + D^{2/D})$, and $0.5 < x_{D,min} < 1$. Moreover

$$f_D(x) < 1 \qquad \text{if and only if} \qquad \delta_D < x < 1$$

where δ_D is the unique solution of the equation $f_D(x) = 1$ in the interval $]0, 1[$. The sequence $\{\delta_D\}_{D\geq 2}$ satisfies the following properties

1. $\delta_{D+1} < \delta_D$;
2. $\delta_D \leq (\log_2 D)/(D + \log_2 D)$.

To summarize we have three bounds, namely (48), (49), and (50). Their combined use gives a bound that is definitively better than the Shannon limit, for any value of $0 < p_1 < 1$. Notice that the width of the interval in which Manstetten's bound (48) holds approaches 0 as D gets large, whereas the total width of (49) and (50) approaches 1.

Theorem 10 *Let p_1 be the probability of the most likely source letter of a discrete source A. The redundancy of the corresponding D-ary, $D \geq 3$, Huffman code is upper bounded by*

$$r_D \leq \sigma_D + p_1 D/(e \ln D) \qquad \text{if } 0 < p_1 \leq \theta_D \tag{51}$$

$$r_D \leq 1 - \mathcal{H}_D(p_1) + (1-p_1)2/D \qquad \text{if } \theta_D < p_1 < 1/2 \tag{52}$$

$$r_D \leq 1 - \mathcal{H}_D(p_1) \qquad \text{if } 1/2 \leq p_1 < 1 \tag{53}$$

where θ_D is the unique solution of the equation $\sigma_D + xD/(\ln D)e = 1 - \mathcal{H}_D(x) + (1-x)2/D$ in the interval $x \in]0, 0.5[$. The sequence $\{\theta_D\}_{D\geq 3}$ satisfies the following inequalities

1. $\delta_D < \theta_D < \gamma_D$;

2. $\theta_D < e(1 + \ln\ln D)/D$.

The bound provided by Theorem 10 can be further improved using inequality (9). Since the minimum of $g_D(x) = 1 - \mathcal{H}_D(x)$ is reached at 0.5, the minimum of $f_D(x)$ satisfies $0.5 < x_{D,min} < 1$, $f_D(1) = g_D(1) = 1$, $f_D(\theta_D) = \sigma_D + \theta_D D/(e \ln D) < 1$, with $\theta_D < 0.5$, and moreover $f_D(x)$ and $g_D(x)$ are convex, there exists a unique solution of the equation $g_D(x) = f_D(\theta_D)$ in the interval $]0.5, 1[$. Denote this solution by η_D.
Suppose that it is known only an upper bound on p_1, say p, and it is unknown its exact value $0 < p_1 \le p < 1$. Any upper bound on the redundancy r_D of the Huffman code for the source A, as a function of p, must be a nondecreasing function. Making use of Theorem 10 we obtain the following upper bound on r_D

$$r_D \le r_{max}(p) \tag{54}$$

where

$$r_{max}(p) = \begin{cases} \sigma_D + pD/(e \ln D), & \text{if } 0 < p \le \theta_D \\ \sigma_D + \theta_D D/(e \ln D), & \text{if } \theta_D < p \le \eta_D \\ 1 - \mathcal{H}_D(p), & \text{if } \eta_D < p < 1. \end{cases}$$

Recalling (9)

$$r_D \le 1 - \mathcal{H}_D(p_1) + (1 - p_1)(r_{max}(p_1/(1 - p_1)) - 1 + 2/D),$$

and using the identity $\mathcal{H}_D(x) + (1 - x)\mathcal{H}_D(x/(1 - x)) = \mathcal{H}_D(2x) + 2x \log_D 2$, inequality (54) and the above reported expression of $r_{max}(p)$ one has

$$r_D \le \begin{cases} 1 - \mathcal{H}_D(p_1) + (1 - p_1)(\sigma_D + \theta_D D/(e \ln D) - 1 + 2/D), \\ \qquad \text{if } 0 < p_1 \le \eta_D/(1 + \eta_D) \\ 1 + 2/D - 2p_1(1/D + \log_D 2) - \mathcal{H}_D(2p_1), \\ \qquad \text{if } \eta_D/(1 + \eta_D) < p_1 < 0.5. \end{cases} \tag{55}$$

We will now combine bound (48), bound (49) and the new derived bounds (55) all together.
First observe that $\eta_D/(1 + \eta_D) > 1/3$, since $\eta_D > 0.5$. Hencefrom we get $\eta_D/(1+\eta_D) > \theta_D$. Set $h_D(x) = 1 - \mathcal{H}_D(x) + (1-x)(\sigma_D + \theta_D D/(e \ln D) - 1 + 2/D)$. $h_D(x)$ is a continuous, convex $\cup$ function of x that evaluated at $x = 0$ gives $h_D(0) > \sigma_D$. Since $\sigma_D + \theta_D D/(e \ln D) < 1$, $h_D(x) < f_D(x)$ for $x < 0.5$, and $\theta_D < 0.5$, at the point $x = \theta_D$ it results $h_D(\theta_D) < f_D(\theta_D) = \sigma_D + \theta_D D/(e \ln D)$. Therefore the equation $h_D(x) = \sigma_D + xD/(e \ln D)$, as a function of x, has a unique solution μ_D in the interval $]0, \gamma_D[$.
Thus bound (55) improves (48) for $\mu_D < p_1 \le \theta_D$, and improves (52) for $\theta_D \le p_1 < 0.5$.
Combining (48) and (55) we can finally state our result in the following

Theorem 11 *Let p_1 be the probability of the most likely source letter of a discrete source A. The redundancy of the corresponding D-ary Huffman code is upper bounded by*

$$r_D \le \sigma_D + p_1 D/(e \ln D) \qquad \textit{if } 0 < p_1 \le \mu_D$$

$$r_D \leq 1 - \mathcal{H}_D(p_1) + (1-p_1)(f_D(\theta_D) - 1 + 2/D) \qquad \textit{if } \mu_D < p_1 \leq \eta_D/(1+\eta_D)$$
$$r_D \leq 1 + 2/D - 2p_1(1/D + \log_D 2) - \mathcal{H}_D(2p_1) \qquad \textit{if } \eta_D/(1+\eta_D) < p_1 \leq 0.5$$
$$r_D \leq 1 - \mathcal{H}_D(p_1) \qquad \textit{if } 0.5 \leq p_1 < 1.$$

Bounds $r_D \leq 1 - \mathcal{H}_D(p_1) + (1-p_1)(\sigma_D + \theta_D D/(e \ln D) - 1 + 2/D)$ and $r_D \leq 1 + 2/D - 2p_1(1/D + \log_D 2) - \mathcal{H}_D(2p_1)$ are new and reported here for the first time. Bound $r_D \leq 1 - \mathcal{H}_D(p_1)$ is tight and it is reached by a source consisting of D letters having probabilities $p_1 = p$ and $p_2 = 1 - p - (D-2)\epsilon$ and $p_i = \epsilon$ for $3 \leq i \leq D$ [19].

4.2 A lower bound as function of p_1

Now we turn our attention to the derivation of a tight lower bound. The bound holds whatever D is.
Let us define

$$\alpha_{D,0} = 1$$
$$\alpha_{D,m} = 1 - \left(\log_D \frac{D^{m+1}-1}{D^m - 1}\right)^{-1} \qquad \text{if } m \geq 1.$$

For $m \geq 0$ it results $\alpha_{D,m+1} < \alpha_{D,m}$; moreover $\lim_{m \to \infty} \alpha_{D,m} = 0$.

Lemma 8 *If a D-ary Huffman code has a minimum codeword length of l, $l \geq 1$, then its redundancy satisfies*

$$r_D \geq l - \mathcal{H}_D(p_1) - (1-p_1)\log_D(D^l - 1).$$

Proof. By hypothesis $n_1 = l$. Let $q_1', q_2', \ldots, q_L'$, $L = D^l$, be the probabilities of nodes $g_1, g_2, \ldots, g_L$ at level l in the Huffman code tree T. We may assume, without loss of generality that $p_1 = q_1'$. Let T_i, $2 \leq i \leq L$, the subtree of the tree T rooted at g_i. Let P_i be the set of source letters that correspond to leaves of T_i. Subtrees $T_2, T_3, \ldots, T_L$ determine a partition $P_2, P_3, \ldots, P_L$ of the source letters $\{a_2, a_3, \ldots, a_L\}$. Let H_i denote the entropy of P_i, and E_i the average codeword length of the Huffman code for subsource P_i. H_i and E_i are related to the entropy $H(p_1, p_2, \ldots, p_N)$ of the source and to the average codeword length E of the Huffman code by the following relations:

$$H(p_1, p_2, \ldots, p_N) = H(q_1', q_2', \ldots, q_L') + \sum_{i=2}^{L} q_i' H_i$$

and

$$E = l + \sum_{i=2}^{L} q_i' E_i.$$

Finally, making use of $E_i \geq H_i$, we get

$$\begin{aligned}
r_D &= E - H(p_1, p_2, \ldots, p_N) \\
&= l - H(q'_1, q'_2, \ldots, q'_L) + \sum_{i=2}^{L} q'_i(E_i - H_i) \\
&\geq l - H(q'_1, q'_2, \ldots, q'_L) \\
&= l - \mathcal{H}_D(p_1) - (1-p_1)H\left(\frac{q'_2}{1-p_1}, \frac{q'_3}{1-p_1}, \ldots, \frac{q'_L}{1-p_1}\right) \\
&\geq l - \mathcal{H}_D(p_1) - (1-p_1)\log_D(D^l - 1).
\end{aligned}$$

□

Theorem 12 *If for some $m \geq 1$,*

$$\alpha_{D,m} < p_1 \leq \alpha_{D,m-1}$$

then the redundancy of any D-ary Huffman code satisfies

$$r_D \geq m - \mathcal{H}_D(p_1) - (1-p_1)\log_D(D^m - 1). \tag{56}$$

The bound is tight.

Proof. Assume that $\alpha_{D,m} < p_1 \leq \alpha_{D,m-1}$. Then since $1/(D^{m+1} - D + 1) \leq \alpha_{D,m} < \alpha_{D,m-1} \leq D/(D^m + D - 1)$ for $m > 1$, Lemma 2 implies that the minimum codeword length of the D-ary Huffman code can be $m-1$, m or $m+1$, if $m > 1$, and 1 or 2 if $m = 1$.
If it is $m-1$, $m > 1$, then from Lemma 8 we get

$$\begin{aligned}
r_D &\geq m - 1 - \mathcal{H}_D(p_1) - (1-p_1)\log_D(D^{m-1} - 1) \\
&\geq m - \mathcal{H}_D(p_1) - (1-p_1)\log_D(D^m - 1)
\end{aligned}$$

since $p_1 \leq \alpha_{D,m-1}$.
If it is m, $m \geq 1$, then the lemma follows from Lemma 8.
If it is $m+1$, $m \geq 1$, then from Lemma 8 we get

$$\begin{aligned}
r_D &\geq m + 1 - \mathcal{H}_D(p_1) - (1-p_1)\log_D(D^{m+1} - 1) \\
&\geq m - \mathcal{H}_D(p_1) - (1-p_1)\log_D(D^m - 1)
\end{aligned}$$

since $\alpha_{D,m} < p_1$.

The lower bound is tight. Indeed it is reached by any source consisting of $D^{m+1} - D + 1$ letters having probabilities $p_1 = p$ and $p_i = (1-p)/(D^{m+1} - D)$ for $2 \leq i \leq D^{m+1} - D + 1$. Notice that a Huffman code for this source has minimum codeword length $n_1 = m$, whereas remaining codeword lengths are $n_i = m + 1$.
□

The bound of Theorem 12 is due to Capocelli and De Santis and appeared in [5], it has been subsequently rediscovered by [19]. For $D = 2$ it provides the lower bound on the redundancy of Huffman codes as function of p_1 obtained by [3] and independently by [21]. Notice that, putting $m = 1$ in Theorem 12, we obtain a bound that corrects Golic and Obradovic's Theorem 2 [14]. Indeed, their bound is greater than our tight lower bound, and then is erroneous. The difference between the two bounds is $(1 - p_1)\log_D(D - 1)$, that is positive for any $p_1 < 1$, $D > 2$.

Let $\phi^-(D, p_1)$ be the lower bound on r_D as stated in Theorem 12. Then it is easily seen that whatever p_1, $1 > p_1 > 0$, it results $\lim_{D\to\infty} \phi^-(D, p_1) = p_1$. On the other hand Theorem 9 states that for large D any upper bound approaches 1. Informally speaking, for large values of D the redundancy of D-ary Huffman codes can be bounded by $p_1 - \epsilon \leq r_D \leq 1$, $\epsilon > 0$, both inequalities being the best possible. More formally, making use of Theorem 9 and of the above mentioned limit $\lim_{D\to\infty} \phi^-(D, p_1) = p_1$, it is easily obtained the following [5]

Theorem 13 *For every $\epsilon > 0$, there exists $D_0 \geq 2$ such that for $D \geq D_0$ the redundancy of D-ary Huffman codes can be bounded by $p_1 - \epsilon \leq r_D \leq 1$. Moreover for every $\epsilon_1 > 0$, $\epsilon_2 > 0$ and $1 > p_1 > 0$ there exist two sources A_1 and A_2, where p_1 is the probability of the most likely letter, such that redundancies $r_D{}'$ and $r_D{}''$ of the corresponding D-ary Huffman codes for A_1 and A_2 satisfy $p_1 - \epsilon < r_D{}' < p_1$ and $1 - \epsilon_2 < r_D{}'' < 1$.*

Figure 2 shows the lower bound on the redundancy of binary Huffman codes versus the probability p_1 of the most likely letter.

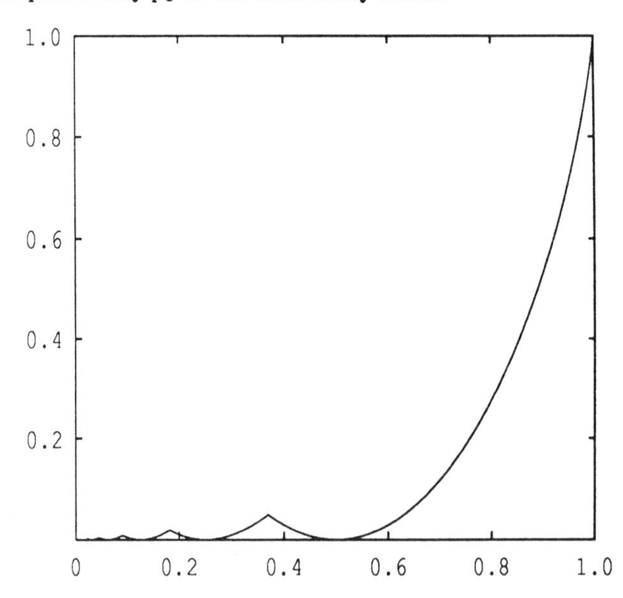

Figure 2. Lower bound of binary Huffman codes as function of p_1.

5 Bounds on the redundancy as function of p_N

In this section we consider binary codes and the case in which only the least likely probability p_N of the source is given. This problem has been considered for the first time by Reza [22] and Horibe [15]. They proved that the redundancy r of binary Huffman codes, as a function of p_N, is upper bounded by

$$r \leq 1 - 2p_N. \tag{57}$$

The bound (57) is not tight if $p_N \neq 1/2$. In the sequel we present both an upper bound and a lower bound as function of p_N that are tight whatever p_N is [6]. To prove the bounds we make use of the following lemmas.

Lemma 9 *Let p_1 and p_N be the probabilities of the most and least likely source letters of a discrete source A, that consists of N letters. Then*

$$\frac{1-p_N}{N-1} \leq p_1 \leq 1-(N-1)p_N.$$

Proof. The lemma follows from $1 - p_N = \sum_{i=1}^{N-1} p_i \leq \sum_{i=1}^{N-1} p_1 = (N-1)p_1$ and $1 - p_1 = \sum_{i=2}^{N} p_i \geq \sum_{i=2}^{N} p_N = (N-1)p_N$. □

Lemma 10 *Let p_1 and p_N be the probabilities of the most and least likely source letters of a discrete source A, that consists of N letters. The redundancy of the corresponding binary Huffman code is*

$$r = 1 - \mathcal{H}(p_N), \qquad \text{if } N = 2$$

$$r = 2 - p_1 - H(p_1, 1-p_1-p_N, p_N), \qquad \text{if } N = 3$$

$$r \leq 2 - H(p_1, 1-p_1-2p_N, p_N, p_N), \qquad \text{if } N = 4 \text{ and } p_N > 1/5.$$

Proof. If $N = 2$ then $L = 1$ and $H = \mathcal{H}(p_1)$. Hence $r = 1 - \mathcal{H}(p_N)$. If $N = 3$ then $L = 2 - p_1$ and $r = L - H = 2 - p_1 - H(p_1, 1-p_1-p_N, p_N)$.
Suppose now that the source A consists of four letters and $p_N > 1/5$. From Lemma 9 we obtain $(1-p_N)/3 \leq p_1 \leq 1 - 3p_N$. Since each codeword of the Huffman code has length 2, then its redundancy is

$$\begin{aligned} r &= 2 - H(p_1, p_2, 1-p_1-p_2-p_N, p_N) \\ &\leq \max_{\{x \epsilon P_1\}} \{2 - H(p_1, x, 1-p_1-x-p_N, p_N)\} \end{aligned}$$

where P_1 is the set of all possible values of p_2, that is $P_1 = \{x | (1-p_1-p_N)/2 \leq x \leq 1-p_1-2p_N\}$.
The function $\phi(x) = 2 - H(p_1, x, 1-p_1-x-p_N, p_N)$ is a convex $\cup$ function of x, and thus it takes the maximum value at an extreme point. At $x = 1-p_1-2p_N$ it is equal to

$$\phi(1-p_1-2p_N) = 2 - H(p_1, 1-p_1-2p_N, p_N, p_N).$$

On the other hand, if $x = (1 - p_1 - p_N)/2$ it is equal to

$$\begin{aligned}\phi\left(\frac{1-p_1-p_N}{2}\right) &= 2 - H(p_1, 1-p_1-p_N, p_N) - (1-p_1-p_N) \\ &\leq 2 - H(p_1, 1-p_1-2p_N, p_N, p_N).\end{aligned}$$

Hence the lemma. □

5.1 An upper bound when only p_N is known

The following theorem improves the upper bound (57). It is easy to see that the bound is the best possible bound expressed as a function only of the probability p_N.

Theorem 14 *Let p_N be the probability of the least likely source letter of a discrete source A. The redundancy of the corresponding binary Huffman code is upper bounded by*

$$r \leq 1 - \mathcal{H}(p_N). \tag{58}$$

Proof. The theorem is easily proved if $N = 2$. Assume $N \geq 3$ and therefore $p_N \leq 1/3$. We distinguish two cases: $p_N > 1/4$ and $p_N \leq 1/4$.

First consider the case $p_N > 1/4$. Then $N = 3$. Making use of Lemmas 8 and 9 we find that $(1 - p_N)/2 \leq p_1 \leq 1 - 2p_N$ and

$$\begin{aligned} r &= 2 - p_1 - H(p_1, 1-p_1-p_N, p_N) \\ &\leq \max_{\{x \epsilon P_2\}} \{2 - x - H(x, 1-x-p_N, p_N)\}\end{aligned}$$

where $P_2 = \{x | (1-p_N)/2 \leq x \leq 1 - 2p_N\}$ is the set of all possible values of p_1. The function to maximize is a convex ∪ function, and thus it takes its maximum value at an extreme point. If $x = (1-p_N)/2$, recalling that $p_N \leq 1/3$, one gets

$$\begin{aligned} 2 - x - H(x, 1-x-p_N, p_N) &= 1/2 + 3p_N/2 - \mathcal{H}(p_N) \\ &\leq 1 - \mathcal{H}(p_N).\end{aligned}$$

Whereas if $x = 1 - 2p_N$ one finds

$$\begin{aligned} 2 - x - H(x, 1-x-p_N, p_N) &= 1 - \mathcal{H}(2p_N) \\ &\leq 1 - \mathcal{H}(p_N).\end{aligned}$$

The theorem is thus proved if $p_N > 1/4$.

Now consider the case $p_N \leq 1/4$. We know from [2] that $r \leq p_1 + \beta - p_N$. If $p_1 < 1/3$ one gets

$$\begin{aligned} r &< 1/3 + \beta - p_N \\ &< 1 - \mathcal{H}(p_N).\end{aligned}$$

Assume $p_1 \geq 1/3$. We prove (58) by induction on p_N. We have shown that (58) is true for $p_N > 1/4$. Suppose that (58) is true for $p_N \geq \eta$, and let $2\eta/3 \leq p_N < \eta$. Thus $p_N/(1-p_1) \geq 3p_N/2 \geq \eta$. From (8) it then follows

$$\begin{aligned} r &\leq 1 - \mathcal{H}(p_1) + (1-p_1)\left(1 - \mathcal{H}\left(\frac{p_N}{1-p_1}\right)\right) \\ &= 2 - p_1 - H(p_1, 1-p_1-p_N, p_N). \end{aligned}$$

The last expression is a convex ∪ function of p_1. The maximum value as a function of p_1 is taken therefore on the boundary. Making use of Lemma 9 one finds $1/3 \leq p_1 \leq 1 - 2p_N$. If $p_1 = 1 - 2p_N$, since $p_N \leq 1/4$, one gets

$$r \leq 1 - \mathcal{H}(2p_N) < 1 - \mathcal{H}(p_N).$$

Whereas, if $p_1 = 1/3$ we get $r \leq 5/3 - \mathcal{H}(p_N) - (1-p_N)\mathcal{H}(1/(3-3p_N))$. From $p_N \leq 1/4$ it follows $3/4 \leq 1 - p_N$ and $\mathcal{H}(1/(3-3p_N)) \geq \mathcal{H}(1/3)$; thus

$$r \leq 5/3 - (3/4)\mathcal{H}(1/3) - \mathcal{H}(p_N) < 0.98 - \mathcal{H}(p_N).$$

This completes the proof of the theorem. □

5.2 A lower bound when only p_N is known

We now prove a lower bound on the redundancy of binary Huffman codes in terms of the probability p_N of the least likely source letter. The bound is the best possible lower bound as function only of p_N. For the proof of it we need the following auxiliary result.

Lemma 11 *Let $p_N > 0$ be the probability of the least likely letter of a finite source A. If the corresponding binary Huffman code has a maximum codeword length of n, then its redundancy is bounded by*

$$r \geq n - \log(2^n - 1) + p_N \log(2^n - 1) - \mathcal{H}(p_N), \quad \text{if } p_N \leq 2^{-n}$$

$$r \geq n - 1 - \log(2^{n-1} - 1) + 2p_N \log(2^{n-1} - 1) - \mathcal{H}(2p_N), \quad \text{if } p_N > 2^{-n}.$$

Proof. Without loss of generality assume that the codewords for the two least likely probabilities, p_{N-1} and p_N, have length n. From the identity

$$p_i n_i = p_i\left(n - H(S_{n-n_i})\right),$$

where S_{n-n_i} is the source with 2^{n-n_i} equiprobable letters, we have

$$\begin{aligned} r &= \sum_{i=1}^{N} p_i n_i - H(p_1, p_2, \ldots, p_N) \\ &= \sum_{i=1}^{N} p_i\left(n - H(S_{n-n_i})\right) - H(p_1, p_2, \ldots, p_N) \\ &= n - \sum_{i=1}^{N} p_i H(S_{n-n_i}) - H(p_1, p_2, \ldots, p_N) = n - H(Q) \end{aligned}$$

where Q is the source consisting of 2^n letters of which 2^{n-n_i} have probability $p_i/2^{n-n_i}$, for $i = 1, 2, ..., N$. Observe that in Q two letters probabilities equal to p_{N-1} and p_N, since $n_{N-1} = n_N = n$. Hence

$$r \geq \min\left\{n - H(y_1, y_2, ..., y_{2^n-2}, y, p_N)\right\} \tag{59}$$

where the minimum is taken over all the possible choices of positive y_i and y that satisfy the constraints $y \geq p_N$ and $\sum y_i + y + p_N = 1$. The right side of (59) can be written as

$$\min\left\{n - H(y, p_N, 1 - y - p_N) - (1 - y - p_N)H\left(\frac{y_1}{1 - y - p_N}, ..., \frac{y_{2^n-2}}{1 - y - p_N}\right)\right\}.$$

Therefrom, since the maximum value of the entropy of a source with $2^n - 2$ letters is $\log(2^n - 2)$, one gets

$$r \geq \min_{y \geq p_N} \left\{n - H(y, p_N, 1 - y - p_N) - (1 - y - p_N)\log(2^n - 2)\right\}. \tag{60}$$

In order to compute above minimum, observe that the function

$$g(x) = -H(x, p_N, 1 - x - p_N) - (1 - x - p_N)\log(2^n - 2)$$

is a convex $\cup$ function of x, with an absolute minimum in $x^* = (1 - p_N)/(2^n - 1)$. From the convexity, one gets that

$$\min_{x \geq p_N} g(x) = \begin{cases} g(x^*), & \text{if } x^* \geq p_N \\ g(p_N), & \text{if } x^* < p_N. \end{cases} \tag{61}$$

Since the condition $x^* \geq p_N$ is equivalent to $p_N \leq 2^{-n}$, substituting (61) into (60) yields

$$r \geq \begin{cases} n - g((1 - p_N)/(2^n - 1)), & \text{if } p_N \leq 2^{-n} \\ n - g(p_N), & \text{if } p_N > 2^{-n}; \end{cases}$$

that, recalling the definition of $g(x)$, proves the lemma. □

Previous Lemma 11 gives a lower bound on the redundancy of Huffman codes expressed in terms of both p_N and the maximum codeword length. We now provide a lower bound expressed only in terms of p_N. To this aim define

$$v(n, x) = n - \log(2^n - 1) + x\log(2^n - 1) - \mathcal{H}(x)$$

and

$$w(n, x) = n - \log(2^n - 1) + 2x\log(2^n - 1) - \mathcal{H}(2x)$$

for $n \geq 1$ and $0 < x \leq 0.5$. The lower bound given by Lemma 11 can be written as $r \geq l(n, p_N)$, where

$$l(n, p_N) = \begin{cases} v(n, p_N), & \text{if } p_N \leq 2^{-n} \\ w(n - 1, p_N), & \text{if } p_N > 2^{-n}. \end{cases} \tag{62}$$

For a fixed value of p_N, minimizing (62) over all possible values of n gives

$$r \geq \min_n l(n, p_N). \tag{63}$$

Notice that, since $v(n,x)$ and $w(n,x)$ are both nonnegative, the minimum in (63) is nonnegative.
A simple algebra gives the following two inequalities for $n \geq 1$ and $0 < x \leq 0.5$

$$v(n,x) > v(n+1,x) \quad \text{iff} \quad x < \gamma_{n+1} \tag{64}$$

$$w(n,x) > w(n+1,x) \quad \text{iff} \quad x < \gamma_{n+1}/2. \tag{65}$$

Where γ_m is defined as follows

$$\gamma_1 = 1$$

$$\gamma_m = 1 - \left(\log \frac{2^m - 1}{2^{m-1} - 1}\right)^{-1} \qquad \text{if } m \geq 2,$$

and it satisfies, for $m \geq 2$, the following inequality [6]

$$\frac{1}{2^{m-1}} > \frac{\log e}{2^m - 2 + \log e} > \gamma_m > \frac{\log e}{2^m - 1 + \log e} > \frac{1}{2^m}. \tag{66}$$

The following Table 3 shows the first few values of γ_m:

m	γ_m	m	γ_m
2	.36907024642854256291	12	.00035222539851554996
3	.18193210089874833082	13	.00017611146664321106
4	.09052518799091549829	14	.00008805542525326059
5	.04516896462115465377	15	.00004402763562020250
6	.02256278256945430729	16	.00002201379855982639
7	.01127616132175507164	17	.00001100689446751098
8	.00563679649606634260	18	.00000550344603067574
9	.00281808006029552038	19	.00000275172271457055
10	.00140896083609551391	20	.00000137586128209376
11	.00070446066341018734	21	.00000068793062224906

Table 3.

We are now ready to prove the lower bound.

Theorem 15 *Let $p_N > 0$ be the probability of the least likely letter of a finite source A. The redundancy of the corresponding binary Huffman code is lower bounded by*

$$r \geq m - \log(2^m - 1) + p_N \log(2^m - 1) - \mathcal{H}(p_N), \quad \text{if } \alpha_{m+1} < p_N \leq 2^{-m} \tag{67}$$

$$r \geq m - \log(2^m - 1) + 2p_N \log(2^m - 1) - \mathcal{H}(2p_N), \quad \text{if } 2^{-m-1} < p_N \leq \alpha_{m+1} \tag{68}$$

where α_m, $m \geq 2$, is the unique zero of the function

$$\psi_m(x) = \mathcal{H}(x) - \mathcal{H}(2x) + x\log(2^{m-1} - 1)$$

for $0 < x < 0.5$. Moreover, for $m \geq 3$ α_m satisfies

$$\frac{1}{2^m} < \alpha_m < \frac{\log e}{2^m - 1 + \log e}.$$

The bound is tight: for any value of p_N, a source exists whose binary Huffman code satisfies (67) and (68) with equality.

Proof. By inequality (63) we know that

$$\min_n l(n, p_N) \tag{69}$$

is a lower bound on the redundancy r. To prove the theorem it is then enough to show that $\min_n l(n, p_N)$ is given by (67) and (68).
For that consider the case $p_N = 2^{-m}$, for some $m \geq 1$. From (62) we get $l(m, p_N) = v(m, 2^{-m}) = 0$. Since $\min_n l(n, p_N)$ cannot be negative, we obtain that $l(m, p_N) = v(m, 2^{-m}) = 0$ gives the minimum (69), proving thus (67).
Let us now suppose that $2^{-m-1} < p_N < 2^{-m}$. First note that in computing the minimum (69) we can restrict our attention only to two possible values for n, namely m and $m+1$; that is

$$\min_n l(n, p_N) = \min_{n \in \{m, m+1\}} l(n, p_N). \tag{70}$$

To see this, let n_1 be a positive integer smaller than m. Since $p_N < 2^{-m} < 2^{-n_1}$, recalling (62) we have

$$l(n_1, p_N) = v(n_1, p_N)$$
$$l(m, p_N) = v(m, p_N).$$

From (66) we also have that $p_N < 2^{-m} < \gamma_m$. Hencefrom making use of (64) and (65) we get $v(n_1, p_N) > v(m, p_N)$ that yields $l(n_1, p_N) > l(m, p_N)$, proving thus that

$$\min_n l(n, p_N) \geq \min_{n \geq m} l(n, p_N).$$

Let n_2 be a positive integer greater than $m+1$. Since $p_N > 2^{-m-1} > 2^{-n_2}$, from (62) we have

$$l(n_2, p_N) = w(n_2 - 1, p_N)$$
$$l(m+1, p_N) = w(m, p_N).$$

From (66), $p_N > 2^{-m-1} > \gamma_{m+1}/2$, hence making use of (64) and (65) we get

$$w(n_2 - 1, p_N) > w(m, p_N)$$

that yields

$$l(n_2, p_N) > l(m+1, p_N),$$

concluding the proof of (70).

Since by hypothesis $2^{-m-1} < p_N < 2^{-m}$, it follows that

$$\min_{n \in \{m, m+1\}} l(n, p_N) = \min\{v(m, p_N),\ w(m, p_N)\}. \tag{71}$$

The difference between the two functions w and v appearing in the previous equation is given by

$$w(m, p_N) - v(m, p_N) = \psi_{m+1}(p_N), \tag{72}$$

where

$$\psi_{m+1}(x) = \mathcal{H}(x) - \mathcal{H}(2x) + x \log(2^m - 1).$$

$\psi_{m+1}(x)$ is a convex $\cup$ function of x that satisfies $\psi_{m+1}(0) = 0$, $\psi_{m+1}(2^{-m-1}) < 0$, for $m \geq 1$, and $\psi_{m+1}((\log e)/(2^{m+1} - 1 + \log e)) > 0$, for $m \geq 2$ [6]. Because of the convexity, $\psi_{m+1}(x)$ has a unique zero, α_{m+1}, in the interval $[2^{-m-1}, (\log e)/(2^{m+1} - 1 + \log e)]$ and

$$\psi_{m+1}(x) \begin{cases} \leq 0, & \text{if } 2^{-m-1} < x \leq \alpha_{m+1} \\ > 0, & \text{if } \alpha_{m+1} < x < 2^{-m}. \end{cases}$$

Recalling (72) we get

$$w(m, p_N) \leq v(m, p_N) \quad \text{if } 2^{-m-1} < p_N \leq \alpha_{m+1}$$

$$w(m, p_N) > v(m, p_N) \quad \text{if } \alpha_{m+1} < p_N < 2^{-m}$$

that together with (70) and (71) prove lower bounds (67) and (68).

To complete the proof of the theorem it remains to show that bounds (67) and (68) are tight. We prove this by exhibiting a source whose redundancy satisfies the bounds with equality, for any value of the minimum letter probability p_N. We distinguish two cases: $\alpha_{m+1} < p_N \leq 2^{-m}$ and $2^{-m-1} < p_N \leq \alpha_{m+1}$. Let U_{n+1} be the code consisting of the codeword 0^n and the n codewords 0^i1 for $i = 0, 1, 2, ..., n-1$. For example $U_4 = \{1, 01, 001, 000\}$.
Let p be in the interval $\alpha_{m+1} < p \leq 2^{-m}$. Consider the source S_{m+1} consisting of $N = m+1$ letters with increasing probabilities p, $(1-p)/(2^m - 1)$, $2(1-p)/(2^m - 1)$, $4(1-p)/(2^m - 1)$, ..., $2^{m-1}(1-p)/(2^m - 1)$. Notice that p is the minimum probability. Using the code U_{m+1} to encode S_{m+1}, we obtain a redundancy that satisfies (67) with equality.
Let p be such that $2^{-m-1} < p \leq \alpha_{m+1}$. Consider the source S'_{m+2} consisting of $N = m+2$ letters with increasing probabilities p, p, $(1-2p)/(2^m - 1)$, $2(1-2p)/(2^m-1)$, $4(1-2p)/(2^m-1)$, ..., $2^{m-1}(1-2p)/(2^m-1)$. Notice that p is the minimum probability, since $\alpha_{m+1} < (\log e)/(2^{m+1} - 1 + \log e) < 1/(2^m + 1)$, $m \geq 2$. Encoding S'_{m+2} with the code U_{m+2} gives an average length that satisfies (68) with equality. □

The following Table 4 shows the first few values of α_m:

m	α_m	m	α_m
2	.3333333333333333333	13	.0001659099719321012
3	.1688049231302295736	14	.0000829551820392400
4	.0847122375438607569	15	.0000414776400291675
5	.0424184376535891141	16	.0000207388322658771
6	.0212233278506082497	17	.0000103694191956252
7	.0106150274029276644	18	.0000051847103634672
8	.0053083352059168857	19	.0000025923553731451
9	.0026543706236411642	20	.0000012961777344251
10	.0013272357765186743	21	.0000006480888791756
11	.0006636304683684728	22	.0000003240444425785
12	.0003318183747184671	23	.0000001620222220369

Table 4.

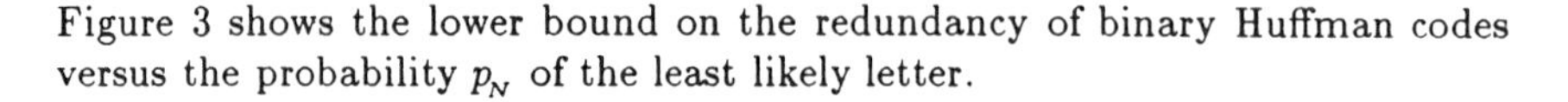

Figure 3 shows the lower bound on the redundancy of binary Huffman codes versus the probability p_N of the least likely letter.

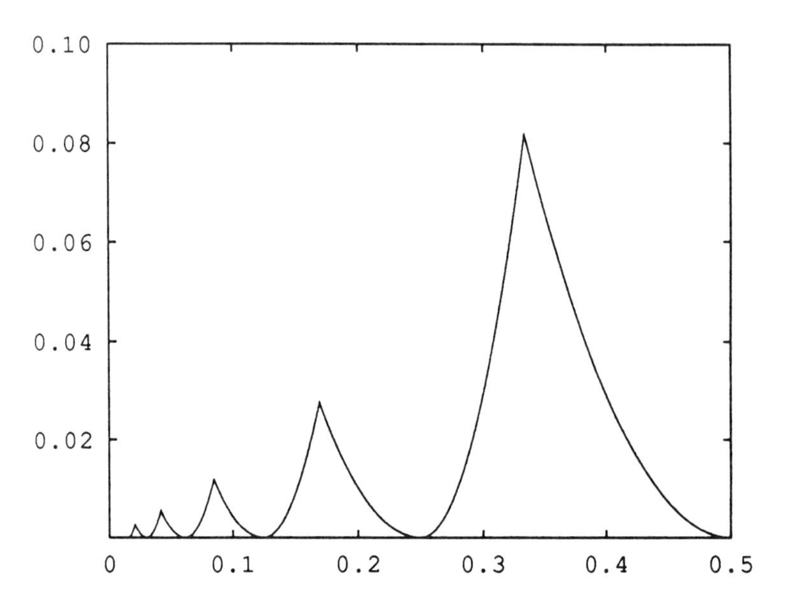

Figure 3. Lower bound of binary Huffman codes as function of p_N.

References

[1] T. Berger and R. Yeung, "Optimum '1'-ended Binary Prefix Codes," *IEEE Trans. Inform. Theory*, vol. IT-36, n. 6, pp. 1435–1441, Nov. 1986.

[2] R. M. Capocelli, R. Giancarlo, and I. J. Taneja, "Bounds on the Redundancy of Huffman codes," *IEEE Trans. Inform. Theory*, vol. IT-32, n. 6, pp. 854–857, Nov. 1986.

[3] R. M. Capocelli and A. De Santis, "Tight Bounds on the Redundancy of Huffman Codes," IEEE International Symposium on Information Theory, Kobe, Japan, June 1988. (Also, IBM Research Report RC–14154.)

[4] R. M. Capocelli and A. De Santis, "Tight Upper Bounds on the Redundancy of Huffman Codes," *IEEE Trans. Inform. Theory*, vol. IT-35, n. 5, Sept. 1989.

[5] R. M. Capocelli and A. De Santis, "A Note on *D*-ary Huffman Codes," *IEEE Trans. Inform. Theory*, vol. IT-37, n. 1, Jan. 1991.

[6] R. M. Capocelli and A. De Santis, "New Bounds on the Redundancy of Huffman Codes," *IEEE Trans. Inform. Theory*, vol. IT-37, n. 4, pp. 1095–1104, July 1991.

[7] R. M. Capocelli, A. De Santis, L. Gargano, and U. Vaccaro, "On the Construction of Statistically Synchronizable Codes," *IEEE Trans. Inform. Theory*, vol. IT-38, n. 2, March 1992.

[8] R. M. Capocelli and A. De Santis, "'1'-ended Binary Prefix Codes," IEEE International Symposium on Information Theory, San Diego, California, Jan. 1990.

[9] R. M. Capocelli and A. De Santis, "Minimum Codeword Length and Redundancy of Huffman Codes," in EUROCODE '90, *Lecture Notes in Computer Science*, vol. 514, pp. 309–317, Springer-Verlag, 1991.

[10] R. M. Capocelli and A. De Santis, "On the Redundancy of Optimal Codes with Limited Word Length," *IEEE Trans. Inform. Theory*, vol. IT-38, n. 2, March 1992.

[11] N. Faller, "An Adaptive System for Data Compression," *Record of 7th Asilomar Conference on Circuits, Systems and Computers*, pp. 593–597, 1973.

[12] R. G. Gallager, "Variations on a Theme by Huffman," *IEEE Trans. Inform. Theory*, vol. IT-24, n. 6, pp. 668–674, Nov. 1978.

[13] E. N. Gilbert, "Codes Based on Inaccurate Source Probabilities," *IEEE Trans. Inform. Theory*, vol. IT-17, n. 3, pp. 304–314, May 1971.

[14] J. Dj. Golic and M. M. Obradovic, "A Lower Bound on the Redundancy of D-ary Huffman Codes," *IEEE Trans. Inform. Theory*, vol. IT-33, no. 6, pp. 910–911, Nov. 1987.

[15] Y. Horibe, "An Improved Bound for Weight-Balanced Trees," *Inform. Contr.*, vol. 24, n. 2, pp. 148–151, 1977.

[16] D. A. Huffman, "A Method for the Construction of Minimum Redundancy Codes," *Proc. IRE*, 40, n. 2, pp. 1098–1101, March 1952.

[17] D. E. Knuth, "Dynamic Huffman Coding," *J. Algorithms,* vol. 6, pp. 163–180, 1985.

[18] O. Johnsen, "On the Redundancy of Huffman Codes," *IEEE Trans. Inform. Theory*, vol. IT-26, n. 2, pp. 220–222, March 1980.

[19] D. Manstetten, "Tight Bounds on the Redundancy of Huffman Codes," *IEEE Trans. Inform. Theory*, vol. IT-38, no. 1, pp. 144–151, Jan. 1992.

[20] B. L. Montgomery and B. V. K. Vijaya Kumar, "On the Average Codeword Length of Optimal Binary Codes for Extended Sources," *IEEE Trans. on Inform. Theory*, vol. IT-33, no. 2, pp. 293–296, Mar. 1987.

[21] B. L. Montgomery and J. Abrahams, "On the Redundancy of Optimal Binary Prefix-condition Codes for Finite and Infinite Sources," *IEEE Trans. Inform. Theory*, vol. IT-33, no. 1, pp. 156–160, Jan. 1987.

[22] F. M. Reza, *Introduction to Information Theory.* McGraw-Hill, 1951.

[23] J. S. Vitter, "Design and Analysis of Dynamic Huffman Codes," *J. of the ACM,* vol. 34, no. 4, pp. 825–845, Oct. 1987.

[24] R. Yeung, "Local Redundancy and Progressive Bounds on the Redundancy of a Huffman Code," *IEEE Trans. Inform. Theory*, vol. IT-37, no. 3, pp. 687–691, May 1991.

9

ON THE CODING DELAY OF A GENERAL CODER

Marcelo J. Weinberger †, Abraham Lempel † †, and Jacob Ziv †

Technion - Israel Institute of Technology

Haifa 32000, Israel

Abstract. We propose a general model for a sequential coder, and investigate the associated coding delay. This model is employed to derive lower and upper bounds on the delay associated with the Lempel-Ziv encoder and decoder.

---

† Department of Electrical Engineering, Technion - Israel Institute of Technology.

† † Department of Computer Science, Technion - Israel Institute of Technology.

1. INTRODUCTION

This paper discusses the concept of coding delay of a general coder. The coding delay of a data compression system has been treated extensively [1,2] but, to the best of our knowledge, no general definitions are available. Instead, the notion of delay is adapted to the specific context of each application. The treatment of delay in [1] is in terms of a code, i.e., the combination of both encoder and decoder, rather than the delay of each coder as a separate entity. In addition, the codes treated in [1] are restricted to have a fixed, predetermined, set of input words, and coding delay is defined as the average length of the input words. This definition suggests a model where the encoder emits a codeword instantaneously after receiving a complete input word, and a decoder receives this codeword at the same time. Thus, the overall delay of the code can be interpreted as the (average) number of input symbols processed by the encoder, beginning with the first symbol of an input word and ending when that symbol is recovered by the decoder. In [2], there is no restriction on the set of codes (except for being finite-state and uniquely decipherable), but the definition of delay still involves the entire encoding/decoding process.

In this work, we propose a general model for a sequential coder and its associated coding delay. We derive lower and upper bounds on the delay associated with the L-Z encoder and decoder [3]. Roughly speaking, coding delay is defined as the maximum over all sequences up to a given length, of the elapsed time between the last input and the last output. It is shown that, with respect to the proposed model, this time is closely related to the length of the buffer that must be installed at the output of the coder in order to prevent any loss of information. Although the question of buffer length has been considered in the past [4,5], the approach proposed here is quite different.

In Section 2 we present the basic definitions (sequential coder, coding delay) of the model, obtain some preliminary results regarding the delay of a general coder, and show how this concept relates to buffer length. In Section 3, lower and upper bounds on the delay of the L-Z encoder are derived, while in Section 4, the delay of a source encoder under a given compression ratio constraint is considered. In Section 5 we deal with the delay associated with the corresponding decoders.

2. DEFINITIONS AND PRELIMINARIES

Given discrete alphabets A and B of α and β letters, let A^* and B^* denote the sets of all finite sequences over the respective alphabets, including the null string λ. We denote by z_i^j the segment $z_i, z_{i+1}, \cdots, z_j$ of a sequence z in A^* or B^*, with $z_i^j \triangleq \lambda$ whenever $j < i$. Assume $\alpha = 2^k$ and $\beta = 2^l$, where k and l are positive integers. Let $\$$ denote an end-of-sequence symbol appended to every sequence in A^* and define $\hat{A} \triangleq A \cup \{ \$ \}$. The convenience of using the additional symbol $\$$ will become clear below.

A function $C(\cdot, \cdot)$ from $A^* \times \hat{A}$ to B^* is called a sequential coder. For a sequence $x_1^n = x_1, x_2, \cdots, x_n$ in A^*, the coded image $\mathbf{C}(x_1^n \$)$ of $x_1^n \$$ under C is defined as the sequence y_1^{n+1}, where $y_i \triangleq C(x_1^{i-1}, x_i)$, $1 \le i \le n$, and $y_{n+1} \triangleq C(x_1^n, \$)$. Note that $x_i \neq \$$ for $1 \le i \le n$ and that $\mathbf{C}(x_1^n \$)$ is not necessarily a prefix of $\mathbf{C}(x_1^{n+1} \$)$. The coder C is said to be uniquely decipherable (UD) if for every pair $x, z \in A^*$, $x \neq z$ implies $\mathbf{C}(x \$) \neq \mathbf{C}(z \$)$.

Let a prefix x_1^i of $x\$$, $x \in A^*$, be fed into a UD coder C, starting at time $t=1$ and ending at time $t(x,i)$. Let τ be the time at which the coder completes to output the last $y_j \neq \lambda$, $j \leq i$, and let $T(y,i) = \max\{t(x,i), \tau\}$. The delay $d_C(x_1^i)$ introduced by the coder for the prefix x_1^i is defined as

$$d_C(x_1^i) \triangleq T(y,i) - t(x,i) \ .$$

The *maximal delay* $D_C(n)$ *of order* n introduced by the coder is defined as

$$D_C(n) \triangleq \max_{\substack{x_1^i \in A^i \\ 0<i\leq n}} d_C(x_1^i \$) \ .$$

Our model for the coder is a synchronous machine that can process one α-ary symbol, i.e., an input of k parallel bits, and output one β-ary symbol, consisting of l parallel bits, per unit of time. Now, the length of $y_i \in B^*$ can be more than one symbol, and we do not allow any loss of information. Consequently, the coder must include enough memory, in the form of a β-ary FIFO buffer so it can process any sequence up to a given length n, without overflow. The shortest feasible length of such a buffer will be denoted by $B(n)$. Notice that every coding scheme requires $B(n) \geq 1$.

We assume that the input symbols arrive sequentially, without interruptions. Thus, $t(x,i) = i$. We further assume that every computation is completed within one clock cycle. Hence, every $y_i \neq \lambda$, is fully in the buffer at time $i+1$. Let $z_i \in B \cup \{\lambda\}$ denote the buffer output at time i. Then, $z_i = \lambda$ if and only if the buffer is empty at time i. Moreover, with an initially empty buffer, $z_1 = \lambda$ while, for $i>1$, $z_i = \lambda$ if and only if $T(y,i-1) = i-1$. Let $N(x,i)$ denote the number of null buffer outputs during the time interval $[1,i]$. We have

$$T(y,i) = L(y_1^i) + N(x,i) ,$$

where $L(y_1^i) \triangleq \sum_{j=1}^{i} length(y_j)$. It follows that

$$d_C(x_1^i) = L(y_1^i) + N(x,i) - i . \quad (1)$$

Now, let $B(x,i)$ denote the queue length of the buffer at time i (note that at this time z_i still resides in the buffer and, therefore, is part of the queue). The above assumptions imply

$$T(y,i) = i + B(x,i+1) , \quad i \geq 1 ,$$

or, equivalently,

$$d_C(x_1^i) = B(x,i+1), \quad i \geq 1 . \quad (2)$$

Lemma 1: $D_C(n) \leq B(n) \leq D_C(n) + 1$.

Proof: By definition, $D_C(n) = d_C(x_1^i\$)$ for some x and i, $0 < i \leq n$. Hence, by (2),

$$D_C(n) = B(x\$,i+2) .$$

To prevent overflow, we need $B(x\$,i+2) \leq B(n)$, implying the left inequality of the lemma. If the maximal queue length is produced by some sequence at time $i+1$, $i \leq n$, then, by (2),

$$B(n) = d_C(x_1^i) \leq d_C(x_1^i\$) + 1 \leq D_C(n) + 1 .$$

Q.E.D.

Note that the extra unit of length beyond $D_C(n)$, is required only when the maximal delay is caused by a sequence $x_1^i\$$ for which $y_{i+1} = \lambda$.

<u>Lemma 2:</u> The delay $d_C(x_1^i)$ for $x_1^i \in A^* \hat{A}$ satisfies the recursion

$$\begin{aligned} d_C(x_1^1) &= length(y_1) \\ d_C(x_1^i) &= length(y_i) + [d_C(x_1^{i-1}) \ominus 1] \ , \ i>1 \ , \end{aligned} \tag{3}$$

where $a \ominus b \triangleq \max(0, a-b)$.

<u>Proof:</u> The lemma is trivial for $i=1$. Clearly, with an initially empty buffer, $B(x,i)$ satisfies the recursion

$$B(x,i+1) = length(y_i) + [B(x,i) \ominus 1] \ , \ i>0 \ . \tag{4}$$

Substituting (2) into (4), we obtain (3). (Note that $B(x,1) = 0$).

Q.E.D.

Since

$$(a \ominus b) \ominus c = a \ominus (b+c)$$

for every $c \geq 0$, we have the following corollary.

<u>Corollary:</u> If $y_j = \lambda$ for $1 \leq m < j < i$, then

$$d_C(x_1^i) = length(y_i) + [d_C(x_1^m) \ominus (i-m)] \ . \tag{5}$$

<u>Example:</u> Block encoder

Let b denote the block length, and assume that $b \mid n$, and that we only consider input sequences where \$ appears after an integer number of blocks. Let l_i denote the length of the i-th codeword. First, assume that the encoder cannot output any part of a codeword until the corresponding input block has been completely received. By the corollary to Lemma 2, the delay d_E introduced by the encoder satisfies

$$d_E(x_1^{bi}) = l_i + [d_E(x_1^{b(i-1)}) \ominus b] \ , \ 1 \le i \le \frac{n}{b} \ , \tag{6}$$

where $d_E(\lambda) \stackrel{\Delta}{=} 0$. Let r denote the length of the longest codeword. By (6), the maximal delay $D_E(n)$ of order n introduced by the encoder is achieved when $l_i = r$ for all i, $1 \le i \le \frac{n}{b}$. Solving the recursion (6), and accounting for the $ symbol in the definition of $D_E(n)$, we obtain

$$D_E(n) = (\frac{n}{b} - 1)(r \ominus b) + r - 1 \ .$$

Now, if we consider an encoder E' that can begin the output of a codeword before receiving all the corresponding input block, it can be readily seen that, by (1), the economy is negligible.

3. THE MAXIMAL DELAY FOR THE L-Z ENCODER

In this section we derive lower and upper bounds on the maximal delay $D_{LZ}(n)$ of order n introduced by the L-Z encoder (Theorem 1 below). We assume $\alpha=\beta=2$ (i.e., $k=l=1$). Let u denote the infinite binary sequence formed by all distinct words of length one, followed by all distinct words of length two, and so forth, with words of the same length appearing, say, in lexicographic order. Let u_1^n denote the sequence formed by the first n symbols of u.

Theorem 1: For every $\varepsilon>0$ and for sufficiently large n we have

$$\frac{2n}{\log n} \le d_{LZ}(u_1^n \$) \le D_{LZ}(n) \le \frac{(3+\varepsilon) \cdot n}{\log n} \ .$$

In order to prove this theorem, we need some further concepts and three lemmas. First, we notice that every integer n can be written in the form $n = \sum_{j=1}^{p} j2^j + \Delta_p$, for some integers p and Δ_p, $p \geq 0$, $0 \leq \Delta_p < 2^{p+1}(p+1)$.

Hence,

$$n = 2^{p+1}(p-1) + 2 + \Delta_p \quad . \tag{7}$$

Let $c(n)$ denote the number of phrases obtained in the incremental L-Z parsing of u_1^n. Clearly,

$$c(n) = \sum_{j=1}^{p} 2^j + q = 2^{p+1} + q - 2 \ , \tag{8}$$

where $q \triangleq \left\lceil \frac{\Delta_p}{p+1} \right\rceil$ (note that the last phrase, which might be incomplete, is taken into account). Let $L(n)$ denote the length of the coded image of $u_1^n\$$, obtained employing an implementation of the L-Z algorithm that uses $1 + \lceil \log i \rceil$ bits to encode the i-th phrase.

Lemma 3:

$$L(n) - n \geq \frac{2n}{p+1} - 3 \ , \tag{9}$$

$$L(n) - n \geq \frac{2n}{\log n} \ , \ p > 2 \ , \tag{10}$$

and

$$L(n) - n \leq \frac{2n}{p-1} \leq \frac{2n}{\log n - \log \log n - 3} \ , \ p > 1 \ . \tag{11}$$

<u>**Proof:**</u> The codewords for the two phrases of length one are of total length 3. It can be easily verified that for phrases of length greater than one, with the possible exception of the very last phrase, the difference between the length of the codeword assigned to a given phrase and the length of the phrase is exactly 2 for all but the first two phrases of each length, for which this difference is 1. For the very last phrase the difference is increased by $q(p+1)-\Delta_p$. Accounting for all the differences, we obtain

$$L(n) = n + q(p+1) - \Delta_p + 2c(n) - (2p+1) - \min\{2,q\}$$

$$= 2^{p+1}(p+1) - 2p + (p+3)q - 3 - \min\{2,q\} \ .$$

Applying (7), we can eliminate q to obtain

$$0 \le L(n) - n - \frac{2\cdot(n + 2^{p+2} - 2)}{p+1} + 2p + 7 \le p + 4 \ . \tag{12}$$

(9) and the first inequality of (11) are easily obtained from (12). For $p > 2$, (9) can be tightened to $L(n) - n \ge \frac{2n}{p+1}$. By (7), $p+1 \le \log n$ whenever $p > 1$, which implies (10). Since $\Delta_p < 2^{p+1}(p+1)$, we have

$$2^{p+2} \ge \frac{n}{p+1} \ ,$$

implying

$$p+2 \ge \log n - \log\log n \ , \tag{13}$$

and the second inequality in (11).

Q.E.D.

Lemma 4: Let j, γ, σ, and η be strictly positive integers satisfying $j = 2^\sigma + \eta < 2^\gamma$, $0 \le \sigma < \gamma$, and $0 \le \eta < 2^\sigma$. Given n such that $p + \sigma - \gamma \ge 1$, we have

$$c\left(\left\lfloor \frac{jn}{2^\gamma} \right\rfloor\right) \le \frac{2^{p+3+\sigma-\gamma} + 2^{p+1-\gamma} j(p-1) + 2^{-\gamma} j \Delta_p}{p+2+\sigma-\gamma} - 1 .$$

Proof: By (7),

$$\frac{jn}{2^\gamma} = n \cdot \frac{2^\sigma + \eta}{2^\gamma} = (2^\sigma + \eta)\, 2^{p+1-\gamma}(p-1) + \frac{2^\sigma + \eta}{2^{\gamma-1}} + \frac{2^\sigma + \eta}{2^\gamma} \cdot \Delta_p$$

$$= 2^{p+1+\sigma-\gamma}(p-1+\sigma-\gamma) + 2^{p+1+\sigma-\gamma}(\gamma-\sigma)$$

$$+ \eta(p-1) \cdot 2^{p+1-\gamma} + \frac{2^\sigma + \eta}{2^{\gamma-1}} + \frac{2^\sigma + \eta}{2^\gamma} \cdot \Delta_p .$$

With $p + \sigma - \gamma \ge 1$, we obtain

$$\left\lfloor \frac{jn}{2^\gamma} \right\rfloor = \sum_{h=1}^{p+\sigma-\gamma} h\, 2^h + \Delta ,$$

where

$$\Delta \triangleq \left\lfloor \frac{2^\sigma+\eta}{2^{\gamma-1}} + \frac{2^\sigma+\eta}{2^\gamma} \cdot \Delta_p + 2^{p+1-\gamma}\eta(p-1) \right\rfloor + 2^{p+1+\sigma-\gamma}(\gamma-\sigma) - 2 . \tag{14}$$

In case $\Delta < (p+1+\sigma-\gamma)\, 2^{p+1+\sigma-\gamma}$, (8) implies

$$c\left(\left\lfloor \frac{jn}{2^\gamma} \right\rfloor\right) = 2^{p+1+\sigma-\gamma} - 2 + \left\lceil \frac{\Delta}{p+1+\sigma-\gamma} \right\rceil \le 2^{p+1+\sigma-\gamma} - 2 + \left\lceil \frac{\Delta + 2^{p+1+\sigma-\gamma}}{p+2+\sigma-\gamma} \right\rceil . \tag{15}$$

In case $\Delta \geq (p+1+\sigma-\gamma)\, 2^{p+1+\sigma-\gamma}$, we have

$$\left\lfloor \frac{jn}{2^\gamma} \right\rfloor = \sum_{h=1}^{p+1+\sigma-\gamma} h\,2^h + \Delta - (p+1+\sigma-\gamma)\, 2^{p+1+\sigma-\gamma}$$

and, hence,

$$c\left(\left\lfloor \frac{jn}{2^\gamma} \right\rfloor\right) \leq 2^{p+2+\sigma-\gamma} - 2 + \left\lceil \frac{\Delta - (p+1+\sigma-\gamma)\, 2^{p+1+\sigma-\gamma}}{p+2+\sigma-\gamma} \right\rceil$$
$$= 2^{p+1+\sigma-\gamma} - 2 + \left\lceil \frac{\Delta + 2^{p+1+\sigma-\gamma}}{p+2+\sigma-\gamma} \right\rceil .$$

Thus, (15) holds for every value of Δ. After some manipulations, Lemma 4 follows from (14) and (15).

Q.E.D.

Let $Z(c)$ denote the length of the (L-Z)-coded image of a sequence which is parsed into c phrases. Let J denote the subset of the integers defined by

$$J \triangleq \{\, j : c(j) = c(j-1) + 1 \,, j \geq 2 \,\} \cup \{0\,,1\} \ ,$$

i.e., J contains 0, 1, and those integers j for which the last phrase in the incremental parsing of u_1^{j-1} is complete. Let $f(w)$, $w \geq 0$, denote a real-valued function of a real variable such that $f(j) = c(j)$ for every $j \in J$, and $f(w)$ is piecewise linear between consecutive elements of J. Finally, given n, let $g_n(w)$ denote the real-valued function defined by

$$g_n(w) \triangleq Z(\lceil f(w) + f(n-w) \rceil + 1) - \frac{2w}{p+1} \,, 0 \leq w \leq n \ .$$

Lemma 5: For every $\varepsilon>0$ and for sufficiently large n we have

$$g_n(w) - n < \frac{(3+\varepsilon)\cdot n}{\log n}$$

Proof: Let $Z_n(w) \triangleq Z(\lceil \phi_n(w) \rceil + 1)$, where $\phi_n(w) \triangleq f(w)+f(n-w)$. First, we show that $Z_n(w)$ is a non-decreasing function of w for $0 \le w \le \frac{n}{2}$, attaining its maximum at $w = \frac{n}{2}$. Clearly, it suffices to prove that this holds for $\phi_n(w)$. Let $j_1 < j_2 < j_3$ be consecutive elements of J. Since the length of the phrases in the sequence u of Theorem 1 is non-decreasing, it follows that $j_3 - j_2 \ge j_2 - j_1$. Hence, $f(w)$ is a convex-$\cap$ function, implying the claimed behavior of $\phi_n(w)$. Thus, $g_n(w) = Z_n(w) - \frac{2w}{p+1}$ is maximum at w_0 belonging to the interval $[0, \frac{n}{2}]$. Now, let $M = \{w_1, w_2, \cdots, w_m\}$ be a partition of this interval, with $0 = w_1 < w_2 < \cdots < w_m = \frac{n}{2}$, and assume $w_0 \in [w_j, w_{j+1}]$ for some j, $1 \le j \le m-1$. Hence,

$$g_n(w) \le g_n(w_0) = Z_n(w_0) - \frac{2w_0}{p+1} \le Z_n(w_{j+1}) - \frac{2w_j}{p+1} \quad . \tag{16}$$

Let I denote the set of the integers and let γ be a given positive integer. Choosing $M = \left\{ \frac{jn}{2^\gamma} : j \in I, 0 \le j \le 2^{\gamma-1} \right\}$, it follows from (16) that

$$g_n(w) \le \max_{1 \le j \le 2^{\gamma-1}} \left\{ Z_n\left(\frac{jn}{2^\gamma} \right) - \frac{(j-1)n}{2^{\gamma-1}(p+1)} \right\} \quad . \tag{17}$$

To obtain an upper bound on $Z_n\left[\frac{jn}{2^\gamma}\right]$, let $j = 2^\sigma + \eta$, where σ and η are integers satisfying $0 \le \sigma \le \gamma - 2$ and $0 \le \eta < 2^\sigma$, or $0 \le \sigma = \gamma - 1$ and $\eta = 0$ (note that these conditions are covered by Lemma 4). By the definition of $f(w)$, for every $w \ge 0$ we have

$$f(w) \le c(\lfloor w \rfloor) + 1 .$$

Hence,

$$\left\lceil \phi_n\left(\frac{jn}{2^\gamma}\right)\right\rceil \le c\left(\left\lfloor \frac{jn}{2^\gamma}\right\rfloor\right) + c\left(\left\lfloor n(1 - \frac{j}{2^\gamma})\right\rfloor\right) + 2 . \tag{18}$$

Since

$$n(1 - \frac{j}{2^\gamma}) = \frac{n}{2^\gamma} \cdot [\, 2^{\gamma-1} + (2^{\gamma-1} - 2^\sigma - \eta)] ,$$

we can upper-bound $c\left(\left\lfloor n(1 - \frac{j}{2^\gamma})\right\rfloor\right)$ using Lemma 4 with σ replaced by $\gamma - 1$ and j replaced by $2^\gamma - j$. We obtain

$$\begin{aligned} c\left(\left\lfloor n(1 - \frac{j}{2^\gamma})\right\rfloor\right) &\le \frac{2^{p+2} + (1 - j\, 2^{-\gamma}) \cdot [\, 2^{p+1}(p-1) + \Delta_p]}{p+1} - 1 \\ &= 2^{p+1} + \frac{\Delta_p}{p+1} - \frac{j\, 2^{-\gamma} \cdot [\, 2^{p+1}(p-1) + \Delta_p]}{p+1} - 1 . \end{aligned} \tag{19}$$

By (18), (19), and Lemma 4, we have

$$\left\lceil \phi_n\left(\frac{jn}{2^\gamma}\right)\right\rceil \le \frac{2^{p+3+\sigma-\gamma} + 2^{p+1-\gamma} j\,(p-1) + 2^{-\gamma} j\,\Delta_p}{p+2+\sigma-\gamma} + 2^{p+1} + \frac{\Delta_p}{p+1}$$

$$- \frac{j\,2^{-\gamma}\cdot[\,2^{p+1}(p-1)+\Delta_p\,]}{p+1}$$

$$= 2^{p+1} + \frac{\Delta_p}{p+1} + \frac{j}{2^\gamma}\cdot\left(\frac{1}{p+2+\sigma-\gamma} - \frac{1}{p+1}\right)$$

$$\cdot\,[\Delta_p + 2^{p+1}(p-1)\,] + \frac{2^{p+3+\sigma-\gamma}}{p+2+\sigma-\gamma}$$

which, by (8) and the inequality $\Delta_p < 2^{p+1}(p+1)$, implies

$$\left\lceil \phi_n\left(\frac{jn}{2^\gamma}\right)\right\rceil + 1 \le c\,(n) + 3 + \frac{jp\,(\gamma-\sigma-1)\cdot 2^{p+2-\gamma}}{(p+1)\,(p+2+\sigma-\gamma)} + \frac{2^{p+3+\sigma-\gamma}}{p+2+\sigma-\gamma}$$

$$\le c\,(n) + 3 + \frac{2^{p+2-\gamma}\,[\,j\,(\gamma-\sigma-1) + 2^{\sigma+1}]}{p+2+\sigma-\gamma} \tag{20}$$

$$\stackrel{\Delta}{=} c\,(n) + e\,(j,\gamma)\ .$$

Now, we are ready to upper-bound $Z_n\left[\frac{jn}{2^\gamma}\right]$. First, note that since $\phi_n(w)$ is maximum at $w = n/2$, (20) implies

$$\left\lceil \phi_n\left(\frac{jn}{2^\gamma}\right)\right\rceil + 1 \le \left\lceil \phi_n\left(\frac{n}{2}\right)\right\rceil + 1 \le c\,(n) + e\,(1,1)$$

$$= c\,(n) + 3 + \frac{2^{p+2}}{p+1} < c\,(n) + 2^{p+2}\ , \tag{21}$$

whenever $p \ge 1$. Since the length of the phrases in the incremental parsing of u_1^n is, at most, $p+1$, it follows, from (21), that each of the first $\left\lceil \phi_n\left(\frac{jn}{2^\gamma}\right)\right\rceil + 1$ phrases of u contains, at most, $p+2$ bits. Hence, as stated in the proof of Lemma 3, the length of the corresponding codewords is, at most, $p+4$. Therefore, by (20),

$$Z_n\left\lceil \frac{jn}{2^\gamma} \right\rceil \leq L(n) + (p+4)\, e(j,\gamma) \ .$$

Thus, (17) yields

$$g_n(w) - n \leq L(n) - n + \max_{1 \leq j \leq 2^{\gamma-1}} \left\{ (p+4)\, e(j,\gamma) - \frac{(j-1)n}{2^{\gamma-1}(p+1)} \right\}$$

which, by the first inequality in (11), the definition of $e(j,\gamma)$, and the fact that $n \geq 2^{p+1}(p-1)$ for $p>1$, implies

$$\begin{aligned} g_n(w) - n &\leq \frac{2n}{p-1} + 3p + 12 + \frac{n\, 2^{1-\gamma}}{p-1} \\ &\quad \cdot \max_{1 \leq j \leq 2^{\gamma-1}} \left\{ \frac{p+4}{p+2+\sigma-\gamma} [\, j(\gamma-\sigma-1) + 2^{\sigma+1}] - \frac{p-1}{p+1} \cdot (j-1) \right\} \\ &= \frac{2n}{p-1} + 3p + 12 + \frac{n\, 2^{1-\gamma}}{p-1} \\ &\quad \cdot \max_{1 \leq j \leq 2^{\gamma-1}} \left\{ \frac{2+\gamma-\sigma}{p+2+\sigma-\gamma} [\, j(\gamma-\sigma-1) + 2^{\sigma+1}] + \frac{2(j-1)}{p+1} \right\} \qquad (22) \\ &\quad + \frac{n\, 2^{1-\gamma}}{p-1} \cdot \max_{1 \leq j \leq 2^{\gamma-1}} \{\, j(\gamma-\sigma-2) + 2^{\sigma+1} + 1\} \\ &\leq \frac{2n}{p-1} + 3p + 12 + \frac{n(\gamma+1)(\gamma+2)}{(p+2-\gamma)(p-1)} + \frac{2n}{(p+1)(p-1)} \\ &\quad + \frac{n\, 2^{1-\gamma}}{p-1} \cdot \max_{1 \leq j \leq 2^{\gamma-1}} \{\, j(\gamma-\sigma-2) + 2^{\sigma+1} + 1\} \ . \end{aligned}$$

Next, we compute $\Gamma \overset{\Delta}{=} \max_{1 \le j < 2^{\gamma-1}} \{ j(\gamma - \sigma - 2) + 2^{\sigma+1} + 1\}$. Since this maximum is computed for $j < 2^{\gamma-1}$, we have $j \le 2^{\sigma+1} - 1$ and $\sigma \le \gamma - 2$. Hence,

$$\Gamma = 2 + \max_{0 \le \sigma \le \gamma-2} \{(2^{\sigma+1} - 1)(\gamma - \sigma - 2) + 2^{\sigma+1} - 1\}$$

$$= 2 + \max_{1 \le \xi \le \gamma - 1} \{(2^{\xi} - 1)(\gamma - \xi)\}$$

It can be readily verified that this maximum is achieved for $\xi = \gamma - 1$, implying $\Gamma = 2^{\gamma-1} + 1$. Now, for the remaining value of j, namely $j = 2^{\gamma-1}$, the same value is obtained. Hence, (22) takes the form

$$g_n(w) - n \le \frac{2n}{p-1} + 3p + 12 + \frac{n(\gamma+1)(\gamma+2)}{(p+2-\gamma)(p-1)} + \frac{2n}{(p+1)(p-1)}$$

$$+ \frac{n \cdot 2^{1-\gamma}(2^{\gamma-1}+1)}{p-1} = \frac{n(3+2^{1-\gamma})}{p-1} + 3p + 12$$

$$+ \frac{n(\gamma+1)(\gamma+2)}{(p+2-\gamma)(p-1)} + \frac{2n}{(p+1)(p-1)} = \frac{n}{p-1}$$

$$\cdot \left[3 + 2^{1-\gamma} + \frac{(3p+12)(p-1)}{n} + \frac{(\gamma+1)(\gamma+2)}{p+2-\gamma} + \frac{2}{p+1} \right] \tag{23}$$

$$\le \frac{n}{\log n} \quad \frac{1}{1 - \dfrac{\log \log n + 3}{\log n}} \cdot \left[3 + 2^{1-\gamma} \right.$$

$$+ \frac{(3 \log n + 9)(\log n - 2)}{n} + \frac{(\gamma+1)(\gamma+2)}{\log n - \log \log n - \gamma}$$

$$\left. + \frac{2}{\log n - \log \log n - 1} \right] ,$$

where the last inequality follows from (13) and from $p+1 \le \log n$. Now, chose γ

such that $2^{1-\gamma} < \frac{\varepsilon}{2}$. Thus, (23) takes the form

$$g_n(w) - n < \frac{n}{\log n} \cdot [\, 3 + \frac{\varepsilon}{2} + \delta(n)] \ ,$$

where $\lim_{n \to \infty} \delta(n) = 0$. For sufficiently large n, we have $\delta(n) < \frac{\varepsilon}{2}$, and the proof is complete.

Q.E.D.

<u>Proof of Theorem 1:</u> Clearly, the number $N(u, n)$ of null buffer outputs for u_1^n is one, so the leftmost inequality follows from (1) and (10), implying the lower bound on $D_{LZ}(n)$.

As for the upper bound, consider a binary sequence x_1^n, and let i denote the last time for which the output z_i of the buffer is null. Consequently, $z_{i+1} \neq \lambda$ is the first symbol of the coded image of a phrase x_m^i for some integers $m \leq i \leq n$. Hence, by (1),

$$d_{LZ}(x_1^n \$) = i + L(y_m^{n+1}) - n - 1 \ , \tag{24}$$

where y_m^{n+1} is the coded output for $x_m^n \$$. We can rewrite (24) as

$$\begin{aligned} d_{LZ}(x_1^n \$) &= L(i) + L(y_m^{n+1}) - [L(i) - i] - n - 1 \\ &\leq L(i) + L(y_m^{n+1}) - \frac{2i}{p+1} - n + 2 \ , \end{aligned} \tag{25}$$

where the inequality follows from (9) and holds for $0 < i \leq n$. Now, the number of phrases in the incremental parsing of x_m^n (including a possibly incomplete last phrase whose coded image is y_{n+1}) is clearly upper bounded by $c(n-i) + 1$. Recalling that $Z(c)$ denotes the length of the (L-Z)-coded image of a sequence which is parsed into c phrases, it follows from (25) that

$$\begin{aligned} d_{LZ}(x_1^n\$) &\leq Z(c(i)+c(n-i)+1)-\frac{2i}{p+1}-n+2 \\ &\leq Z(\lceil f(i)+f(n-i)\rceil+1)-\frac{2i}{p+1}-n+2 \\ &= g_n(i)-n+2 \ . \end{aligned} \tag{26}$$

By Lemma 5, and for sufficiently large n,

$$g_n(i)-n < \frac{(3+\frac{\varepsilon}{2})\cdot n}{\log n}$$

which, by (26), implies

$$d_{LZ}(x_1^n\$) \leq \frac{(3+\frac{\varepsilon}{2}+\frac{2\log n}{n})\, n}{\log n} \ . \tag{27}$$

Since the right-hand side of (27) is a function of n only, independent of the sequence x, and since $\frac{2\log n}{n} < \frac{\varepsilon}{2}$ for sufficiently large n, the claimed upper bound on $D_{LZ}(n)$ follows.

Q.E.D.

The sequence u used to lower-bound the maximal delay is not the worst case. For a given n, consider the sequence v_1^n obtained by reordering the phrases of u_1^n so that, for every point in the new order, the phrase appearing in that point is preceded by all its prefixes and is the longest phrase with the said property (in case of a tie, apply a lexicographic ordering). For example, with $p=3$ and $\Delta_p=0$,

$$v_1^{34} = 0\,,00\,,000\,,001\,,01\,,010\,,011\,,$$
$$1\,,10\,,100\,,101\,,11\,,110\,,111\ .$$

Clearly, the same code length $L(n)$ corresponds to v_1^n and to u_1^n. By the corollary to Lemma 2 and the structure of the sequence v, the number $N(v,n)$ of null buffer outputs can be expressed as a function of p and Δ_p. To state this function we need some further definitions. Let μ denote the length of the longest phrase in u_1^n and let ν denote the number of complete phrases of length μ. Thus, in case $\Delta_p \geq p+1$, we have $\mu = p+1$ and $\nu = \left\lfloor \frac{\Delta_p}{p+1} \right\rfloor$ while, for $\Delta_p < p+1$, we have $\mu \triangleq p$ and $\nu = 2^p$ (note that we always have $0 < \nu \leq 2^\mu$). Then, it can be shown that

$$N(v,n) = 2^{\mu-2} + 1 - G(\mu,\nu)\ ,$$

where

$$G(\mu,\nu) \triangleq \begin{cases} 0 & \text{if } 2^{\mu-3} \leq \nu \leq 2^\mu \\ 1 + \lceil \log(2^{\mu-3} - \nu) \rceil & \text{if } 2^{\mu-4} \leq \nu < 2^{\mu-3} \\ 1 + \lfloor \log \nu \rfloor + 2 \cdot (2^{\mu-4} - \nu) & \text{if } 0 < \nu < 2^{\mu-4}\ . \end{cases}$$

It can be further shown that, in any case,

$$N(v,n) > \frac{n}{8 \log n}$$

and, hence,

$$D_{LZ}(n) > \frac{(2 + \frac{1}{8}) \cdot n}{\log n}\ ,$$

thus obtaining a lower bound tighter than that of Theorem 1. In addition, it can be shown that for $2^{\mu-4} \leq \nu < 2^{\mu-3}$ and sufficiently large n,

$$N(v,n) > \frac{2n}{9 \log n} . \tag{28}$$

Moreover, there exist infinitely many values of n for which the multiplying constant in (28) is $\frac{4}{17}$, so the multiplying constant in the upper bound of Theorem 1 cannot be reduced below $2 + \frac{4}{17}$.

4. THE DELAY OF A SOURCE ENCODER UNDER A GIVEN COMPRESSION RATIO CONSTRAINT

In this section we restrict our discussion to source encoders. The definition of maximal delay stated in Section 2 implies a maximization over all possible input sequences up to a given length n. When we deal with source encoders, we are interested in "compressible" sequences, i.e., sequences x_1^n for which the normalized compression ratio $R(x_1^n)$ achieved by the encoder satisfies

$$R(x_1^n) \triangleq \frac{l \cdot L(y_1^{n+1})}{nk} \leq \rho ,$$

where ρ is a given constant satisfying $0<\rho<1$. (Note that, in this context, the concept of compressibility refers to the pair {sequence, encoder}, and not only to the sequence as in the original definition of [5]). Roughly speaking, the concept of coding delay under a compression ratio constraint is obtained as the result of maximizing the coding delay over the prefixes of sequences x_1^n for which $R(x_1^n)$ is approximately ρ. We derive lower and upper bounds for this quantity for block and L-Z encoders. Note that the above constraint is a function of both ρ and the specific encoder. In fact, a more interesting question in this context is one where the maximum delay of an individual sequence is a function of the sequence only, just as its compressibility. However, such formulation is outside the scope of this work.

Let ρ and $\Delta\rho$ denote given constants, with $0 \le \rho - \Delta\rho < \rho$ (although typically $\rho < 1$, we do not require this condition), and define the range $\rho_- \stackrel{\Delta}{=} [\rho - \Delta\rho, \rho]$. Let E denote a source encoder. The *ρ-maximal delay $D_E(n, \rho_-)$ of order n* introduced by E is defined as

$$D_E(n, \rho_-) \stackrel{\Delta}{=} \max_{x_1^n \in S(n, \rho_-)} \max_{0 < i \le n} d_E(x_1^i \$) ,$$

where $S(n, \rho_-) \stackrel{\Delta}{=} \left\{ x_1^n \in A^n : \frac{(\rho - \Delta\rho) nk}{l} \le L(y_1^{n+1}) \le \rho \frac{nk}{l} \right\}$.

For the block encoder considered in Section 2, we further denote by s the length of the shortest codeword. We can assume $\frac{lr}{kb} \ge \rho > \frac{ls}{kb}$, and $s < b < r$, for otherwise the problem is trivial. Then, it can be shown that the associated ρ-maximal delay of order n is $n[F + O(n^{-1})]$, where $F \stackrel{\Delta}{=} \frac{(r-b)(\rho kb - ls)}{lb(r-s)} > 0$. The maximum is attained by a sequence formed by a segment with blocks that have a coded image of length s, followed by a segment with blocks whose coded image has length r, where the length of each segment is chosen such that the sequence is in $S(n, \rho_-)$.

As for the L-Z encoder, we derive lower and upper bounds on the associated ρ-maximal delay of order n. We assume, again, $\alpha = \beta = 2$ (i.e., $k = l = 1$). Let U denote the infinite binary sequence formed by all distinct words of length one except 0, followed by all distinct words of length two except 00, and so forth where, for each length, we exclude the all-zero word, and with words of the same length appearing, say, in lexicographic order. Note that U is obtained from u by removing the all-zero words. Let U_1^n denote the sequence formed by the first n symbols of U, and let $L'(n)$ denote the length of the coded image of $U_1^n\$$, obtained by an implementation of the L-Z algorithm that uses $1 + \lceil \log i \rceil$ bits to encode the i-th phrase.

Lemma 6: For every $n > 1$ we have

$$L'(n) - n > \frac{2n}{\log n} - 4 \; .$$

The proof of Lemma 6 is similar to that of Lemma 3. For $n > 48$, the inequality in Lemma 6 can be tightened to $L'(n) - n \geq \frac{2n}{\log n}$.

Lemma 7: Let K be a positive real constant, and let ξ and ψ denote real variables such that $\psi \geq 2$ and $\xi > e^{K+1}$. If

$$\xi \leq \psi \, (1 + \frac{K}{\log \psi}) \tag{29}$$

then

$$\psi > \xi \, (1 - \frac{K}{\log \xi}) \; .$$

Proof: If $\psi \geq \xi$, there is nothing to prove. Therefore, we assume $\psi = \xi(1-\delta)$ for some δ, $0<\delta<1$. By (29),

$$\delta \leq \frac{K(1-\delta)}{\log\,[(1-\delta)\xi]} \; .$$

Using the well-known inequality $\log x \leq (x-1) \log e$ with $x=(1-\delta)^{-1}$, we obtain

$$\delta \leq \frac{K(1-\delta)}{\log \xi} \cdot \frac{1}{1-\frac{\delta}{(1-\delta)\ln \xi}} \quad . \tag{30}$$

Now, since $\psi \geq 2$ and $\xi > e^{K+1}$, we have $\xi > e^{\frac{K}{\log \psi}+1}$ which, by (29), yields

$$\psi \ln \xi > \psi\,(1+\frac{K}{\log \psi}) \geq \xi \ ,$$

Consequently, by the definition of δ,

$$(1-\delta)\ \ln\xi > 1 \tag{31}$$

Finally, by (30) and (31),

$$\delta < \frac{K}{\log \xi} \ ,$$

and the proof is complete.

Q.E.D.

Theorem 2: For every $\varepsilon>0$, for sufficiently large n, and every $\rho>0$, the ρ-maximal delay $D_{LZ}(n,\rho\_)$ of order n introduced by the L-Z encoder satisfies

$$\frac{(2-\varepsilon)\cdot \rho n}{\log n} \leq D_{LZ}(n,\rho\_) \leq \frac{(3+\varepsilon)\cdot \rho n}{\log n} \quad .$$

Proof: First, we prove the upper bound. Consider a prefix x_1^i of a binary sequence $x_1^n \in S(n,\rho\_)$, and let $i-j$ denote the last time in the interval $[1,i]$ for which the output z_{i-j} of the buffer is null. Consequently, $z_{i-j+1} \neq \lambda$ is the first symbol of the coded image of a phrase x_m^{i-j} for some integers $1 \leq m < i-j \leq i$. Hence, by (1),

$$d_{LZ}(x_1^i\,\$) = L(y_1^i v) - (s+j) - 1 \ , \tag{32}$$

where $y_1^i v$ is the coded image of $x_1^i\,\$$ (v denotes the output corresponding to $\$$)

and $s \stackrel{\Delta}{=} L(y_1^{m-1})$. Next, we lower-bound $s+j$. Let $s+j = \sum_{h=1}^{r} h 2^h + \Delta_r$, for some integers r and Δ_r, $r \geq 0$, $0 \leq \Delta_r < 2^{r+1}(r+1)$. By (9),

$$\begin{aligned} L(y_1^i v) = s + L(y_m^i v) \leq L(s) + L(y_m^i v) - \frac{2s}{r+1} + 3 \\ < Z(c(s)+c(j)+1) - \frac{2s}{r+1} + 3 \\ \leq Z(\lceil f(s)+f((s+j)-s) \rceil + 1) - \frac{2s}{r+1} + 3 = g_{s+j}(s) + 3 \ . \end{aligned} \tag{33}$$

By Lemma 5, given $\varepsilon>0$, there exists a constant $K(\varepsilon)$ such that, for every $s+j > K(\varepsilon)$, we have

$$g_{s+j}(s) - (s+j) < \frac{(3+\varepsilon/2)\cdot(s+j)}{\log(s+j)} \ . \tag{34}$$

In order to show that, for sufficiently large n, we can indeed assume $s+j > K(\varepsilon)$, we note first that the minimum value of $s+j$ is achieved when x_1^i is the all-zero sequence. It can be readily seen that this implies

$$s+j \geq Z(\sqrt{2i} - 3) \ .$$

It follows that (34) holds for sufficiently large i. We can assure a sufficiently large i by selecting a sufficiently large n, for otherwise $d_{LZ}(x_1^i \$)$ is upper-bounded by a constant, independent of n. Hence, by (33),

$$L(y_1^i v) - 3 < (s+j)\left[1 + \frac{3+\varepsilon/2}{\log(s+j)}\right] \ . \tag{35}$$

By Lemma 7, (35) yields

$$s+j > [L(y_1^i v) - 3]\left[1 - \frac{3+\varepsilon/2}{\log[L(y_1^i v) - 3]}\right] .$$

Substituting this bound on $s+j$ into (32), we obtain

$$d_{LZ}(x_1^i \$) < \frac{[L(y_1^i v) - 3]\cdot(3+\varepsilon/2)}{\log[L(y_1^i v) - 3]} + 2 . \tag{36}$$

Now, since $x_1^n \in S(n, \rho\_)$, we have $L(y_1^i v) \le L(y_1^{n+1}) \le \rho n$. In addition, $w/\log w$ is an increasing function for $w > e$. Hence, for sufficiently large $i \le n$, and any $x_1^n \in S(n, \rho\_)$, (36) implies

$$d_{LZ}(x_1^n \$) < \frac{\rho n\cdot(3+\varepsilon/2)}{\log(\rho n)} + 2 = \frac{\rho n}{\log n}\left[\frac{3+\varepsilon/2}{1+\dfrac{\log\rho}{\log n}} + \frac{2\log n}{\rho n}\right]$$

$$< \frac{\rho n\cdot(3+\varepsilon)}{\log n}$$

This proves the upper bound on $D_{LZ}(n, \rho\_)$.

As for the lower bound, given an integer i, $0 \le i \le n$, consider a binary sequence $x(i)_1^n$ defined as follows. The sequence is formed by two segments, $x(i)_1^i$ and $x(i)_{i+1}^n$. If $i=0$, the first segment is null; if $i=1$ then $x(i)_1 = 1$ and, if $i>1$, then $x(i)_j = 0$, $1 \le j < i$, and $x(i)_i = 1$. Clearly, every phrase of $x(i)_1^n$ that begins in $x(i)_1^i$ also ends in $x(i)_1^i$. Let $x(i)_m^i$ denote the last phrase in $x(i)_1^i$, with $m \stackrel{\Delta}{=} 1$ whenever $i=0$, and with $x(i)_1^0 \stackrel{\Delta}{=} \lambda$. In case $x(i)_m^i$ is not a phrase of U_1^{n-i} (the sequence defined just before Lemma 6), the second segment is $x(i)_{i+1}^n \stackrel{\Delta}{=} U_1^{n-i}$. Otherwise, $x(i)_{i+1}^n$ is defined as the prefix of length $n-i$ of the sequence U' obtained by deleting the phrase $x(i)_m^i$ from U_1^{n-i}. Thus, in any case, every complete phrase of U_1^{n-i} appears among the phrases of $x(i)_1^n$ that belong to $x(i)_m^n$.

The coded image of a segment $x(i)_j^h$ is denoted by $y(i)_j^h$. Clearly,

$$L(y(n)_1^{n+1}) \leq Z(\lceil \sqrt{2n}\, \rceil + 1) < (\sqrt{2n} + 2) \log 4(\sqrt{2n} + 2) \ .$$

Hence, for $\rho>0$ and for sufficiently large n, we have $L(y(n)_1^{n+1}) < \rho n$. Let a be the least integer in the range $[0,n]$ satisfying $L(y(a)_1^{n+1}) < \rho n$. Obviously, we can assume $n-a > 1$. Now, it can be easily verified that the number of phrases in the incremental parsing of $x(i)_1^n$ is, at most, one more than that of $x(i+1)_1^n$. Let c denote the number of phrases in $x(a)_1^n$. By the definition of a, we have

$$Z(c) = L(y(a)_1^{n+1}) \leq \rho n \ , \tag{37}$$

and

$$Z(c+1) > \rho n \ . \tag{38}$$

In addition,

$$Z(c+1) - Z(c) = \lceil \log 2(c+1) \rceil < 2 + \log (c+1)$$

which, by (37) and (38), implies

$$L(y(a)_1^{n+1}) > \rho n - \log (c+1) - 2 \ . \tag{39}$$

By (37) and Stirling's inequality, we also have

$$\rho n \geq Z(c) \geq c + \sum_{j=1}^{c} \log c = c + \log (c!) > c \log \frac{2c}{e} \ .$$

It follows that $\rho n > c+1$ for every $c > 3$ which, by (39), implies

$$L(y(a)_1^{n+1}) > n(\rho - \frac{\log(4\rho n)}{n}) \ . \tag{40}$$

Hence, for sufficiently large n, we have $x(a)_1^n \in S(n, \rho_-)$.

Next, we prove that the delay introduced by the L-Z encoder for the sequence $x(a)_1^n$ is lower-bounded as claimed. By the definition of the sequence, we have

$$d_{LZ}(x(a)_1^n \$) \geq L(y(a)_m^{n+1}) + a - n - 1$$

which, by (40), takes the form

$$d_{LZ}(x(a)_1^n \$) > \rho n - \log(4\rho n) - L(y(a)_1^{m-1}) - (n-a) - 1 \ , \tag{41}$$

where $L(y(a)_1^0) \triangleq 0$. Now, we upper-bound $n-a$. We have

$$\rho n - L(y(a)_1^{m-1}) \geq L(y(a)_m^{n+1}) \geq L'(n-a) \ .$$

(Recall that $L'(n-a)$ denotes the length of the (L-Z)-coded image of $U_1^{n-a}\$$).
Since $n-a > 1$ we obtain, by Lemma 6,

$$\rho n - L(y(a)_1^{m-1}) + 4 > (n-a)(1 + \frac{2}{\log(n-a)}) \ . \tag{42}$$

Clearly, $\rho n \geq L(y(a)_m^{n+1}) > n-a$. Thus, by (42),

$$n-a < \frac{\rho n - L(y(a)_1^{m-1}) + 4}{1 + \dfrac{2}{\log(\rho n)}} \ .$$

Replacing $n-a$ in (41) by this upper bound yields

$$d_{LZ}(x(a)_1^n \$) > [\,\rho n - L(y(a)_1^{m-1}) + 4\,] \left[1 - \frac{1}{1 + \frac{2}{\log(\rho n)}} \right]$$

$$- \log(4\rho n) - 5$$

$$> \frac{2 \cdot [\rho n - L(y(a)_1^{m-1})]}{\log(4\rho n)} - \log(4\rho n) - 5 \; .$$

Since $x(a)_j = 0$ for every $0 < j < m$, it follows that

$$L(y(a)_1^{m-1}) < \sqrt{2m}\, \log(4\sqrt{2m}\,) = \sqrt{\frac{m}{2}}\, \log(32n) < \sqrt{\frac{n}{2}}\, \log(32n) \; .$$

Hence,

$$d_{LZ}(x(a)_1^n \$) > \frac{2\rho n}{\log(4\rho n)} \cdot \left[1 - \frac{\log(32n)}{2\rho\sqrt{2n}} - \frac{\log^2(4\rho n)}{2\rho n} - \frac{5\log(4\rho n)}{2\rho n} \right] = \frac{\rho n}{\log n} \cdot (2 - \delta(n)) \; ,$$

where $\lim_{n \to \infty} \delta(n) = 0$. For sufficiently large n, we have $\delta(n) < \varepsilon$. Since $x(a)_1^n \in S(n, \rho\_)$, the claimed lower bound on $D_{LZ}(n, \rho\_)$ follows.

Q.E.D.

5. DECODING DELAY

5.1 The Maximal Delay

In this section we derive bounds on the delay associated with the decoders for the data compression algorithms considered in the previous sections. The notation, definitions, and assumptions stated in Section 2 for a general coder remain in force.

In particular, we assume that the decoder input symbols arrive sequentially, without interruptions. This is equivalent to the assumption that the beginning of decoding lags behind the beginning of encoding by at least the incurred encoding delay. Note that, with the definitions of Section 2, the delay of a decoder is determined with respect to the length n of the coded sequence, rather than that of the original sequence. Note also that adherence to the notation of Section 2 for alphabets and sequences, requires an obvious adjustment due to the interchange of encoder/decoder roles: for instance, now the input alphabet A denotes the alphabet of the compressed image, and the output alphabet B stands for the source alphabet. We assume that the end-of-sequence symbol \$ appears only at the end of the complete coded image of some β-ary sequence.

For the block decoder, it is easy to see that under the assumption $s \mid n$, the maximal delay $D_D(n)$ of order n is achieved when the encoded input is a concatenation of n/s codewords of length s. Proceeding as with the block encoder, we obtain

$$D_D(n) = (\frac{n}{s} - 1)(b \ominus s) + b - 1 \ .$$

As for the L-Z decoder, we assume α=β=2 (i.e., k=l=1). The length n of the input (encoded) sequence must equal $Z(c) \triangleq c + \sum_{j=1}^{c} \lceil \log j \rceil$, where c denotes the number of phrases in the incremental parsing of the decoded sequence. Let b_i denote the length of the i-th phrase in the decoded sequence. Again, by the corollary to Lemma 2, the delay d_{LZD} introduced by the decoder satisfies

$$d_{LZD}(x_1^{Z(i)}) = b_i + [d_{LZD}(x_1^{Z(i-1)}) \ominus (1+\lceil \log i \rceil)] \quad , 1 \le i \le c \quad ,$$

where $d_{LZD}(\lambda) \stackrel{\Delta}{=} 0$. Since $b_i \le i$, it can be readily seen that the maximal delay $D_{LZD}(n)$ is achieved when $b_i = i$ for all i, $1 \le Z(i) \le n$, namely, when the decoded sequence is the all-zero (or the all-one) sequence of length $\frac{c(c+1)}{2}$. Clearly, in this case, $d_{LZD}(x_1^{Z(i-1)}) \ge 1+\lceil \log i \rceil$ whenever $i \ge 3$. Hence, assuming $c \ge 3$ (i.e., $n \ge 6$) and solving the recursion, we obtain

$$d_{LZD}(x_1^n) = \sum_{i=4}^{c} i - \sum_{i=4}^{c} (1+\lceil \log i \rceil) + d_{LZD}(x_1^{m(3)}) = \frac{c(c+1)}{2} - n + 3 \quad ,$$

implying

$$D_{LZD}(n) = \frac{c(c+1)}{2} - n + 2 \quad . \tag{43}$$

Theorem 3 below states tight lower and upper bounds on $D_{LZD}(n)$ as a function of n only.

<u>**Theorem 3:**</u> For every $\varepsilon > 0$ and for sufficiently large n, the maximal delay $D_{LZD}(n)$ of order n introduced by the L-Z decoder satisfies,

$$\frac{(1-\varepsilon) \cdot n^2}{2 \log^2 n} < D_{LZD}(n) < \frac{(1+\varepsilon) \cdot n^2}{2 \log^2 n} \quad .$$

<u>**Proof:**</u> By (40),

$$n = c + \sum_{i=1}^{c} \lceil \log i \rceil = Z(c) > c \log \frac{2c}{e} \quad . \tag{44}$$

Next, we prove that (44) implies

$$c < \frac{n}{\log n} \cdot (1+\delta(n)) \ , \tag{45}$$

where $\delta(n) \triangleq \dfrac{\log(\frac{e}{2}\log n)}{\log \frac{2n}{\log n}}$, $n>1$. If $c < \dfrac{n}{\log n}$ there is nothing to prove, so assume $c = \dfrac{n}{\log n} \cdot (1+\gamma(n))$ for some function $\gamma(n)$. Thus, (44) takes the form

$$1 > [1+\gamma(n)] \cdot [1 - \frac{\log(\frac{e}{2}\log n)}{\log n} + \frac{\log(1+\gamma(n))}{\log n}]$$

$$\geq 1 + \gamma(n) - \frac{(1+\gamma(n))\log(\frac{e}{2}\log n)}{\log n} + \frac{\gamma(n)\log e}{\log n} \ ,$$

where we used the well-known inequality $x \log x \geq (x-1)\log e$. After some manipulations we obtain $\gamma(n) < \delta(n)$, and (45) follows.

By (43), (45), and assuming $c \geq 3$,

$$D_{LZD}(n) = \frac{c(c+1)}{2} - n + 2 < \frac{c^2}{2} < \frac{n^2}{2\log^2 n} \cdot (1+\delta(n))^2$$

$$< \frac{(1+\varepsilon)\cdot n^2}{2\log^2 n} \ ,$$

where the last inequality holds for sufficiently large n.

On the other hand, $n = Z(c) < c \log 4c$ and $n > 4c$ whenever $c > 15$, implying $c > \dfrac{n}{\log n}$. Hence, by (43),

$$D_{LZD}(n) > \frac{c^2}{2} - n > \frac{n^2}{2\log^2 n} \cdot (1 - \frac{2\log^2 n}{n})$$

$$> \frac{(1-\varepsilon)\cdot n^2}{2\log^2 n} ,$$

for sufficiently large n.

Q.E.D.

5.2 The Delay Under a Given Compression Ratio Constraint

By Section 5.1, we see that for both, the block and the L-Z decoder, the maximal delay is achieved for the coded images of the most compressible sequences. With the L-Z algorithm, the asymptotic compressibility of this sequence is zero. Such cases are of minor importance so, as in Section 4, we consider the concept of decoding delay under a compression ratio constraint. Let ρ and $\Delta\rho$ denote given positive constants, with $\rho < 1$, and define the range $\rho_+ \overset{\Delta}{=} [\rho, \rho + \Delta\rho]$. Let D denote a source decoder. The *ρ-maximal delay $D_D(n, \rho_+)$ of order n* introduced by D is defined as

$$D_D(n, \rho_+) \overset{\Delta}{=} \max_{x_1^n \in S'(n,\rho_+)} \max_{\substack{0 < i \leq n \\ x_1^i \in C(x,i)}} d_D(x_1^i \$) ,$$

where $C(x,i)$ is the set containing all the prefixes of x_1^n that are the coded image of some β-ary sequence under C, and $S'(n, \rho_+) \overset{\Delta}{=} \left\{ x_1^n \in C(x,i) : \frac{nk}{(\rho+\Delta\rho)l} \leq L(y_1^{n+1}) \leq \frac{nk}{\rho l} \right\}$. I.e., we consider only those input (coded) sequences of length n that are images of a β-ary (source) sequence whose compression ratio is in the range ρ_+, and all their prefixes that are the coded image of some source sequence.

For the block encoder we can assume, as in Section 4, $s < b < r$ and $kr > \rho lb > ks$, for otherwise the problem is trivial. If we further assume that there exists a codeword of length $s+1$, then it can be shown that the associated ρ-maximal delay of order n is $n[F' + O(n^{-1})]$, where $F' \triangleq \frac{(b-s)(kr-\rho lb)}{\rho lb(r-s)} > 0$.

As for the L-Z decoder we assume, again, $\alpha = \beta = 2$ and $0<\rho<1$.

Theorem 4: For every $\varepsilon>0$ and for sufficiently large n, the ρ-maximal delay $D_{LZD}(n,\rho_+)$ of order n introduced by the L-Z decoder satisfies

$$(\frac{1}{\rho} - 1) \cdot n \leq D_{LZD}(n,\rho_+) < (\frac{1}{\rho} - 1 + \varepsilon) \cdot n \quad .$$

Before presenting the proof, we discuss the meaning of Theorem 4. While the upper bound applies to the delay of every sequence in $S'(n,\rho_+)$, the lower bound may apply only to a subset of sequences in this set. However, by (1), we have $d_{LZD}(x_1^n\$) \geq (\frac{1}{\rho+\Delta\rho} - 1) \cdot n$ for *every* sequence x_1^n in the set.

Proof of Theorem 4: First, we prove the upper bound. Consider a binary sequence $x_1^n \in S'(n,\rho_+)$, and let $x_1^i \in C(x,i)$, $0<i\leq n$. As in the proof of Theorem 2, let $i-j$ denote the last time in the interval $[1,i]$ for which the symbol z_{i-j} at the output of the buffer is null. Consequently, $z_{i-j+1} \neq \lambda$ is the first symbol of the decoded image of a codeword x_m^{i-j} for some integers $1 \leq m < i-j \leq i$. Hence, by (1),

$$d_{LZD}(x_1^i\$) = L(y_1^i) - (s+j) - 1 \quad , \tag{46}$$

where y_1^i is the decoded image of $x_1^i\$$ (since the code is instantaneous [6, Definition A.3, p.453], and $x_1^i \in C(x,i)$, the output for \$ is null), and $s \triangleq L(y_1^{m-1})$. It follows from the above definitions that x_1^{m-1} is the coded image of a sequence of length s. In addition, the length $i-j-m+1$ of the codeword x_m^{i-j} is $1+\lceil \log(c+1) \rceil$ symbols, where c denotes the number of codewords in x_1^{m-1}. Hence,

$$i-j < L(s) + \log[4(c+1)] < s + [L(s)-s] + \log 4i \quad . \tag{47}$$

From the proof of Lemma 3, it can be seen that, for every $s>1$,

$$L(s) - s < \frac{4s}{\log s} \quad . \tag{48}$$

By (47), (48), the inequality $s<i-j\leq i$, and with $x/\log x$ being an increasing function for $x>e$, we have

$$j+s > i - \frac{4i}{\log i} - \log 4i \quad . \tag{49}$$

Hence, by (46) and (49), we obtain

$$d_{LZD}(x_1^i\$) \leq L(y_1^i) - i + \frac{4i}{\log i} + \log 2i \quad . \tag{50}$$

Now, let $L(y_1^i) \triangleq M$ and $L(y_1^n) \triangleq N$, and consider the binary sequence $w_1^N \triangleq y_1^n$. By definition, x_{i+1}^n is the coded image $\mathbf{C}(w_{M+1}^N\$)$ of $w_{M+1}^N\$$. Hence,

$$n = i + \mathit{length}(\mathbf{C}(w_{M+1}^N\$)) = L(i) + \mathit{length}(\mathbf{C}(w_{M+1}^N\$)) - [L(i)-i] \quad .$$

Proceeding as in the proof of Theorem 2 ((33) through (35)), we obtain

$$n \leq g_{i+N-M}(i) + 3$$

and, for sufficiently large n,

$$i+N-M > (n-3)\,(1-\frac{4}{\log\,(n-3)})\ .$$

Hence, by (50),

$$d_{LZD}\,(x_1^i\,\$) \le N-n+\frac{4(n-3)}{\log\,(n-3)}+\frac{4i}{\log i}+\log 2i+3\ .$$

Since $x_1^n \in S'(n,\rho_+)$, we have $N \le \frac{n}{\rho}$, implying the claimed upper bound for sufficiently large n.

As for the lower bound, consider an input sequence x_1^n, $n=c+\sum_{j=1}^{c}\lceil \log j \rceil$, formed by the coded image of u_1^s with a complete last phrase, followed by the coded image of an all-zero sequence with $c-c(s)$ phrases, the first of which contains one more zero than the longest all-zero phrase in u_1^s. Noting that the number of null buffer outputs is greater than $Z(c(s)+1)-s$ we have, by (1) and Lemma 3,

$$\begin{aligned} d_{LZD}\,(x_1^n\$) &> L\,(y_1^{n+1})+Z\,(c\,(s)+1)-s-n-1 \\ &= L\,(y_1^{n+1})+\lceil \log\,(c\,(s)+1)\rceil+L\,(s)-s-n \qquad (51) \\ &> L\,(y_1^{n+1})+\lceil \log\,(c\,(s)+1)\rceil+\frac{2s}{\log s}-3-n\ , \end{aligned}$$

Now, choose $c(s)$ as the least number of phrases such that $L\,(y_1^{n+1}) \le \frac{n}{\rho}$. Since $\rho<1$, this is always possible, for $c(s)=c$ implies $L\,(y_1^{n+1}) \le n \le \frac{n}{\rho}$. Obviously, since $\rho>0$, for sufficiently large n, we can assume $c(s) \ge 8$. Let w_1^n

denote the sequence obtained with $c(s)-1$ codewords resulting from u, followed by $c-c(s)+1$ codewords resulting from the all-zero sequence. It can be readily verified that the output for w_1^n contains, at most, $c-c(s)+1$ symbols more than that for x_1^n. Hence,

$$L(y_1^{n+1}) > \frac{n}{\rho} - c + c(s) - 1 \ . \tag{52}$$

Clearly, the number $c-c(s)$ of phrases in the all-zero part of the decoded sequence is upper bounded by $\sqrt{2(L(y_1^{n+1})-s)}$. Consequently, (52) implies

$$L(y_1^{n+1}) > \frac{n}{\rho} - \sqrt{2L(y_1^{n+1})} - 1 > \frac{n}{\rho} - \sqrt{\frac{2n}{\rho}} - 1 \ . \tag{53}$$

Hence, for sufficiently large n, we have $x_1^n \in S'(n, \rho_+)$. By (51) and (53), we obtain

$$D_{LZD}(n, \rho_+) \geq d_{LZD}(x_1^n \$) > n(\frac{1}{\rho} - 1) + \frac{2s}{\log s} - \sqrt{\frac{2n}{\rho}} \ .$$

It remains to be shown that for sufficiently large n

$$\frac{2\rho s^2}{\log^2 s} \geq n \ .$$

To this end we have

$$n = Z(c) = L(s) + \sum_{j=c(s)+1}^{c} \lceil \log 2j \rceil < L(s) + [c - c(s)] \log 4c$$

$$= L(s) + [c - c(s)] [\log 4(c - c(s)) + \log \frac{c}{c - c(s)}]$$

$$\leq L(s) + [c - c(s)] \log 4(c - c(s)) + c(s) \log e$$

$$< L(s) + [c - c(s)] \log (c - c(s)) + 2c$$

$$< L(s) + \sqrt{\frac{n}{2\rho}} \log \frac{2n}{\rho} + 2c \quad ,$$

where we have used the inequality $\log x \leq (x-1)\log e$ with $x = \frac{c}{c - c(s)}$, and

the upper bound $\sqrt{\frac{2n}{\rho}}$ on $c - c(s)$ as in (53). By (45), we obtain

$$L(s) > n [1 - \frac{2(1+\delta(n))}{\log n} - \frac{\log (2n/\rho)}{\sqrt{2\rho n}}] \quad . \tag{54}$$

It follows from (48), (54), and from $s < n/\rho$ that

$$s > n [1 - \frac{2(1+\delta(n)}{\log n} - \frac{\log 2n/\rho}{\sqrt{2\rho n}} - \frac{4}{\rho \log n}] \triangleq n (1 - \gamma(n)) \quad ,$$

where $\lim_{n \to \infty} \gamma(n) = 0$. Hence, for sufficiently large n ,

$$\frac{2\rho s^2}{\log^2 s} > \frac{2\rho n^2 (1 - \gamma(n))^2}{\log^2 (n/\rho)} > n \quad .$$

Q.E.D.

References

[1] R.E. Krichevsky and V.K. Trofimov, "The performance of Universal Encoding", IEEE Trans. Infor. Theory, vol. IT-27, pp. 199-207, March 1981.

[2] S. Even, "On information Lossless Automata of Finite Order", IEEE Trans. Electronic Computers, vol. EC-14, pp. 561-569, August 1965.

[3] A. Lempel and J. Ziv, "Compression of Individual Sequences via Variable Rate Coding", IEEE Trans. Infor. Theory, vol. IT-24, pp. 530-536, September 1978.

[4] A.D. Wyner, "On the Probability of Buffer Overflow under an Arbitrary Bounded Input-Output Distribution", SIAM J. Appl. Math., vol. 27, pp. 544-570, December 1974.

[5] N. Merhav, "Universal Coding with Minimum Probability of Codeword Length Overflow", IEEE Trans. Infor. Theory, vol. IT-37, pp. 556-563, May 1991.

[6] F. Jelinek, Probabilistic Information Theory, New York: McGraw-Hill, 1968.

10

Finite State Two-Dimensional Compressibility

Dafna Sheinwald

IBM Scientific Center, Haifa Israel

Abstract

Combining results from [6, 9, 11], a generalized, parametric quantity $q(I)$, measurable on any given individual picture (two dimensional array of data) I, is proved to be a lower bound on the compression ratio attainable for I by any information lossless finite state encoder. Quantity $q(I)$ is shown to be a tight lower bound, asymptotically attainable, and thus an inherent measure of the picture. Different measures of sequence and picture complexity are related to $q(I)$. Other measurable quantities, also serving as tight lower bounds on the attainable compression, are referred to.

1 Introduction

Early works in the field of complexity, or randomness, of given individual sequences tied the measure of complexity with algorithms for producing the sequences. Kolmogorov [4] defined the complexity of a given sequence x as the length of the shortest input for which there exists a Turing machine that outputs x. Although very intuitive and appealing, this measure is not provable for all given sequences. Chaitin [1] showed that no formal verification exists for a complexity exceeding some function of the number of axioms and induction rules of the formal theory.

In [5], Lempel and Ziv suggested a complexity measure that is determined for any given sequence by an iterative process of parsing the sequence into subsequences, such that no subsequence has a copy thereof preceding it in the sequence. In [10], the authors introduced a universal compression algorithm that attains for any sequence given to it a compression ratio correlated with the sequence's complexity that was defined in [5].

In [11], Ziv and Lempel studied the best compression ratio attainable for any given individual sequence by the class of information lossless [2, 3] finite state encoders, named the *finite state compressibility* of the given sequence. The authors derived a quantity $q(x)$, measurable on any given sequence x by parsing growing prefixes thereof into distinct subsequences, and proved it the finite state compressibility of x. Although there exist sequences with high $q(x)$ but low Kolmogorov complexity, still sequences with high $q(x)$ stand most of the "standard" randomness tests and information theoretic entropy measures. Quantity $q(x)$ may, therefore, effectively serve as a complexity measure for x.
The keystone in proving $q(x)$ a lower bound on the attainable compression ratio for x is a derivation of an *upper* bound on the number of *distinct* subsequences of x on which any given information lossless finite state encoder may produce a given output. Compression ratio $q(x)$ was proved attainable by a universal compression algorithm that iteratively parses its input sequence into growing distinct subsequences, and encodes each subsequence into a reference to a previously parsed one.

In [6], Lempel and Ziv extended the encoder model by adding a moving reading head, yielding a finite state model for a picture encoder. They studied the *finite state compressibility* of any given individual picture, namely, the best compression ratio attainable for the picture by any information lossless finite state encoder. The main result of [6] is that the finite state compressibility of the *sequence* read from the picture, when scanned by the Pieano-Hilbert curve, asymptotically approaches the picture compressibility. Employing the optimal sequential compression algorithm from [11] to the above sequence may therefore constitute an asymptotically optimal compression algorithm for the picture. In [6], the keystone of the proof is a derivation of an upper bound on the number of distinct square subpictures, all of the same size, on which any given information lossless finite state encoder may produce a given output. Similarly to the one dimensional case, picture compressibility may very well serve as a measure for its complexity. Specifically, it was proved to asymptotically approach the Shannon entropy of the picture, determined by the relative frequencies in the picture of growing square subpictures.

In this paper we further investigate the compressibility of individual pictures. Results of [11] with respect to one dimensional data are extended, in a more direct, natural way than in [6], to the case of two dimensional data; an extension previously believed to be impossible [7]. We generalize the upper bound proved in [6] on the number of distinct subpictures on which a given information lossless finite state encoder can produce a given output. We cover the case of subpictures of different shapes, different sizes, and even non continuous subpictures. This leads us to a generalized, parametric quantity $q(I)$, analogous to $q(x)$ of [11], which we prove to be the finite state compressibility of picture I. Quantity $q(I)$ is measurable on any given individual picture, and may thus serve as an effective measure for its complexity.

2 The Encoder Model

We view the given picture I as if traced out on a two dimensional array of *cells* of size 1×1 each, one picture symbol in each cell. Each cell has eight *neighbors*, and its location in the picture is specified by a pair of coordinates. The picture is infinite to the right and downward, and we investigate growing top left $N \times N$ portions I_N thereof. Before applying an encoder to I_N, two rows and two columns of cells containing special *edge* symbols (distinct from all picture symbols) are appended to I_N on its top, bottom, left, and right, respectively.

As in [6], our encoder model is a natural extension of the sequential encoder of [11] that was already introduced in [8]. An encoder E is the septuple $E = (S, A, B, D, f, g, t)$ where S is a finite set of states one of which is the starting state s_1; A is a finite input alphabet of α picture symbols and the extra edge symbol; B is a finite set of output binary words including the empty word of length 0; D is the set of incremental displacement vectors that determine the move of the reading head, consisting of the eight moves to any of the eight neighboring cells and the empty move whereby the reading head stays in place; $f : S \times A \to B$ is the output function; $g : S \times A \to S$ is the next state function; and $t : S \times A \to D$ is the incremental displacement function.

When encoder $E = (S, A, B, D, f, g, t)$ processes a picture, it reads a sequence of input letters $x_1 x_2 x_3 \cdots$, $x_i \in A$ from cells in locations $\vec{\Delta}_1$ (= upper left cell of the picture), $\vec{\Delta}_2, \vec{\Delta}_3 \cdots$, respectively; outputs a sequence of words $y_1 y_2 y_3 \cdots$, $y_i \in B$; goes through a sequence of states $z_1 (= s_0) z_2 z_3 \cdots$, $z_i \in S$; and moves its reading head according to a sequence of incremental moves $\vec{d_1} \vec{d_2} \vec{d_3} \cdots$, $\vec{d_i} \in D$. The i-th step of the processing, $i \geq 1$, starts with the encoder being in state z_i, observing input symbol x_i in cell $\vec{\Delta}_i$. The encoder then outputs word $y_i = f(z_i, x_i)$, goes to state $z_{i+1} = g(z_i, x_i)$, and finally moves its reading head to location $\vec{\Delta}_{i+1} = \vec{\Delta}_i + \vec{d_i} = \vec{\Delta}_i + t(z_i, x_i)$.

The subsequence $w_i w_{i+1} \cdots w_j$ of a sequence $w_1 w_2 w_3 \cdots$ is denoted by w_i^j.

An encoder E is said to be *Information Lossless (IL)* if for all $z_1 \in S$ and $x_1^n \in A^n$, $n \geq 1$, the triple $(z_1, f(z_1, x_1^n), g(z_1, x_1^n))$, where $f(z_1, x_1^n)$ is the concatenation of the output words emitted by E when applied to x_1^n in state z_1, and $g(z_1, x_1^n)$ – the state E reached upon completing that application, uniquely determines x_1^n, and hence also z_1^n and $\vec{d}_1^n$.

The following compression ratio definitions are also as in [6]. Given an $N \times N$ picture I_N and an encoder E, the *compression ratio* attained by E for I_N is given by

$$\rho_E(I_N) \triangleq \frac{L(y_1^n)}{N^2 \log(\alpha)}$$

where n ($\geq N^2$) is the number of moves needed for E to complete the scan of I_N, and $L(y_1^n) = \sum_1^n l(y_i)$ with $l(y_i)$ being the length in bits of the i-th output word y_i. Throughout this paper, $\log(w)$ stands for $\log_2(w)$. Let $E(s)$ denote the class of information lossless encoders with no more than s states, and define

$$\rho_s(I_N) \triangleq \min_{E \in E(s)} \{\rho_E(I_N)\}.$$

For the infinite picture I we define

$$\rho_s(I) \triangleq \limsup_{N \to \infty} \rho_s(I_N),$$

and

$$\rho(I) \triangleq \lim_{s \to \infty} \rho_s(I).$$

$\rho(I)$ is called the *finite state compressibility* of I.

For further discussion we need some more definitions, all related to subpictures.

1. The number of cells composing a subpicture T of I_N is called the *area* of T, and denoted by $|T|$.

2. The *perimeter* of T is the set of all cells of T neighboring one or more cells of I_N not in T. The number of cells in the perimeter of T is called the *length of perimeter* of T, denoted $p(T)$.

3. A sequence of cells $c_1 c_2 \cdots c_n$ where c_i is a neighbor of c_{i+1}, $1 \leq i < n$, is called a *path* leading from c_1 to c_n via $c_2, c_3, \ldots, c_{n-1}$. The number n of cells in that sequence is the *length* of the path.

4. A subpicture T is *continuous* if for every pair of cells c_1 and c_2 in T there exists a path leading from c_1 to c_2 via cells of T only.

5. With each subpicture T, associated is the sequence $C(T) = P_1, P_2, \ldots, P_m$ of maximal continuous nonoverlapping subpictures P_i which together cover T. In a raster scan of I_N, a cell of P_i is encountered before any cell of P_{i+1}, $1 \leq i < n$.

6. The *shape* $S(P)$ of a continuous subpicture P is the two dimensional form covered by the cells of P. We say that $S(P) = S(Q)$ iff the two dimensional form covered by the cells of P is congruent to the form covered by the cells of Q, *without* any rotation. The shape $S(T)$ of any subpicture T is defined as the sequence of shapes of the respective continuous subpictures in $C(T)$. That is, if $C(T_1) = P_1, P_2, \ldots, P_n$ and $C(T_2) = Q_1, Q_2, \ldots, Q_m$ then $S(T_1) = S(T_2)$ iff $m = n$ and $S(P_i) = S(Q_i)$ for all $1 \leq i \leq n$.

7. $T_1 \neq T_2$ iff either $S(T_1) \neq S(T_2)$ or $S(T_1) = S(T_2)$ and there exists a pair of cells $c_1 \in T_1$ and $c_2 \in T_2$ such that the relative location of c_1 in $S(T_1)$ is the same as the relative location of c_2 in $S(T_2)$, but the contents of c_1 and c_2 (the input symbols written in them) are different.

Given an infinite picture I, our first goal is to prove a lower bound on $\rho_s(I_N)$. To this end, we partition I_N into subpictures and compute a function of the subpicture set. This partition is only an analysis tool, and by no means restricts the operation of the encoders discussed. In Section 3, we point at a special subset of the subpicture set and prove two auxiliary lemmas about it. In Section 4, we complete the lower bound proof, and investigate it for growing N and s. In Sections 5, 6, and 7, we discuss algorithms that asymptotically attain the proved lower bound.

3 Good Sets

Fix parameters $b_1 > 0$, $0 \leq b_2 < 1$, $b_3 \geq 0$, and $b_4 \geq 0$.
Let $\Pi(I_N) = \{T_1, T_2, \ldots\}$ be a partition of I_N into subpictures and let $G \subseteq \Pi(I_N)$ be a set satisfying the following **good set** condition:

1. the subpictures in G are distinct,
2. for every $T \in G$, $p(T) \leq b_1|T|^{b_2}$, and
3. the number of distinct shapes of subpictures $T \in G$ of area a does not exceed $b_3 a^{b_4}$.

In the sequel, we consider the output produced by any IL finite state encoder E, applied to I_N, while visiting cells of G. We will show that the good set condition implies that not too many members of G can cause E to produce the same total output. Accumulating the lengths of the various total outputs we receive a lower bound on the compression ratio attainable for I_N.

Let g denote the cardinality of G, G_a – the subset of G consisting of all $T \in G$ with $|T| = a$, and g_a – the cardinality G_a. Let m denote the maximal area of any subpicture in G; that is, $\sum_{a=1}^{m} g_a = g$.

The following lemma shows that g is small compared to N^2. More specifically, g/N^2 tends to 0 as N goes to infinity.

Lemma 1

$$g \leq \frac{N^2 \log(\alpha)}{(1-\epsilon_N)\log(N^2)} \qquad \text{where} \qquad \lim_{N\to\infty} \epsilon_N = 0.$$

Proof: This proof is similar to that of Theorem 2 in [5], and that of Lemma 1 in [9]. The aim is to maximize $g = \sum_a g_a$ and still comply with the good set condition, by which g_a can not exceed $b_3 a^{b_4} \alpha^a$, and the constraint by which $\sum_a a g_a \leq N^2$. To this end, we overestimate g_a to its upper bound, $b_3 a^{b_4} \alpha^a$, for $a = 1, 2, 3, \ldots, b$, where b is the maximal integer possible, namely, $S_b = \sum_{a=1}^{b} a b_3 a^{b_4} \alpha^a \leq N^2 < \sum_{a=1}^{b+1} a b_3 a^{b_4} \alpha^a$. We then take $g_b = (N^2 - S_b)/b$, and $g_a = 0$ for $a > b$. Putting all these ingredients into some tedious calculations, we derive

$$\frac{g}{N^2} \leq \frac{\log(\alpha)}{(1-\epsilon_N)\log(N^2)}$$

with

$$\epsilon_N = \frac{3\log(\alpha) + \log(b_3) + (b_4+1)\log(\log(N^2))}{\log(N^2)}.$$

□

Lemma 2

$$g\log(g) \le \sum_{a=1}^{m} g_a \log(g_a) + \delta_N N^2 \qquad \text{with} \qquad \lim_{N\to\infty} \delta_N = 0.$$

For $x = 0$, we take $x\log(x) = 0$.
For proof see Lemma 2 in [9], which is the same as this lemma. It is worthy of noticing that the lemma is valid for any set of integers $\{g_a\}$ and their sum $g = \sum_a g_a$, which depend on N such that $\lim_{N\to\infty} g/N^2 = 0$.

4 Lower Bound

Let E be any IL finite state encoder with s states, and let y_1^n be the total output produced by E in the process of encoding I_N. Let subpicture T be any member of G. Suppose that in the course of compressing I_N, there are r transitions from a cell outside T into a cell belonging to T. That is, T is processed by E in r passes where each pass corresponds to a sequence of transitions inside T. Let $y_{u_j}^{v_j}$, $1 \le j \le r$ denote the substring of y_1^n which is produced by E during the j-th pass through T. By the IL property of E, T is fully recoverable if $S(T)$ is known and for each pass through T given are the entry state z_{u_j}, the output $y_{u_j}^{v_j}$, the exit state z_{v_j+1}, and the coordinates Δ_{u_j} of the entrance cell measured relative to the perimeter of T, $1 \le j \le r$.

While processing I_N, E produces a total output (possibly null) on each and every member of G. Yet, the above conclusion from the IL property of E and the fact that G is a good set imply that not "too many" members $T \in G$ can cause E to produce the same total output. Now, even if the largest possible number of members of G cause E to produce each total output, still, in order to associate a total output with each and every member of G, long outputs must be associated too, and the total length of all the outputs must grow. This is how we arrive at a lower bound on $\rho_s(I_N)$. Following is a formalization of this argument.

Let $h(a,r,l)$ denote the number of members $T \in G_a$ processed by E in r passes while producing a total output of l bits. $h(a,r,l)$ cannot exceed the product $h_1(a) \cdot h_2(a,r) \cdot h_3(r,l)$, where the three functions h_i are as follows.

$h_1(a) \leq b_3 a^{b_4}$ is an upper bound on the number of distinct shapes of subpictures $T \in G_a$.

$h_2(a,r) \leq \binom{p_a \cdot s}{r} \cdot s^r$ is an upper bound on the number of choices of r triples $(z_{in}, \Delta, z_{out})$ indicating entry state, entrance cell (whose coordinates are measured relative to the perimeter of T), and exit state of the r passes of E through T where $p_a = b_1 a^{b_2}$ is an upper bound on the perimeter length of each $T \in G_a$. Note that a pair (z_{in}, Δ) is never chosen more than once if E is not to enter an endless loop.

$h_3(r,l) \leq (l+1)^r 2^l$ is an upper bound on the number of ways to associate r (possibly null) binary substrings of total length $= l$ to r distinct triples $(z_{in}, \Delta, z_{out})$.

Altogether, we can write:

$$h(a,r,l) \leq 2^l b_3 a^{b_4} \binom{p_a \cdot s}{r} (s(l+1))^r.$$

Since the number of passes can not exceed the number of distinct pairs (z_{in}, Δ), it follows that $r \leq p_a s$. Therefore, denoting by $h(a,l)$ the number of $T \in G_a$ for which E produces a total output of l bits, we have

$$h(a,l) = \sum_{r=1}^{p_a s} h(a,r,l) \leq 2^l b_3 a^{b_4} (s(l+1)+1)^{p_a s} \triangleq h_{max}(a,l).$$

Next, let $L(a)$ denote the total output length produced by E in all passes through all members of G_a. Then,

$$L(a) = \sum_l l h(a,l).$$

In order to derive a lower bound on $L(a)$, it is clear that we may overestimate $h(a,l)$ for small values of l, to its upper bound $h_{max}(a,l)$, at the expense of $h(a,l)$ for large values of l, provided that the sum of $h(a,l)$ over all values of l does not exceed g_a. Using calculations similar to those in the proof of Lemma 1, we arrive at the following lower bound:

$$L(a) \geq g_a \left[\log(g_a) - \log(b_3) - b_4 \log(a) - b_1 a^{b_2} s \log(sa \log(\alpha)) - 3\right].$$

Finally, the total output length produced by E while processing all the members of G is given by

$$\begin{aligned} L &= \sum_{a=1}^{m} L(a) \\ &\geq \sum_{a=1}^{m} g_a \log(g_a) - b_4 \sum_{a=1}^{m} g_a \log(a) \\ &\quad - b_1 s \sum_{a=1}^{m} g_a a^{b_2} \log(sa \log(\alpha)) - (\log(b_3) + 3)g. \end{aligned}$$

Using the convexity of $a^{b_2} \log(a)$ – for a greater than a constant depending on b_2, the convexity of x^{b_2}, and $\log(x)$, the fact that $\sum_{a=1}^{m} a g_a \leq N^2$, Lemma 1, and Lemma 2, we arrive, with some more tedious calculations, at the following result:

$$\begin{aligned} \rho_E(I_N) &\geq \frac{L}{N^2 \log(\alpha)} && (1) \\ &\geq \frac{g \log(g)}{N^2 \log(\alpha)} - \psi(N,s) \quad \text{with} \quad \lim_{N \to \infty} \psi(N,s) = 0. && (2) \end{aligned}$$

Note that the dependency on E is reflected in (2) in terms of the number of states s. Thus, (2) is effective for all $E \in E(s)$, and we have

$$\rho_s(I_N) \geq \frac{g \log(g)}{N^2 \log(\alpha)} - \psi(N,s) \quad \text{with} \quad \lim_{N \to \infty} \psi(N,s) = 0. \tag{3}$$

Let $c(I_N)$ denote the maximal cardinality of a good set, where the maximization is taken over all partitions of I_N, and for each partition – over all good sets of subpictures of the partition. (3) is valid for every cardinality of a good set, in particular $c(I_N)$. We therefore complete the proof of the following theorem:

Theorem 1 (Converse-to-Coding Theorem)

$$\rho_s(I_N) \geq \frac{c(I_N) \log(c(I_N))}{N^2 \log(\alpha)} - \psi(N,s) \quad \text{with} \quad \lim_{N \to \infty} \psi(N,s) = 0. \tag{4}$$

Corollary 1

$$\rho_s(I) = \limsup_{N\to\infty} \rho_s(I_N) \geq \limsup_{N\to\infty} \frac{c(I_N)\log(c(I_N))}{N^2\log(\alpha)},$$

and

$$\rho(I) = \lim_{s\to\infty} \rho_s(I) \geq \limsup_{N\to\infty} \frac{c(I_N)\log(c(I_N))}{N^2\log(\alpha)}.$$

Since the derivation of the lower bound on $\rho_s(I_N)$ does not depend on the continuous moving of the encoder outside the subpictures of I_N, it gives rise to the following corollary too:

Corollary 2

$$\begin{aligned}\rho_s(I) \;\geq\; & \limsup_{p\to\infty} \frac{1}{p^2M^2\log(\alpha)} \\ & \sum_{i=1}^{p}\sum_{j=1}^{p} c(I_M(i,j))\log(c(I_M(i,j))) - \eta_s(M) \\ & \qquad \text{with} \quad \lim_{M\to\infty} \eta_s(M) = 0,\end{aligned}$$

where $I_M(i,j)$ is the (i,j)-th block I_M in the partition of I_{pM} into p^2 squares of area M^2 each.

5 Algorithm Framework

The quantity

$$q(I) = \limsup_{N\to\infty} \frac{c(I_N)\log(c(I_N))}{N^2\log(\alpha)}$$

is analogous to quantity

$$q(x) = \limsup_{n\to\infty} \frac{c(x_1^n)\log(c(x_1^n))}{n\log(\alpha)},$$

where $c(x_n)$ is the maximal number of distinct phrases into which the prefix x_1^n of x can be parsed. Quantity $q(x)$ was proved in [11] to be a lower bound on the compression ratio attainable for the infinite sequence x by any IL finite state (one dimensional) encoder.

In [11], Ziv and Lempel introduced their incremental parsing algorithm, ZL, which asymptotically achieves compression $q(x)$ for any sequence given to it, proving $q(x)$ the finite state compressibility of x. Algorithm ZL iteratively parses its input x_1^n into distinct phrases, and encodes them. The algorithm maintains a dictionary D whose initial contents are the empty word of length 0. Each parsed phrase is added to dictionary D upon the completion of its encoding. Step i of the algorithm starts with the empty word and the set of $i-1$ phrases parsed in steps $1, 2, \ldots, i-1$ stored in D, and with the prefix $x_1^{j_{i-1}}$ of length j_{i-1} of x_1^n completely processed. The algorithm then finds the minimal k such that $x_{j_{i-1}+1}^k$ is *not* in D. Denoting this value of k by j_i, it follows that the prefix $x_{j_{i-1}+1}^{j_i-1}$ *is* in D. $x_{j_{i-1}+1}^{j_i}$ is then encoded into the index of its prefix in D, using $\lceil \log(i) \rceil$ bits, and the index of its last letter x_{j_i} in the input alphabet – as a sequence of $\lceil \log(\alpha) \rceil$ bits. Finally, phrase $x_{j_{i-1}+1}^{j_i}$ is added to the dictionary, and the algorithm is about to start its $i+1$-st step at x_{j_i+1}. The process continues until the whole of x_1^n has been processed. The phrase determined at the last step of the algorithm may be equal to a member of the dictionary effective at the beginning of the last encoding step. However, we omit the discussion of this case which does not change the results. Let $p(x_1^n)$ denote the number of steps it takes algorithm ZL to encode x_1^n in entirety, and let $D(x_1^n)$ denote the dictionary at the end of the encoding process. Note that $D(x_1^n)$ includes $p(x_1^n)$ distinct phrases. The total output length produced by the algorithm while processing x_1^n is thus upper bounded by the sum $\sum_{i=1}^{p(x_1^n)} \lceil \log(i) \rceil + p(x_1^n) \lceil \log(\alpha) \rceil \leq p(x_1^n) \log(p(x_1^n)) + bp(x_1^n)$, with $b = 1 + \lceil \log(\alpha) \rceil$. Now, since the members of $D(x_1^n)$ are distinct, $p(x_1^n) \leq c(x_1^n)$ by definition of $c(x_1^n)$. Hence, when n goes to infinity, the compression ratio attained by the Ziv Lempel algorithm tends to:

$$\limsup_{n \to \infty} \rho_{ZL}(x_1^n) \leq \limsup_{n \to \infty} \frac{p(x_1^n) \log(p(x_1^n))}{n \log(\alpha)} \leq q(x). \tag{5}$$

An infinitely long input sequence is fed to algorithm ZL finite segment by finite segment, so that each segment is separately processed, using a finite dictionary. The length of the segments is chosen to satisfy the desired approach of the obtained compression ratio to $q(x)$.

As $q(I)$ is analogous to $q(x)$, we would look for a compression algorithm working in analogy with algorithm ZL; that is, an algorithm of the following framework. The algorithm iteratively parses subpictures T from the picture I_N given to it, and encodes them. Step i of the algorithm begins with the empty subpicture of area 0 and the set of the $i-1$ subpictures parsed in steps $1, 2, \ldots, i-1$ stored in dictionary D. First, a subpicture T_i is parsed from I_N. T_i is such that there exists a subpicture $T_j \in D$, $j < i$, for which $C(T_j)$ is equal to a major part of $C(T_i)$. T_i is then encoded into a sequence of $\lceil \log(i) \rceil$ bits expressing j in the binary notation, and a description of the part of $C(T_i)$ which is not covered by $C(T_j)$. Finally, T_i is added to D, and step i of the algorithm is complete.
Let $p(I_N)$ denote the number of subpictures that have been parsed by the algorithm once the processing of I_N is complete. The total output length produced by the algorithm is

$$\sum_{i=1}^{p(I_N)} \lceil \log(i) \rceil + L(residues), \tag{6}$$

where $L(residues)$ is the total length of the descriptions of the residual parts of the subpictures. The algorithm satisfies:
f1: $p(I_N) \leq c(I_N)$ and
f2: the total length of the descriptions of the residues is negligible compared to N^2.
Combining with (6) and with Theorem 1, the algorithm is asymptotically optimal.

The existence of an actual algorithm that fits this framework was questioned in the past [7]. Note that the algorithm of [6] compresses a *sequence* read from the picture employing an incremental parsing thereof, whereas our algorithm iteratively parses the *picture* given to it into subpictures.

In the following sections we introduce two compression algorithms that fit the above framework. Being such, they prove the lower bound of Theorem 1 tight. Hence, $q(I)$ *is* the compressibility of I.

6 Compression Algorithm A1

Algorithm A1 scans the input picture on a predetermined path, denoted SP, independent of the contents of the picture. The scanning path visits each and every cell of I_N exactly once. Given a picture I_N, let $A(l,m)$ denote the subpicture covered by the subpath of SP which extends from the l-th cell visited by SP to the m-th one.

Step $i \geq 1$ of the algorithm starts with dictionary D including i subpictures (one of which is the empty subpicture), and the l_i-th cell of SP is being observed, where $l_i - 1$ is the total number of cells visited by SP in the first $i-1$ steps. In the course of the i-th step, the scan iteratively proceeds cell by cell on SP, from its l_i-th cell on. When visiting the m-th cell of SP, $m \geq l_i$, A1 checks the membership of $A(l_i, m)$ in D. Thus, as the scan proceeds, m_i satisfying

$$m_i = \min\{m | l_i \leq m \leq N^2, A(l_i, m) \notin D\}$$

is found. The m_i-th cell of SP is the last cell visited in the i-th step of the algorithm. By the definition of m_i and the fact that the empty subpicture is a member of D, it is clear that $A(l_i, m_i - 1) \in D$. Once m_i is determined, A1 encodes $A(l_i, m_i)$ into a binary string composed of two substrings: the first is the binary representation, using $\lceil \log(i) \rceil$ bits, of the index of $A(l_i, m_i - 1)$ in D, and the second is the binary representation, using $\lceil \log(\alpha) \rceil$ bits, of the picture symbol in the m_i-th cell of SP. Finally, $A(l_i, m_i)$ is added to D as its i-th member.

Let $p_{A1}(I_N)$ denote the number of steps it takes A1 to complete the processing of I_N. The total output length produced by A1 during that processing does not exceed

$$\sum_{1=1}^{p_{A1}(I_N)} \lceil \log(i) \rceil + p_{A1}(I_N) \lceil \log(\alpha) \rceil. \tag{7}$$

For proving that A1 fits the framework described above, all that is left to show is that $p_{A1}(I_N) \leq c(I_N)$ for all values of N and pictures I_N. The reason is, that $p_{A1}(I_N) \leq c(I_N)$ will obviously satisfy **f1**. Combining

Lemma 1 with the fact that the residue description lengths accumulate to $p_{A1}(I_N)\lceil\log(\alpha)\rceil$ (see (6)), **f2** will be satisfied as well. One way of proving that $p_{A1}(I_N) \leq c(I_N)$ is to show that the set of $p_{A1}(I_N)$ subpictures, into which I_N is parsed by A1, is a good set. The first part of the good set condition is easily seen to be complied with: the subpictures are distinct by their definition. For complying with the second and third part, we must assume some properties of SP. In what follows, two such properties are specified, a scanning path possessing them is shown, and the way the properties imply the complying with the good set condition is pointed at.

We assume $N = 2^k$. A square subpicture J of I_{2^k} is called a *legal square* if either $J = I_{2^k}$ or J is a quadrant of a legal square of I_{2^k}.

We require that SP possesses the following two properties:

Property 1 (The Legal Square Property) SP covers I_{2^k} legal square by legal square. That is, having entered a legal square, SP leaves it only after visiting each and every cell thereof.

Property 2 (The Square Scan Property) SP scans each legal square in one of a fixed number, denoted $t(SP)$, of different scans.

An example of a path possessing the above two properties is the Peano-Hilbert scan, also used in [6], where a definition thereof can be found. For the Peano-Hilbert scan, $t(SP) = 4$.

Given a subpicture A which is covered by a continuous subpath of SP, we say that J is a *maximal square* of A if J is a legal square included in A but not in a larger legal square that is included in A. The set $Q(A)$ of all the maximal squares of A forms a partition of A. By the legal square property, SP covers the members of $Q(A)$ one by one, and thus we can write these members in a sequence $S(A) = S_1(A), S_2(A), \ldots, S_s(A)$ in accordance with the order SP induces on them.

Lemma 3 If $|S_l(A)| < |S_{l-1}(A)|$, for some $1 < l \leq s$, then for $l \leq i \leq s$, $|S_i(A)| \leq |S_{i-1}(A)|$.

Proof: Assume the contrary of the lemma, and let k be the minimal index in the range $l < k \leq s$, for which $|S_k(A)| > |S_{k-1}(A)|$. By the definition of k it follows that $|S_{k-2}(A)| \geq |S_{k-1}(A)| < |S_k(A)|$. Let J denote the legal square of which $S_{k-1}(A)$ is a quadrant. By the definition of maximal squares, J is not included in A. Also by the definition of l and k, $S_{l-1}(A)$, J, and $S_k(A)$ are pairwise non overlapping. Now, by the definition of $S(A)$, it follows that SP covers the whole of $S_{l-1}(A)$, later it enters J, and before completing the scanning thereof, it leaves and enters $S_k(A)$. This contradicts the legal square property of SP. □

¿From Lemma 3 it follows that the sequence $a(A) = a_1(A), a_2(A), \ldots, a_s(A)$, where $a_i(A)$ is the area $|S_i(A)|$ of $S_i(A)$, either monotonically increases, or monotonically decreases, or first monotonically increases and then monotonically decreases (not necessarily strictly increases or decreases).

Lemma 4 Let b be a non maximal element of $a(A)$. Then, in any of the (at most two) monotonic sections of $a(A)$, there are at most three (successive) appearances of b.

Proof: Assume, to the contrary, that in a monotonically increasing section of $a(A)$, b appears four times or more and let $h > b$ be the element following the last appearance. Let $S_k(A), S_{k+1}(A), S_{k+2}(A), S_{k+3}(A)$, and $S_{k+4}(A)$ be the elements of $S(A)$ that correspond to the last four appearances of b and the following h in $a(A)$. Let J be the legal square of which $S_{k+3}(A)$ is a quadrant. J and S_{k+4} are not overlapping. By the definition of maximal squares, not all of $S_k(A), \ldots, S_{k+3}(A)$ are quadrants of J. It follows that SP starts out of J, at $S_k(A)$, then enters J when entering either of S_{k+1}, S_{k+2}, or S_{k+3}, and before completing the scan of J, SP leaves for S_{k+4}, contradicting the legal square property of SP. The proof for the case of a monotonically decreasing section of $a(A)$ is analogous. □

Lemma 5 Let b be the maximal element of $a(A)$, then there are at most six (successive) appearances of b in $a(A)$.

Proof: Suppose that there were seven or more appearances of b in $a(A)$. If not four of the corresponding maximal squares of A are included in the same legal square of area $4b$, then they are included in at least three such legal squares, none of which is included in A, and SP visits the first of the three, then enters the second, and before completing its scan (since it is not in A), SP leaves for the third. This contradicts the legal square property of PS. The inclusion of four of the maximal squares of area b in the same legal square of area $4b$ contradicts their being maximal squares. □

Lemma 6 The perimeter length $p(A)$ of A does not exceed $48|A|^{1/2}$.

Proof: The length of the perimeter of A does not exceed the total length of the perimeters of all its maximal squares. By the definition of legal square, the area of any maximal square is of the form $2^k \cdot 2^k$ for some non negative integer k, and its perimeter length is 4×2^k. By Lemma 4 and Lemma 5, there are at most six maximal squares of A of each area a, $1 \leq a \leq |A|$, which is a power of 4. Thus, for $\beta = \lfloor \frac{1}{2} \log(|A|) \rfloor$, the perimeter length $p(A)$ of A is upper bounded by

$$\sum_{i=0}^{\beta} 6 \times 4 \times 2^i \leq 48 \times 2^\beta \leq 48|A|^{1/2}. \qquad \square$$

Two subpaths of SP of length l each are said to be *equivalent* if there exists a vector (p, q) such that for $1 \leq i \leq l$ the coordinates $\vec{\Delta}_{i,1}$ of the i-th cell visited by the first subpath, and the coordinates $\vec{\Delta}_{i,2}$ of the i-th cell visited by the second subpath satisfy $\vec{\Delta}_{i,2} = \vec{\Delta}_{i,1} + (p, q)$. Since equivalent subpaths cover subpictures of equal shapes, but not necessarily vice versa, the upper bound r_l on the number of non-equivalent subpaths of SP of length l also bounds from above the number s_l of distinct shapes that subpictures of area l, parsed by algorithm A1, may have.

Lemma 7 $s_l \leq (3t(SP) + 8(t(SP))^2)l$.

Proof: As discussed above, it is enough to show that $r_l \leq t(SP)(3l - 3) + (t(SP))^2(l-1)$. Let m^2 be the area of the smallest legal square that can contain a path of length l. That is, m is a power of 2 satisfying $l \leq m^2 \leq 4(l-1)$. Every subpath of length l is included in one legal square of area m^2, or it starts in one legal square of area m^2 and ends in an adjacent legal square of the same size. Now, the number of non-equivalent subpaths of length l which are included in one $m \times m$ legal square is upper bounded by the product of the number of scans of the legal square (see Property Scan Square above) and the number of starting cell of the subpath within the square. That is, $t(SP)(m^2 - l + 1) \leq t(SP)(4(l-1) - l + 1) = t(SP)(3l - 3)$. The number of non-equivalent subpaths of length l which start in one $m \times m$ legal square and end in an adjacent one is upper bounded by the product of the number of scans of the two legal squares, the number of starting cell for the subpath within the first square, and the number of ways the two legal squares are positioned with respect to one another. That is, $8(t(SP))^2(l-1)$. Altogether, $r_n \leq t(SP)(3l-3)+8(t(SP))^2(l-1) \leq (3t(SP)+8(t(SP))^2)l$. $\square$

Lemma 8 The set of subpictures parsed by algorithm A1 while processing I_N is a good set.

Proof: The subpictures are distinct by their definition. By Lemma 6, part 2 of the good set condition is satisfied with $b_1 = 48$ and $b_2 = 1/2$. By Lemma 7, part 3 of the condition is satisfied with $b_3 = 3t(SP)+8(t(SP))^2$ and $b_4 = 1$. $\square$

¿From Lemma 8 it follows that $p_{A1}(I_N) \leq c(I_N)$, and we conclude that A1 fits our framework, and as such is asymptotically optimal. Namely, for all finite values of s,

Lemma 9

$$\lim_{N \to \infty} \rho_{A1}(I_N) \leq \lim_{N \to \infty} \rho_s(I_N).$$

In analogy with the Ziv Lempel algorithm [11], an infinitely large picture is fed to A1 "big" square subpicture by "big" square subpicture, so that each square subpicture is separately processed by A1, using a finite dictionary. The size of the square subpictures is chosen to satisfy the desired approach of the obtained compression ratio to its lower bound $q(I)$. Formally,

Theorem 2 (Coding Theorem) Let j and m be positive integers. For all $\epsilon > 0$, $q = 2^j$ and $N = q \cdot m$, there exists an IL finite state encoder E with $s(q)$ states, that processes its input in $q \times q$ squares, and attains compression ratio $\rho_E(I_N, q)$ for the top left corner I_N of the infinite picture I that satisfies

$$\limsup_{m \to \infty} \rho_E(I_N, q) \leq \rho(I) + \delta_\epsilon(q), \quad \text{with} \quad \lim_{q \to \infty} \delta_\epsilon(q) = \epsilon.$$

The theorem and its proof are as in [6]; a use is made of Corollary 2.

7 Compression Algorithm A2

Algorithm A2 is discussed in [9] where its asymptotic optimality is proved in a way which is a special case of our proof here. In this section, we briefly describe it and show that our generalized approach of good sets applies to the proof of its optimality. The main difference between algorithm A1 and algorithm A2 is that in step i of A2, a subpicture T is looked for, such that the value of a many to one function f, evaluated on the components $C(T)$, is not yet in the dictionary. The dictionary, thus, includes the values of function f evaluated on the subpictures parsed by the algorithm, not the subpictures themselves. In addition, the range of f is restricted, and thus not too many pictures of a given area are parsed by the algorithm, and in rather early steps it parses subpictures of large area. Since the total output length produced by any algorithm that fits our framework is correlated with the number of steps it takes the algorithm to complete the processing of the picture, it follows that A2 has the potential to better compress finite pictures. As A1 was proved asymptotically optimal, for infinite picture, A2 can not exhibit any gain in compression relative to A1.

Given a picture I_N, algorithm A2 processes, one by one, growing continuous rectangular subpictures thereof, called the *basic rectangles* of I_N.

Step $i \geq 1$ of algorithm A2 starts with dictionary D including i rectangular subpictures, one of which is the empty subpicture. A2 then grows a small subrectangle R_i inside the current basic rectangle. The growing continues as long as R_i matches some member of D. R_i grows by keeping its base fixed inside the basic rectangle, and iteratively raising its top, by a few lines of cells at a time. When R_i no longer matches any member of D, it is encoded into the index in D of the member that matches the whole but the top of it, using $\lceil \log(i) \rceil$ bits, and a description of the top portion thereof, using $\lceil \log(\alpha) \rceil$ bits for each picture symbol in that portion. Finally, R_i is added to D, where its index is set to be i.

In order for the rectangles R_i not to have a too long perimeter, the length b_i of their base, measured in cells, and their and height h_i, also measured in cells, satisfy the relation

$$\sqrt{b_i} \leq h_i < b_i^2. \tag{8}$$

Thus, in step i, subrectangle R_i can not grow unrestrictedly; once it reaches the maximal allowed height (given the length of its base) the roles of its base and hight exchange, and R_i continues to grow (as long as it matched some member of D) in the perpendicular direction.

The initial value of R_i should not have a "too long" base, since, by (8), in such a case its hight would be "rather large" too, and at the event that this value of R_i is not a member of D, describing it using $\lceil \log(\alpha) \rceil$ bits per letter would have a bad affect on the attained compression.

Therefore, once the current basic rectangles become large enough, R_i does not start with its base coinciding with the basic rectangle's base. Rather, a subrectangle is cut out of the basic rectangle, and R_i starts growing therein. It might occur that following a small rectangle R_i, the next rectangle R_{i+1} grows much larger. In such a case, the remaining of the basic rectangle, having determined R_i, may not fit one continuous big rectangle, and some "cutting and pasting" is needed in order to prepare a room for a large rectangle to grow further on the area of the basic rectangle. The aforementioned function f is, in fact, that process of forming a rectangle from several continuous subpictures of I_N matched together.

The careful elaboration of all these details is described in [9]. A (long) proof is given there to the property that the residues (the top portions of the rectangles R_i) are of total area that vanishes with N. Also proved there is the property that the "cutting and pasting" of subrectangles of basic rectangles does not cause any rectangle R_i to be formed of more than seven continuous subrectangles of I_N.

Therefore, in order to show that A2 fits our framework, all that is left to show is that $p_{A2}(I_N) \leq c(I_N)$, where $p_{A2}(I_N)$ is the number of steps it takes A2 to complete the processing of I_N. To this end, it is enough to show that the set of subpictures R_i is a good set.

Lemma 10 The set of subpictures parsed by algorithm A2 is a good set.

proof: As in the case of A1, the distinctness of the subpictures is by definition. Let b_i and h_i be, respectively, the base length and the height of the i-th parsed subpicture R_i. ¿From (8) it follows that $b_i, h_i \leq (b_i \cdot h_i)^{2/3}$. Hence, the total length of the perimeters of the (at most 7) continuous components of R_i does not exceed $7 \times 4 \times |R_i|^{2/3}$, and part 2 of the good set condition is satisfied with $b_1 = 28$ and $b_2 = 2/3$. Since only rectangular shapes are allowed, there are no more than a distinct shapes of area a, as the length of the base determines the shape. A bit more elaboration is needed in order to carry this argument over to the continuous components of the subpictures R_i, but as their number is limited to 7, the result is not dramatically affected. Specifically, with some tedious calculation we find out that the number of distinct shapes of the subpictures of area a, parsed by A2, is upper bounded by $7^4 \cdot 28a^2$, and part 3 of the good set condition is satisfied as well. □

¿From Lemma 10 it follows that $p_{A2}(I_N) \leq c(I_N)$, and we conclude that A2 fits our framework, and as such is asymptotically optimal. Namely, for all finite values of s,

Lemma 11

$$\lim_{N\to\infty} \rho_{A2}(I_N) \leq \lim_{N\to\infty} \rho_s(I_N).$$

As in the case of A1, an infinitely large picture is fed to A2 "big" square subpicture by "big" square subpicture, so that each square subpicture is separately processed by A2, using a finite dictionary, and the size of the square subpictures is chosen to satisfy the desired approach of the obtained compression ratio to its lower bound $q(I)$. The concluding Coding Theorem (Theorem 2) of A1 applies to A2 too.

Acknowledgement

I thank A. Lempel and J. Ziv for many enlightening discussions in the course of the research led to this work.

References

[1] G.J. Chaitin, *Information Theoretic Computational Complexity*, IEEE Trans. Inform. Theory, vol IT-20, pp 10-15, 1974.

[2] S. Even, *On Information Lossless Automata of Finite Order*, IEEE Trans. Electronic Computers, vol EC-1, pp 561-569, 1965.

[3] D.A. Huffman, *Canonical Forms for Information Lossless Finite State Logical Machines*, IRE Trans. Circuit Theory, vol CT-6, pp. 41-59, 1959.

[4] A.N. Kolmogorov, *Three aproaches to the Quantitative Definition of Information*, Probl. Inform. Transmission, vol 1, pp 1-7, 1965.

[5] A. Lempel and J. Ziv, *On the Complexity of Finite State Sequences*, IEEE Trans. Inform. Theory, vol IT-22, pp. 75-81, 1976.

[6] A. Lempel and J. Ziv, *Compression of Two Dimensional Data*, IEEE Trans. Inform. Theory, vol IT-32, pp. 1-8, 1986.

[7] J. Rissanen, *A Universal Data Compression System*, IEEE Trans. Inform. Theory, vol IT-29, pp. 656-664, 1983.

[8] C.E. Shannon, *A Mathematical Theory of Cummunication*, Bell System Journal, vol 27, pp 379-423, 1948.

[9] D. Sheinwald, A. Lempel, and J. Ziv, *Two Dimensional Encoding by Finite State Encoders*, IEEE Transactions on Communications, vol 38, pp. 341-347, 1990.

[10] J. Ziv and A. Lempel, *A Universal Algorithm for Sequential Data Compression*, IEEE Trans. Inform. Theory, vol IT-23, pp. 337-343, 1977.

[11] J. Ziv and A. Lempel, *Compression of Individual Sequences via Variable Rate Coding*, IEEE Trans. Inform. Theory, vol IT-24, pp. 530-536, 1978.

Data Compression Bibliography

James A. Storer
Computer Science Department
Brandeis University
Waltham, MA 02254

This bibliography consists of some of the data compression references that the editor has encountered in his research. It represents only a very small fraction of the existing data compression research, but should provide researchers a "jumping off" point into the literature on image and text compression. Many important references may be missing (either because I was unaware of them or because of oversights when compiling the bibliography). Authors are encouraged to contribute additional entries (or corrections exisiting ones) for future revisions of the bibliography.

E. Abaya and G. L. Wise [1982]. "On the existence of optimal quantizers", *IEEE Transactions on Information Theory* IT-28:6, 937-940.

D. J. Abel and J. L. Smith [1983]. "A Data Structure and Algorithm Based on a Linear Key for a Rectangle Retrieval Problem", *Computer Graphics Image Processing* 24:1, 1-13.

J. P. Abenstein and W. J. Tompkins [1982]. "A New Data-Reduction Algorithm for Real Time ECG Analysis", *IEEE Trans. on Biomedical Engineering, BME-29:1*, 43-48.

D. M. Abrahamson [1989]. "An adaptive dependency source model for data compression", *Commun. ACM 32:1*, 77-83.

N. Abramson [1963]. *Information Theory and Coding*, McGraw-Hill, New York, NY.

H. Abut, R. M. Gray, and G. Rebolledo [1982]. "Vector quantization of speech and speech like waveforms", *IEEE Trans. on Asoustics, Speech, and Signal Processing* ASSP-30, 423-435.

H. Abut, editor [1990]. *Vector Quantization*, IEEE Reprint Collection IEEE Press, Piscataway, NJ.

J. Aczel and Z. Daroczy [1975]. *On Measures of Information and Their Characterizations*, Academic Press.

E. H. Adelson, E. Simoncelli, and R. Hingorani [1987]. "Orthogonal pyramid transforms for image coding", *Proceedings of SPIE.*

S. C. Ahalt, A. K. Krishnamurthy, P. Chen, and D. E. Nelon [1990]. "Competitive Learning Aogorithms for Vector Quantization", *Neural Networks 3*, 277-290.

N. Ahmed and K. R. Rao [1975]. *Orthogonal Transforms for Digital Signal Processing*, Springer-Verlag, NY.

N. Ahmed, P. J. Milne and S. G. Harris [1975]. "Electrocardiographic Data Compression via Orthogonal Transfomrs", *IEEE Transactions on Biomedical Engineering BME-22:6*, 484-487.

N. Ahmed, T. Natarajan, and K. R. Rao [1974]. "Dicrete Cosine Transform", *IEEE Trans. on Computers C-23*, 90-93.

M. Ai-Suwaiyel and E. Horowitz [1984]. "Algorithms for Trie Compaction", *ACM Transactions on Database Systems* 9:2, 243-263.

K. Aizawa, H. Harashima, and H. Miyakawa [1986]. "Adaptive discrete cosine transform coding with vector quantization for color images", *Proc. of ICASSP*, 985-988.

K. Aizawa, H. Harashimia, and H. Miyakawa [1987]. "Adaptive vector quantization of picture signals in discrete cosine transform domain", *Electronics and Communications in Japan, Part 1*, 70.

K. Aizawa, H. Harashima, and T. Saito [1989]. "Model-based analysis synthesis image coding (MBASIC) system for a person's face", *Image Communications 1:2*, 117-152.

V. R. Algazi [1966]. "Useful approximations to optimum quantization", *IEEE Trans. Commun. Technol.* COM-14, 297-301.

P. Algoet and T. Cover [1988]. "A Sandwitch Proof of the Shannon-McMillan-Breiman Theorem", *Annals of Probability 16*, 899-909.

M. Ali, C. Papadopoulos, and T. Clarkson [1992]. "The Use of Fractal Theory in a Video Compression System", *Proceedings IEEE Data Compression Conference*, Snowbird, Utah, 259-268.

E. W. Allender [1987]. "Some Consequences of the Existence of Pseudorandom Generators", *Proceedings Nineteenth Annual ACM Symposium on the Theory of Computing*, New York City, NY, 151-159.

P. A. Alsberg [1975]. "Space and Time Savings Through Large Data Base Compression and Dynamic Restructuring", *Proceedings of the IEEE* 63:8, 1114-1122.

A. Amir and G. Benson [1992]. "Efficient Two-dimensional Compressed Matching", *Proceedings IEEE Data Compression Conference*, Snowbird, Utah, 279-288.

M. R. Anderberg [1973]. *Cluster Analysis for Applications*, Academic Press, New York, NY.

W. M. Anderson [1992]. "A Dynamic Dictionary Compression Method with Genetic Algorithm Optimization", *Proceedings IEEE Data Compression Conference*, Snowbird, Utah, 393.

J. B. Anderson and J. B. Bodie [1975]. "Tree Encoding of Speech", *IEEE Trans. on Info. Theory IT-21*, 379-387.

G. B. Anderson and T. S. Huang [1971]. "Piecewise Fourier Transformation for Picture Bandwidth Compression", *IEEE Trans. on Comm. Tech. COM-19:2*, 133-140.

R. Anderson, J. Bowers, W. C. Fang, D. Johnson, J. Lee, and R. Nixon [1991]. "A Very High Speed Noiseless Data Compression Chip for Space Imaging Applications", *Proceedings IEEE Data Compression Conference*, Snowbird, Utah, 462.

H. C. Andrews [1970]. *Computer Techniques in Image Processing*, Academic Press, NY.

C. A. Andrews, J.M. Davies, and E. Schwarz [1967]. "Adaptive Data Compression", *Proceedings of IEEE 55:3.*

P. Ang, P. Ruetz, and D. Auld [1991]. "Video Compression Makes Big Gains", *IEEE Spectrum*, 16-19.

R. Ansari et. al. [1991]. "Wavelet Construction using Lagrange Halfband Filters", *IEEE Transactions on Circuits and Systems 38:9*, 1116-1118.

E. Antonidakis and D. A. Perreault [1992]. "Time-Space Compression Method Using Simultaneous Program Execution", *Proceedings IEEE Data Compression Conference*, Snowbird, Utah, 439.

A. Apostolico [1979]. "Linear Pattern Matching and Problems of Data Compression", *Proceedings IEEE International Symposium on Information Theory.*

A. Apostolico [1985]. "The Myriad Virtues of Subword Trees", *Combinatorial Algorithms on Words*, Springer-Verlag (A. Apostolico and Z. Galil, editors), 85-95.

A. Apostolico and A. S. Fraenkel [1985]. "Robust Transmission of Unbounded Strings Using Fibonacci Representations", TR-CS545, Purdue University.

A. Apostolico and R. Giancarlo [1986]. "The Boyer-Moore-Galil String Searching Strategies Revisited", *SIAM Journal on Computing* 15:1.

A. Apostolico and E. Guerrieri [1983]. "Linear Time Universal Compression Techniques Based on Pattern Matching", *Proceedings Twenty-First Allerton Conference on Communication, Control, and Computing*, Monticello, Ill., 70-79.

A. Apostolico, C. Iliopoulos, G. M. Landau, B. Schieber, and U. Vishkin [1987]. "Parallel Construction of a Suffix Tree with Applications", *Technical Report*, Department of Computer Science, Purdue University, West Lafayette, IN.

R. Aravind and A. Gersho [1987]. "Image Compression based on Vector Quantization with Finite Memory", *Optical Engineering* 26:7, 570-580.

J. F. Arnold and M. C. Cavenor [1981]. "Improvements to the Constant Area Quantization Bandwidth Compression Scheme", *IEEE Trans. on Comm. COM-29:12*, 1818-1823.

D. S. Arnstein [1975]. "Quantization Error in Predictive Coders", *IEEE Transactions on Communications COM-23*, 423-429.

J. Aronson [1977]. "Data Compression-A Comparison of Methods", *National Bureau of Standards PB*, 269-296.

R. B. Arps [1974]. "Bibliography on Digital Graphic Image Compression and Quality", *IEEE Trans. on Info. Theory, IT-20:1*, 120-122.

R. B. Arps [1980]. "Bibliography on Binary Image Compression", *Proc. of the IEEE, 68:7*, 922-924.

R. B. Arps, T. K. Truong, D. J. Lu, R. C. Pasco, and T. D. Friedman [1988]. "A multi-purpose VLSI chip for adaptive data compression of bilevel images", *IBM J. Res. Develop. 32*, 775-795.

A. Artieri and O. Colavin [1990]. "A Real-Time Motion Estimation Circuit", *Advanced Imaging*, 43-44.

Ash [1965]. *Information Theory*, John Wiley and Sons, New York, NY.

B. S. Atal and S. L. Hanauer [1971]. "Speech Analysis and Synthesis by Linear Prediction of the Speech Wave", *J. Acoust. Soc. Amer.* 50:2, 637-655; also in *Bell System Tech. Journal*, 1973-1986.

M. Auslander, W. Harrison, V. Miller, and M. Wegman [1985]. "PCTERM: A Terminal Emulator Using Compression", *Proc. IEEE Globcon '85*, IEEE Press, New York, 860-862.

E. Ayanoglu and R. M. Gray [1986]. "The design of predictive trellis waveform coders using the generalized Lloyd algorithm", *IEEE Trans. Comm. , COM-34*, 1073-1080.

F. L. Bacon and D. J. Houde [1986]. "Data Compression Apparatus and Method", U.S. Patent No. 4,612,532.

M. J. Bage [1986]. "Interframe predictive coding of images using hybrid vector quantization", *IEEE Trans. on Commun. COM-34:4*, 411-415.

L. R. Bahl and H. Kobayashi [1974]. "Image Data Compression by Predictive Coding II: Encoding Algorithms", *IBM Journal of Research and Development 18:2*, 172-179.

R. L. Baker [1984]. *Vector Quantization of Digital Images*, Ph.D. dissertation, Stanford Univ., Stanford, CA.

R. L. Baker and H. H. Shen [1987]. "A finite-state vector quantizer for image sequence coding", *International Conference on Acoustics, Speech, and Signal Processing, vol. 2*, Dallas, TX, 760-3.

R. L. Baker and R. M. Gray [1982]. "Image compression using non-adaptive spatial vector quantization", in *Proc. Conf. Rec. Sixteenth Asilomar Conf. Circuits, Syst., Comput.*, 55-61.

R. L. Baker and J. L. Salinas [1988]. "A motion compensated vector quantizer with filtered prediction", *International Conference on Acoustics, Speech, and Signal Processing, vol. 2*, New York, NY, 1324-1327.

J. L. Balcazar and R. V. Book [1987]. "Sets with Small Generalized Kolmogorov Complexity", *Acta Informatica.*

F. Banchilon, P. Richard, M. Scholl [1982]. "On Line Processing of Compacted Relations", *Proceedings of the Eighth International Conference on Very Large Data Bases*, September 1982, 263-269.

N. Baran [1990]. "Putting the squeeze on graphics", *Byte*, Dec., 289-294.

J. Barba, N. Scheinberg, D. L. Schilling, J. Garodnick, and S. Davidovici [1981]. "A Modified Adaptive Delta Modulator", *IEEE Trans. on Comm. COM-29:12*, 167-1785.

G. A. Barnard [1955]. "Statistical Calculation of Word Entropies for Four Western Languages", *IEEE Trans. Information Theory 1*, 49-53.

C. F. Barnes and R. L. Frost [1990]. "Necessary conditions for the optimality of residual vector quantizers", *Proceedings IEEE International Symposium on Information Theory.*

M. Barnsley [1988]. *Fractals Everywhere*, Academic Press.

M. F. Barnsley [1989]. *The Desktop Fractal Design Handbook*, Academic Press.

M. F. Barnsley and S. Demko [1985]. "Iterated Function Systems and the Global Construction of Fractals", Proceedings of the Royal Society of London A399, 243-275.

M. F. Barnsley and A. Jacquin [1988]. "Application of recurrent iterated function systems to images", *SPIE Vis. Comm. and Image Processing 1001*, 122-131.

M. F. Barnsley and A. D. Sloan [1988]. "A Better Way to Compress Images", *Byte Magazine* 215-223.

M. F. Barnsley and A. D. Sloan [1990]. "Methods and apparatus for image compression by iterated function systems", *U. S. patent no. 4941193.*

M. F. Barnsley, V. Ervin, D. Hardin, and J. Lancaster [1986]. "Solution of an Inverse Problem for Fractals and Other Sets", Proceedings of the National Academy of Science 83, 1975-1977.

L. K. Barrett and R. K. Boyd [1992]. "Minimization of Expected Distortion through Simulated Annealing of Codebook Vectors", *Proceedings IEEE Data Compression Conference*, Snowbird, Utah, 417.

I. J. Barton, S.E. Creasey, M.F. Lynch, and M.J. Snell [1974]. "An Information-Theoretic Approach to Text Searching in Direct Access Systems", *Comm. of the ACM 17:*, 345-350.

M. A. Bassiouni [1984]. "Data Compression in Scientific and Statistical Databases", *IEEE Transactions on Software Engineering 11:10*, 1047-1058.

M. A. Bassiouni and B. Ok [1986]. "Double Encoding- A Technique for Reducing Storage Requirements of Text", *Information Systems 11:2*, 177-184.

M. Bassiouni, A. Mukherjee, and N. Tzannes [1991]. "Experiments on Improving the Compression of Special Data Types", *Proceedings IEEE Data Compression Conference*, Snowbird, Utah, 433.

M. Bassiouni, N. Ranganathan, and A. Mukherjee [1988]. "Software and Hardware Enhancement of Arithmetic Coding", *Proceedings Fourth International Conference on Statistical and Scientific Database Management*, Roma, Italy.

D. Basu [1991]. "Text Compression Using Several Huffman Trees", *Proceedings IEEE Data Compression Conference*, Snowbird, Utah, 452.

R. H. T. Bates and M. J. McDonnell [1986]. *Image Restoration and Reconstruction*, Oxford University Press, New York, NY.

J. W. Bayless, S. J. Campanella, and A. J. Goldberg [1973]. "Voice Signals: Bit-by-Bit", *IEEE Spectrum 10:10*, 28-34.

J. M. Beaulieu and M. Goldberg [1989]. "Hierarchy in picture segmentation:A stepwise optimization approach", *IEEE Transactions on Pattern Analysis and Machine Intelligence 11:2*, 150-163.

C. D. Bei and R. M. Gray [1986]. "Simulation of vector trellis encoding systems", *IEEE Trans. Comm. , COM-34*, 214-18.

T. C. Bell [1985]. "Better OPM/L Text Compression", *IEEE Transactions on Communications* 34:12, 1176-1182.

T. C. Bell [1987]. "A Unifying Theory and Improvements for Existing Approaches to Text Compression", *Ph. D. Thesis. Department of Computer Science*, University of Calgary, Calgary, Alberta, Canada.

T. C. Bell and A. M. Moffat [1989]. "A Note on the DMC Data Compression Scheme", *Computer J. 32:1*, 16-20.

T. C. Bell, J. G. Cleary, and I. H. Witten [1990]. *Text Compression*, Prentice Hall, Englewood Cliffs, NJ.

T. C. Bell, I. H. Witten, and J. G. Cleary [1989]. "Modelling for Text Compression", *ACM Computing Surveys* 21:4, 557-591.

R. W. Bemer [1960]. "Do it by the Numbers- Digital Shorthand", *Communications of the ACM 3:N180*, 530-536.

W. R. Bennett [1948]. "Spectra of quantized signals", *Bell Syst. Tech. J.* 27, 446-472.

W. R. Bennett [1976]. *Scientific and Engineering Problem-Solving with the Computer* (Chapter 4), Prentice-Hall, Englewood Cliffs, NJ.

J. L. Bentley [1975]. "Multidimensional binary search trees used for associative searching", *Communications of the ACM* 18:9, 509-517.

J. L. Bentley and A. C. Yao [1976]. "An Almost Optimal Algorithm for Unbounded Searching", *Information Processing Letters* 5, 1976.

J. L. Bentley, D. D. Sleator, R. E. Tarjan, and V. K. Wei [1986]. "A Locally Adaptive Data Compression Scheme", *Communications of the ACM 29:4*, 320-330.

H. A. Bergen and J. M. Hogan [1992]. "Data Security in Arithmetic Coding Compression Algorithms", *Proceedings IEEE Data Compression Conference*, Snowbird, Utah, 390.

T. Berger [1971]. *Rate Distortion Theory: A Mathematical Basis for Data Compression*, Prentice-Hall, Englewood Cliffs, NJ.

T. Berger [1972]. "Optimum quantizers and permutation codes", *IEEE Transactions on Information Theory* IT-18, 759-765.

T. Berger [1982]. "Minimum entropy quantizers and permutation codes", *IEEE Trans. Inform. Theory* IT-28:2, 149-157.

T. Berger and R. Yeung [1986]. "Optimun '1'-ended binary prefix codes", *IEEE Trans. Inform. Theory IT-36:6*, 1435-1441.

J. Berstel and D. Perrin [1985]. *Theory of Codes*, Academic Press, New York, NY.

T. Bially [1969]. "Space-Filling Curves: Their Generation and Their Application to Bandwidth Reduction", *IEEE Transactions on Computers* 15:6, 658-664.

T. P. Bizon, W. A. Whyte, and V. R. Marcopoli [1992]. "Real-Time Demonstration Hardware for Enchanced DPCM Video Compression Algorithm", *Proceedings IEEE Data Compression Conference*, Snowbird, Utah, 429.

R. E. Blahut [1972]. "Computation of channel capacity and rate-distortion functions", *IEEE Transactions on Information Theory* IT-18, 460-473.

R. E. Blahut [1983]. *Theory and Practice of Error Control Codes*, Addison-Wesley, Reading, MA.

R. E. Blahut [1983b]. *Fast Algorithms for Digital Signal Processing*, Addison-Wesley, Reading, MA.

R. E. Blahut [1987]. *Principles and Practice of Information Theory*, Addison-Wesley, Reading, MA.

H. Blasbalg and R. Van Blerkom [1972]. "Message Compression", *IRE Transactions on Space Electronics and Telemetry*, 228-238.

M. Bleja and M. Domański [1990]. "Image data compression using subband coding", *Annales des Télécommunications*, 477-486.

R. Blitstein [1991]. "A Design to Increase Media-Independent Data Integrity and Availability Through the Use of Robotic Media Management Systems", *Proceedings IEEE Data Compression Conference*, Snowbird, Utah, 429.

J. J. Bloomer [1992]. "Compression for Medical Digital Fluoroscopy", *Proceedings IEEE Data Compression Conference*, Snowbird, Utah, 386.

L. I. Bluestein [1964]. "Asymptotically optimum quantizers and optimum analog to digital converters", *IEEE Transactions on Information Theory* IT-10, 242-246.

M. Blum [1967]. "On the Size of Machines", *Information and Control* 11, 257-265.

A. Blumer [1985]. "A Generalization of Run-Length Encoding", *Proceedings IEEE Symposium on Information Theory.*

A. Blumer [1987]. "Min-Max Universal Noisless Coding with Unifilar and Markov Sources", *IEEE Transactions on Information Theory.*

A. Blumer [1988]. "Arithmetic Coding with Non-Markov Sources", *Technical Report*, Tufts University, Medford, MA.

A. Blumer and J. Blumer [1987]. "On-Line Construction of a Complete Inverted File", *Technical Report*, Dept. of Mathematics and Computer Science, University of Denver, Denver, CO.

A. C. Blumer and R. J. McEliece [1988]. "The Renyi redundancy of generalized Huffman codes", *IEEE Trans. Inform. Theory IT-34*, 1242-1249.

A. Blumer, J. Blumer, A. Ehrenfeuchtr, D. Haussler, and R. McConnell [1984]. "Building the Minimal DFA for the Set of all Subwords of a Word in Linear Time", *Lecture Notes in Computer Science* 172, Springer-Verlag, New York, NY, 109-118.

A. Blumer, J. Blumer, A. Ehrenfeuchtr, D. Haussler, and R. McConnell [1984b]. "Building a Complete Inverted File for a Set of Text Files in Linear Time", *Proceedings Sixteenth Annual ACM Symposium on the Theory of Computing*, Washington, DC, 349-358.

A. Blumer, A. Ehrenfeucht, and D. Haussler [1989]. "Average size of suffix trees and DAWGS", *Discrete Applied Mathematics 24*, 37-45.

A. Blumer, A. Ehrenfeuchtr, D. Haussler, and M. Warmuth [1986]. "Classifying Learnable Geometric Concepts with the Vapnik-Chervonekis Dimension", *Proceedings Eighteenth Annual ACM Symposium on the Theory of Computing*, Berkeley, CA, 273-282.

H. Bodlaender, T. Gonzales, and T. Kloks [1991]. "Complexity Aspects of Map Compression", *Proceedings IEEE Data Compression Conference*, Snowbird, Utah, 287-296.

A. Bookstein and G. Fouty [1976]. "A Mathematical Model for Estimating the Effectiveness of Bigram Coding", *Information Processing Management 12*. 11-116.

A. Bookstein and S. T. Klein [1991]. "Flexible Compression for Bitmap Sets", *Proceedings IEEE Data Compression Conference*, Snowbird, Utah, 402-410.

A. Bookstein and S. T. Klein [1992]. "Models of Bitmap Generation: A Systematic Approach to Bitmap Compression", *Journal of Information Processing and Management.*

A. Bookstein and J. A. Storer [1992]. "Data Compression", *Journal of Information Processing and Management.*

A. Bookstein, S. T. Klein, and T. Raita [1992]. "Model Based Concordance Compression", *Proceedings IEEE Data Compression Conference*, Snowbird, Utah, 82-91.

T. E. Bordley [1983]. "Linear Predictive Coding of Marine Seismic Data", *IEEE Trans. on Acoustics, Speech, and Signal Processing, ASSP-31:4*, 828-835.

L. Boroczky, F. Fazekas, and T. Szabados [1990], "Theoretical and experimental analysis of a pel-recursive Wiener-based motion estimation algorithm", 9-10, 471-476.

G. Bostelman [1974]. "A Simple High Quality DPCM-Codec for Video Telephony Using and Mbit Per Second", *NTZ 3*, 115-117.

C. P. Bourne and D. F. Ford [1961]. "A Study of Methods for Systematically Abbreviating English Words and Names", *J. Association for Computing Machinery 8*, 538-552.

R. S. Boyer [1977]. "A Fast String Searching Algorithm", *Communications of the ACM* 20:10, 762-772.

P. Bratley and Y. Choueka [1982]. "Processing Truncated Terms in Document Retrieval Systems", *Inf. Processing and Management 18*, 257-266.

R. P. Brent [1987]. "A Linear Algorithm for Data Comrpession", *The Australian Computer Journal 19:2*, 64-68.

D. J. Brown and P. Elias [1976]. "Complexity of Acceptors for Prefix Codes", *IEEE Transactions on Information Theory* 22:3, 357-359.

J. A. Bucklew [1981]. "Upper bounds to the asymptotic performance of block quantizers", *IEEE Transactions on Information Theory* IT-27:5, 577-581.

J. A. Bucklew [1984]. "Two results on the asymptotic performance of quantizers", *IEEE Transactions on Information Theory* IT-30:2, 341-348.

J. A. Bucklew and G. L. Wise [1982]. "Multidimensional Asymptotic Quantization Theory with rth Power Distortion Measures", *IEEE Transactions on Information Theory IT-28*, 239-247.

J. Buhmann and H. Kuhnel [1992]. "Complexity Optimized Vector Quantization: A Neural Network Approach", *Proceedings IEEE Data Compression Conference*, Snowbird, Utah, 12-21.

S. Bunton and G. Borriello [1992]. "Practical Dictionary Management for Hardware Data Compression", *CACM 35:1*, 95-104.

B. L. Burrows and K. W. Sulston [1991]. "Measurement of Disorder in Non-Periodic Sequences", *J. Phys. A: Math Gen. 24*, 3979-3987.

P. J. Burt and E. H. Adelson [1983]. "The laplacian pyramid as a compact image code", *IEEE Trans. Communications COM-31:4*, 532-540.

P. J. Burville and J. F. C. Kingman [1973]. "On a Model for Storage and Search", *Journal of Applied Probability* 10, 697-701.

A. R. Butz [1969]. "Convergence with Hilbert's Space Filling Curve", *Journal of Computer and System Sciences* 3, 128-146.

A. R. Butz [1971]. "Alternate Algorithm for Hilbert's Space-Filling Curve", *IEEE Transactions on Computers* (April), 424-426.

A. Buzo, A. H. Gray, R. M. Gray, and J. D. Markel [1980]. "Speech Coding Based upon Vector Quantization", *IEEE Trans. on Acoustics, Speech, and Signal Processing ASSP-28:5*, 562-574.

A. Buzo, A. H. Gray, R. M. Gray, and J. D. Markel [1982]. "Speech coding based upon vector quantization", *IEEE Trans. Inform. Theory IT-28*, 205-210.

C. Cafforio and F. Rocca [1983]. "The Differential Model for Motion Estimation", *Image Sequence Processing and Dynamic Scene Analysis* (T. S. Huang, editor), Springer-Verlag, New York, NY.

P. Camana [1979]. "Video Bandwidth Compression: A Study in Tradeoffs", *IEEE Spectrum*, 24-29.

R. D. Cameron [1988]. "Source encoding using syntatic information source models", *IEEE Transactions on Information Theory 34:4*, 843-850.

S. J. Campanella and G. S. Robinson [1971]. "A Comparison of Orthogonal Transformations for Digital Speech Processing", *IEEE Trans. on Comm. Tech. COM-19:6*, 1045-1050.

G. Campbell, T. A. DeFanti, J. Frederiksen, S. A. Joyce, A. L. Lawrence, J. A. Lindberg, and D. J. Sandin [1986]. "Two bit/pixel full color encoding", *Computer Graphics* 20:4, 215-223.

R. M. Capocelli and A. De Santis [1988]. "Tight upper bounds on the redundancy of Huffman codes", *Proceedings IEEE International Symposium on Information Theory*, Kobe, Japan; also in *IEEE Trans. Inform. Theory IT-35:5* (1989).

R. M. Capocelli and A. De Santis [1990]. "'1'-ended binary prefix codes", *Proceedings IEEE International Symposium on Information Theory*, San Diego, CA.

R. M. Capocelli and A. De Santis [1991]. "A note on D-ary Huffman codes", *IEEE Trans. Inform. Theory IT-37:1.*

R. M. Capocelli and A. De Santis [1991]. "New Bounds on the Redundancy of Huffman Code", *IEEE Trans. Inform. Theory IT-37*, 1095-1104.

R. M. Capocelli and A. De Santis [1992]. "Variations on a Theme by Gallager", *Image and Text Compression*, Kluwer academic Press, Norwell, MA.

R. M. Capocelli and A. De Santis [1992b]. "On the construction of statistically synchronizable codes", *IEEE Trans. Inform. Theory IT-38:2.*

R. M. Capocelli, R. Giancarlo, and I. J. Taneja [1986]. "Bounds on the redundancy of Huffman codes", *IEEE Trans. Inform. Theory IT-32:6*, 854-857.

V. Cappellini [1985]. *Data Compression and Error Control Techniques with Applications*, Academic Press, New York.

P. Cappello, G. Davidson, A. Gersho, C. Koc, and V. Somayazulu [1986]. "A systolic vector quantization processor for real-time speech coding", *Proc. ICASSP*, 2143-2146.

B. Carpentieri and J. A. Storer [1991]. "The Complexity of Aligning Vectors", Technical Report, Brnadeis University, Waltham, MA 02254.

B. Carpentieri and J. A. Storer [1992]. "A Split-Merge Parallel Block Matching Algorithm for Video Displacement Estimation", *Proceedings IEEE Data Compression Conference*, Snowbird, Utah, 239-248.

J. B. Carroll [1966]. "Word-Frequency Studies and the Lognormal Distribution", *Proc. Conference on Language and Language Behavior* (E. M. Zale, editor), Appleton Century-Crofts, New York, 213-235.

J. B. Carroll [1967]. "On Sampling from a Lognormal Model of Word-Frequency Distribution", *Computational Analysis of Present-Day American English* (H. Kucera and W. N. Francis, editors), Brown University Press, Providence, RI, 406-424.

A. E. Cetin and V. Weerackody [1988]. "Design vector quantizers using simulated annealing", *Proceedings of the International Conference on Acoustics, Speech, and Signal Processing*, 1550.

G. J. Chaitin [1966]. "On the Length of Programs for Computing Finite Binary Sequences", *Journal of the ACM* 13:4, 547-569.

G. J. Chaitin [1969]. "On the Length of Programs for Computing Finite Binary Sequences; Statistical Considerations", *Journal of the ACM* 16:1, 145-159.

G. J. Chaitin [1969b]. "On the Simplicity and Speed for Computing Infinite Sets of Natural Numbers", *Journal of the ACM* 16:3, 407-422.

G. J. Chaitin [1975]. "A Theory of Program Size Formally Identical to Information Theory", *Journal of the ACM* 22:3, 329-340.

G. J. Chaitin [1975b]. "Randomness and Mathematical Proof", *Scientific American*, May, 47-52.

G. J. Chaitin [1976]. "Information-Theoretic Characterizations of Recursive Infinite Strings", *Theoretical Computer Science* 2, 45-48.

G. J. Chaitin [1987]. *Algorithmic Information Theory*, Cambridge University Press, New York, NY.

W. Y. Chan and A. Gersho [1991]. "Constrained constrained storage quantization of multiple vector sources by codebook sharing", *IEEE Trans. Comm. , COM-38(12)* 11-13.

D. K. Chang [1991]. "Exact Data Commpression Using Hierarchical Dictionaries", *Proceedings IEEE Data Compression Conference*, Snowbird, Utah, 431.

H. K. Chang and S. H. Chen [1992]. "A New, Locally Adaptive Data Compression Scheme Using Multilist Structure", *Proceedings IEEE Data Compression Conference*, Snowbird, Utah, 421.

P. C. Chang and R. M. Gray [1986]. "Gradient Algorithms for Designing Predictive VQ's", *IEEE ASSP-34*, 957-971.

M. Chang and G. G. Langdon [1991]. "Effects of Coefficient Coding on JPEG Baseline Image Compression", *Proceedings IEEE Data Compression Conference*, Snowbird, Utah, 430.

S. Y. Chang and J. J. Metzner [1991]. "Universal Data Compression Algorithms by Using Full Tree Models", *Proceedings IEEE Data Compression Conference*, Snowbird, Utah, 458.

P. C. Chang, R. M. Gray, and J. May [1987]. "Fourier transform vector quantization for speech coding", IEEE Trans. Comm. , COM-35, 1059-1068.

W. K. Chau, S. K. Wong, X. D. Yang, and S. J. Wan [1991]. "On the Selection of Color Basis for Image Compression", *Proceedings IEEE Data Compression Conference*, Snowbird, Utah, 441.

D. T. S. Chen [1977]. "On two- or more dimensional optimum quantizers", *Proc. IEEE Int. Conf. Acoust., Speech, Signal Processing*, 640-643.

H. H. Chen and A. Gersho [1987]. "Gain-adaptive vector quantization with application to speech coding", *IEEE Trans. on Communications COM-35:9.*

C. C. Chen and J. H. Hsieh [1992]. "Singular Value Decomposition for Texture Compression", *Proceedings IEEE Data Compression Conference*, Snowbird, Utah, 434.

H. H. Chen and T. S. Huang [1988]. "A Survey of Construction and Manipulation of Octrees", *Computer, Vision, Graphics, and Image Processing 43*, 409-431.

W. Chen and W. K. Pratt [1984]. "Scene adaptive coder", *IEEE Trans. Comm. 32*, 225.

K. Chen and T. V. Ramabadran [1991]. "An Improved Hierarchical Interpolation (HINT) Method for the Reversible Compression of Grayscale Images", *Proceedings IEEE Data Compression Conference*, Snowbird, Utah, 436.

M. T. Chen and J. Seiferas [1985]. "Efficient and Elegant Subword-Tree Construction", *Combinatorial Algorithms on Words*, Springer-Verlag (A. Apostolico and Z. Galil, editors), 97-110.

O. T. C. Chen, B. J. Sheu, W. C. Fang [1992]. "Image Compression on a VLSI Neural-Based Vector Quantizer", *Journal of Information Processing and Management.*

O. Chen, Z. Zhang, and B. Sheu [1992]. "An Adaptive High-Speed Lossy Data Compression", *Proceedings IEEE Data Compression Conference*, Snowbird, Utah, 349-358.

Y. Cheng and P. Fortier [1992]. "A Parallel Algorithm for Vector Quantizer Design", *Proceedings IEEE Data Compression Conference*, Snowbird, Utah, 423.

D. Cheng and A. Gersho [1986]. "A fast codebook search algorithm for nearest neighbor pattern matching", *Proc. ICASSP*, 265-268.

J. M. Cheng and G. Langdon [1992]. "QM-AYA Adaptive Arithmetic Coder", *Proceedings IEEE Data Compression Conference*, Snowbird, Utah, 428.

S. C. Cheng and W. H. Tsai [1992]. "Image Compression by Moment-Preserving Edge Detection", *Proceedings IEEE Data Compression Conference*, Snowbird, Utah, 405.

D. Chevion, E. D. Karnin, and E. Walach [1991]. "High Efficiency, Multiplication Free Approximation of Arithmetic Coding", *Proceedings IEEE Data Compression Conference*, Snowbird, Utah, 43-52.

W. Chou [1983]. *Computer Communications*, Prentice-Hall, Englewood Cliffs, NJ.

W. K. Chou and D. Y. Yun [1991]. "A Uniform Model for Parallel Fast Fourier Transform (FFT) and Fast Discrete Cosine Transform (FDCT)", *Proceedings IEEE Data Compression Conference*, Snowbird, Utah, 457.

P. A. Chou, T. Lookabaugh, and R. M. Gray [1989]. "Entropy-constrained vector quantization", *IEEE Trans. Acoust., Speech, Signal Process.* 37:1, 31-42.

P. A. Chou, T. Lookabaugh, and R. M. Gray [1989b]. "Optimal pruning with applications to tree-structured source coding and modeling", *IEEE Transactions on Information Theory* IT-35:2, 299-315.

P. A. Chou, T. Lookabaugh, and R. M. Gray [1989]. "Entropy constrained vector quantization", *IEEE Trans. Acoust. Speech Signal Process.* , 31-42.

Y. Choueka, A. S. Fraenkel, and Y. Perl [1981]. "Polynomial construction of optimal prefix tables for text compression", *Proceedings Nineteenth Annual Allerton Conferences on Communication, Control, and Computing*, 762-768.

E. Cinlar [1975]. *Introduction to Stochastic Processes*, Prentice-Hall.

A. C. Clare, E. M. Cook, and M. F. Lynch [1972]. "The Identification of Variable-Length Equifrequent Character Strings in a Natural Language Data Base", *Computer Journal 15:3*, 259-262.

R. J. Clarke [1985]. *Transform Coding of Images*, Academic Press, London.

Clavier, Panter, and Grieg [1947]. "PCM distortion analysis", *Elec. Eng.* 66, 1110-1122.

J. G. Cleary and I. H. Witten [1984]. "Data Compression Using Adaptive Coding and Partial String Matching", *IEEE Transactions on Communications* 32:4, 396-402.

J. G. Cleary and I. H. Witten [1984b]. "A Comparison of Enumerative and Adaptive Codes", *IEEE Trans. Information Theory IT-30:2*, 306-315.

F. Cohen [1973]. "A Switched Quantizer for Markov Sources Applied to Speech Signals", *NTZ 26:11*, 520-522.

M. Cohn and C. Kozhukhin [1992]. "Symmetric-Context Coding Schemes", *Proceedings IEEE Data Compression Conference*, Snowbird, Utah, 438.

M. Cohn [1992]. "Ziv-Lempel Compressors with Deferred-Innovation", *Image and Text Compression*, Kluwer academic Press, Norwell, MA.

J. M. Combes. A. Grossmann, and P. Tchamitchian, editors [1989]. *Wavelets: Time-Frequency Methods and Phase Space*, Springer-Verlag, New York, NY.

D. Comer, R. Sethi [1977]. "The Complexity of Trie Index Construction", *Journal of the ACM* 24:3, 428-440.

J. B. Connell [1973]. "A Huffman-Shannon-Fano code", *Proceedings of the IEEE 61:7*, 1046-1047.

D. J. Connor, R. F. W. Pease, and W. G. Scholes [1971], "Television coding using two-dimensional spatial prediction", *Bell Syst. Tech. J.* 50, 1049-1061.

J. F. Contla [1985]. "Compact coding of syntatically correct source programs", *Software Practice and Experience 15:7*, 625-636.

J. H. Conway and N. J. A. Sloane [1982]. "Voronoi regions of lattices, second moments of polytopes, and quantization", *IEEE Transactions on Information Theory* IT-28:2, 211-226.

J. H. Conway and N. J. A. Sloane [1982b]. "Fast quantizing and decoding algorithms for lattice quantizers and codes", *IEEE Transactions on Information Theory* IT-28:2, 227-232.

J. H. Conway and N. J. A. Sloane [1983]. "A fast encoding method for lattice codes and quantizers", *IEEE Transactions on Information Theory* IT-29, 820-824.

J. H. Conway and N. J. A. Sloane [1985]. "A lower bound on the average error of vector quantizers", *IEEE Transactions on Information Theory* IT-31:1, 106-109.

D. Cooper and M. F. Lynch [1982]. "Text Compression Using Variable to Fixed-Length Encodings", *Journal of the American Society for Information Science 33:1*, 18-31.

N. Coppisetti, S. C. Kwatra, and A. K. Al-Asmari [1992]. "Low Complexity Subband Encoding for HDTV Images", *Proceedings IEEE Data Compression Conference*, Snowbird, Utah, 415.

R. M. Cormack [1971]. "A review of classification", *J. of the Royal Statistical Society, Series A* 134, 321-367.

G. V. Cormack [1985]. "Data Compression on a Database System", *Communications of the ACM* 28:12, 1336-1342.

G. V. Cormack and R. N. S. Horspool [1987]. "Data Compression Using Dynamic Markov Modelling", *Comput. J. 30:6*, 541-550.

G. V. Cormack and R. Nigel Horspool [1984]. "Algorithms for Adaptive Huffman Codes", *Information Processing Letters* 18, 159-165.

D. Cortesi [1982]. "An Effective Text-Compression Algorithm", *Byte 7:1*, 397-403.

P. C. Cosman, E. A. Riskin, and R. M. Gray [1991]. "Combining Vector Quantization and Histogram Equalization", *Proceedings IEEE Data Compression Conference*, Snowbird, Utah, 113-118; also in *Journal of Information Processing and Management.*

N. Cot [1977]. "Characterization and Design of Optimal Prefix Codes", *Ph.D. Thesis*, Computer Science Dept. Stanford University, Stanford, CA.

T. M. Cover [1973]. "Enumerative Source Encoding", *IEEE Transactions on Information Theory* 19:1, 73-77.

T. M. Cover and R. C. King [1978]. "A Convergent Gambling Estimate of the Entropy of English", *IEEE Trans. Information Theory, IT-24:4*, 413-421.

L. H. Croft and J. A. Robinson [1992]. "Subband Image Coding Using Watershed and Watercourse Lines", *Proceedings IEEE Data Compression Conference*, Snowbird, Utah, 437.

K. Culik and S. Dube [1991]. "Using Fractal Geometry for Image Compression", *Proceedings IEEE Data Compression Conference*, Snowbird, Utah, 459.

K. Culik and I. Fris [1985]. "Topological Transformations as a Tool in the Design of Systolic Networks", *Theoretical Computer Science* 37, 183-216.

V. Cuperman and A. Gersho [1985]. "Gain-Adaptive Vector Quantization with Applications to Speech Coding", *Proc. of IEEE Trans. Commun. COM-35*, 918-930.

C. C. Cutler [1952]. "Differential quantization of communication signals", U.S. Patent 2 605 361.

R. P. Daley [1973]. "An Example of Information and Computation Trade-Off", *Journal of the ACM* 20:4, 687-695.

R. P. Daley [1974]. "The Extent and Density of Sequences Within the Minimal-Program Complexity Hierarchies", *Journal of Computer and System Sciences* 9, 151-163.

R. P. Daley [1976]. "Noncomplex Sequences: Characterizations and Examples", *Journal of Symbolic Logic* 41:3, 626-638.

S. Daly [1992]. "Incorporation of Imaging System and Visual Parameters into JPEG Quantization Tables", *Proceedings IEEE Data Compression Conference*, Snowbird, Utah, 410.

D. L. Dance and U. W. Pooch [1976]. "An Adaptive On Line Data Compression System", *Computer Journal* 19:3, 216-224.

J. C. Darragh and R. L. Baker [1989]. "Fixed distortion subband coding of images for packet-switched networks", *IEEE J. on Selected Areas in Commun.* 5:5, 789-800.

R. T. Dattola [1969]. "A fast algorithm for automatic classification", *Journal of Library Automation* 2, 31-48.

I. Daubechies [1988]. "Orthonormal basis of compactly supported wavelets", *Comm. on Pure and Appl. Math. 41*, 909-996.

I. Daubechies [1990]. "The Wavelet Transform, Time-frequency Localization and Signal Analysis", *IEEE Transactions on Information Theory 36:5*, 961-1005.

J. G. Daugman [1988]. "Complete Discrete 2-D Gabor Trandforms by Neural Networks for Image Analysis and Compression", *IEEE Trans. Acoust. Speech, Signal Processing ASSP-36*, 1169-1179.

W. B. Davenport and W. L. Root [1958]. *An Introduction to the Theory of Random Signals and Noise*, McGraw-Hill, NY.

G. A. Davidson and A. Gersho [1988]. "Systolic Architectures for Vector Quantization", *IEEE Trans. on ASSP 36:10*, 1651-1664.

G. A. Davidson, P. Capello, and A. Gersho [1988]. "Systolic Architectures for Vector Quantization", *IEEE Trans. ASSP*, 1651-1664.

G. Davis [1989]. "Method and System for Storing and Retrieving Compressed Data", *US Patent no. 4,868,570.*

G. Davis [1992]. "Collision String Repopulation of Hash Tables", *Proceedings IEEE Data Compression Conference*, Snowbird, Utah, 384.

M. D. Davis and E. J. Weyuker [1983]. *Computability, Complexity, and Languages*, Academic Press, New York, NY.

L. D. Davisson [1966]. "Theory of Adaptive Data Compression", *Advances in Communications Systems* (A.V. Balakrishinan, editor), Academic Press, New York, NY, 173-192.

L. D. Davisson [1967]. "An Approximate Theory of Prediction for Data Compression", *IEEE Trans. IT-13:2*, 274-278.

L. D. Davisson [1968]. "Data Compression Using Straight Line Interpolation", *IEEE Trans. on Info. Theory, IT-14:3*, 390-394.

L. D. Davisson [1968b]. "The Theoretical Analysis of Data Compression Systems", *Proc. IEEE 56:2*, 176-186.

L. D. Davisson [1973]. "Universal Noiseless Coding", *IEEE Transactions on Information Theory* 19, 783-795.

L. D. Davisson [1983]. "Min-Max Noisless Universal Coding for Markov Sources", *IEEE Transactions on Information theory.*

L. D. Davisson and R. M. Gray [1975]. "Advances in Data Compression", in *Advances in Communication Systems* 4, Academic Press, New York, 199-228.

L. D. Davisson and R. M. Gray, editors [1976]. *Data Compression*, Dowden, Hutchinson, and Ross, Stroudsburg, PA.

L. D. Davisson, McLiece, Pursley, and Wallace [1981]. "Efficient Universal Noisless Source Codes", *IEEE Transactions on Information Theory.*

S. De Agostino and J. A. Storer [1992]. "Parallel Algorithms for Optimal Compression Using Prefix Property Dictionaries", *Proceedings IEEE Data Compression Conference*, Snowbird, Utah, 52-61.

P. A. D. De Main, K. Kloss, and B.A. Marron [1967]. "The SOLID System III. Alphanumeric Compression", *US Government Printing Office, NBS Technical Note 413*, Washington.

P. A. D. De Maine, T. Rotwitt, Jr. [1971]. "Storage Optimization of Tree Structured Files Representing Descriptor Sets", *Proceedings ACM SIGFIDET Workshop on Data Description, Access and Control*, November 1971, 207-217.

A. De Santis and G. Persiano [1991]. "An Optimal Algorithm for the Construction of Optimal Prefix Codes", *Proceedings IEEE Data Compression Conference*, Snowbird, Utah, 297-306.

P. Delogne [1990]. "Picture coding for video telephony, television and high-definition television", *Annales des Télécommunications*, 519-527.

D. J. Delp and O. R. Mitchell [1979]. "Image compression using block truncation coding", *IEEE Transactions on Communications* 27:9, 1335-1342.

A. Desoky and Y. You [1991]. "Performance Analysis of a Vector Quantization Algorithm of Image Data", *Proceedings IEEE Data Compression Conference*, Snowbird, Utah, 443.

R. A. Devore, B. Jawerth, and B. J. Lucier [1991]. "Data Compression Using Wavelets: Error, Smoothness, and Quantization", *Proceedings IEEE Data Compression Conference*, Snowbird, Utah, 186-195.

R. Dianysian and R. L. Baker [1987]. "A VLSI chip set for real time vector quantization of image sequences", *Proc. ISCS.*

R. Dionysian [1992]. "Variable Precision Representation for Efficient VQ Codebook Storage", *Proceedings IEEE Data Compression Conference*, Snowbird, Utah, 319-328.

Y. Dishon [1977]. "Data Compression in Computer Systems", *Computer Design*, 85-90.

V. L. Doucette, K. M. Harrison, and E. J. Schuagref [1977]. "A Comparative Evaluation of Fragment Dictionaries for the Compression of French, English, and German Bibliographic Data Bases", *Proceedings Third International Conference in the Humanities* (S. Lusignan and J. S. North, editors), University of Waterloo Press, Waterloo, Ontario, Canada, 297-305.

Y. P. Drobyshev, V. V. Pukhov [1979]. "Analysis of the Influence of a System on Objects as a Problem of Transformation of Data Tables", in *Modeling and Optimization of Complex Systems* (Proceedings IFIP-TC 7 Working Conference, Novosibirisk, 1978), Lecture Notes in Control and Information Science, 18, Springer-Verlag, New York, NY, 187-197.

W. J. Duh and J. L. Wu [1992]. "Parallel Image and Video Coding Schemes in Multi-Computers", *Proceedings IEEE Data Compression Conference*, Snowbird, Utah, 424.

J. G. Dunham [1978]. "A Note on the Abstract Alphabet Block Source Coding with a Fidelity Criterion Theorem", *IEEE Transactions on Information Theory IT-24*, 760.

M. O. Dunham and R. M. Gray [1985]. "An algorithm for design of labeled-transition finite-state vector quantizers", *IEEE Trans. Commun.* COM-33:1, 83-89.

S. J. Eggers and A. Shoshani [1980]. "Efficient Access of Compressed Data", *Proceedings Very Large Data Bases International Conference vol. 6*, 205-211.

L. Ehrman [1967]. "Analysis of Some Redundancy Removal Bandwidth Compression Techniques", *Proc. IEEE 55:3*, 278-287.

E. B. Eichelberger, W. C. Rodgers, E. W. Stacy [1968]. "Method for Estimation and Optimization of Printer Speed Based on Character Usage Statistics", *IBM Journal of Research and Development* 12:2, March 1968, 130-139.

G. Einarsson [1991]. "An Improved Implementation of Predictive Coding Compression", *IEEE Trans. Commun. COM-39:169*, 169-171.

P. Elias [1955]. "Predictive Coding", *IRE Trans. Info. Theory IT-1:1*, 16-33.

P. Elias [1970]. "Bounds on performance of optimum quantizers", *IEEE Transactions on Information Theory* IT-16, 172-184.

P. Elias [1970b]. "Bounds and asymptotes for the performance of multivariate quantizers", *Ann. Math. Statist.* 41, 1249-1259.

P. Elias [1975]. "Universal Codeword Sets and Representations of the Integers", *IEEE Transactions on Information Theory* 21:2, 194-203.

P. Elias [1987]. "Interval and Recency Rank Source Coding: Two On-Line Adaptive Variable Length Schemes", *IEEE Transactions on Information Theory* 33:1, 3-10.

J. A. Elliott, P. M. Grant, and G. G. Sexton [1991]. "Concurrent Techniques for Developing Motion Video Compression Algorithms", *Proceedings IEEE Data Compression Conference*, Snowbird, Utah, 426.

P. Elmer [1989]. "The Design of High Bit Rate HDTV Codec", *Proceedings Third International Workshop on HDTV*, Torino, Italy.

S. E. Elnahas and J. G. Dunham [1987]. "Entropy coding for low-bit-rate visual communication", *IEEE J. Selected Areas in Commun. SAC-5*, 1175-1182.

H. Enomoto and K. Shibata [1971]. "Orthogonal Transform Coding System for Television Signals", *IEEE Trans. Electromag. Compat. EMC-13*, 11-17.

Y. Ephraim and R. M. Gray [1988]. "A unified approach for encoding clean and noisy sources by means of waveform and autoregressive model vector quantization", *IEEE Transactions on Information Theory* IT-34, 826-834.

B. Epstein, R. Hingorani, J. Shapiro, and M. Czigler [1992]. "Multispectral KLT-Wavelet Data Compression for Landsatematic Mapper Images", *Proceedings IEEE Data Compression Conference*, Snowbird, Utah, 200-208.

W. Equitz [1987]. "Fast algorithms for vector quantization picture coding", *Proc. Int. Conf. Acoust., Speech, and Signal Processing*, 725-728.

W. Equitz [1989]. "A new vector quantization clustering algorithm", *IEEE Trans. on Asoustics, Speech, and Signal Processing* 37:10, 1568-1575.

S. Even and M. Rodeh [1978]. "Economical Encoding of Commas Between Strings", *Communications of the ACM* 21:4, 315-317.

S. Even, D. Lichtenstein, and Y. Perl [1979]. "Remarks on Ziegler's Method for Matrix Compression", draft, Technion University, Haifa, Israel.

N. Faller [1973]. "An Adaptive System for Data Compression", *Conference Record of the Seventh IEEE Asilomar Conference on Circuits and Systems*, 593-597.

W. C. Fang, C. Y. Chang, and B. J. Sheu [1990]. "Systolic Tree Searched Vector Quantizer for Real Time Image Compression", *Proceedings VLSI Signal Processing IV* (H. S. Moscovitz, K. Yao and R. Jain, editors), IEEE Press.

W. Fang, B. Sheu, and O. T. Chen [1991]. "A Neural Network Based VLSI Vector Quantizer for Real-Time Image Compression", *Proceedings IEEE Data Compression Conference*, Snowbird, Utah, 342-351.

R. M. Fano [1949]. *Ph.D. Thesis*, Massachusetts Institute of Technology, Cambridge, MA.

R. M. Fano [1961]. *Transmission of Information*, Wiley, New York, NY.

P. M. Farrelle [1990]. *Recursive Block Coding for Image Data Compression*, Springer-Verlag, New York, NY.

N. Farvardin and J. W. Modestino [1984]. "Optimum quantizer performance for a class of non-Gaussian memoryless sources", *IEEE Transactions on Information Theory* IT-30:3, 485-497.

N. Farvardin and J. W. Modestino [1984]. "Optimum quantizer performance for a class of non-gaussian memoryless sources", *IEEE Trans. Inform. Theory*, 485-497.

T. Feder and D. H. Greene [1988]. "Optimal algorithms for approximate clustering", *Proc. of the 20th Annual ACM Symp. Theory Computing*, 434-444.

C. Fenimore and B. F. Field [1992]. "An Analysis of Frame Interpolation in Video Compression and Standards Conversion", *Proceedings IEEE Data Compression Conference*, Snowbird, Utah, 381.

E. R. Fiala and D. H. Green [1988]. "Data compression with finite windows", *Communications of the ACM 32:4*, 490-505.

W. A. Finamore and W. A. Pearlman [1980]. "Optimal encoding of discrete-time continuous-amplitude memoryless sources with finite output alphabets", *IEEE Transactions on Information Theory* IT-26:2, 144-155.

B. J. Fino [1972]. "Relations Between Haar and Walsh/Hadamard Transforms", *Proc. IEEE 60:5*, 647-648. A. Habibi [1977]. "Survey of Adaptive Image Coding Techniques", *IEEE Trans. on Comm. COM-25:11*, 1275-1284.

B. J. Fino and V. R. Algazi [1974]. "Slant-Haar transform", *Proc. IEEE* 62, 653-654.

M. J. Fischer [1980]. "Optimal Tree Layout", *Proceedings Twelfth Annual ACM Symposium on the Theory of Computing*, 177-189.

T. R. Fischer [1984]. "Quantized control with data compression constraints", *Optimal Control Applications and Methods, 5*, 39-55.

T. R. Fischer [1986]. "A pyramid vector quantizer", *IEEE Transactions on Information Theory* IT-32, 568-583.

T. R. Fischer [1989]. "Geometric source coding and vector quantization", *IEEE Transactions on Information Theory* IT-35:1, 137-145.

T. R. Fischer and R. M. Dicharry [1984]. "Vector quantizer design for memoryless Gaussian, Gamma, and Laplacian sources", *IEEE Transactions on Communications* COM-32:9, 1065-1069.

T. R. Fischer and M. Wang [1991]. "Entropy-Constrained Trellis Coded Quantization", *Proceedings IEEE Data Compression Conference*, Snowbird, Utah, 103-112.

Y. Fisher, E. W. Jacobs, and R. D. Boss [1992]. "Fractal Image Compression Using Iterated Transforms", *Image and Text Compression*, Kluwer academic Press, Norwell, MA.

B. Fitingof and Z. Waksman [1988]. "Fused trees and some new approaches to source encoding", *IEEE Transactions on Information Theory 34:3*, 417-424.

J. L. Flanagan [1972]. *Speech Analysis. Synthesis and Perception*, Springer-Verlag, New York, NY.

J. K. Flanagan, D. R. Morrell, R. L. Frost, C. J. Read, and B. E. Nelson [1989]. "Vector Quantization Codebook Generation Using Simulated Annealing", *Proc. IEEE ICASSP 3*, 1759-1762.

J. L. Flanagan, M. R. Schroeder, B. Atal, R. E. Crochiere, N. S. Jayant, and J. M. Tribolet [1979]. "Speech Coding", *IEEE Trans. on Comm. , vol. COM-27:4*, 710-737.

P. Fleischer [1964]. "Sufficient conditions for achieving minimum distortion in a quantizer", *IEEE Int. Conv. Rec.* , 104-111.

G. D. Forney and W.Y. Tao [1976]. "Data Compression Increases Throughout", *Data Communications.*

J. Foster, R. M. Gray, and M. O. Dunham [1985]. "Finite-state vector quantization for waveform coding", *IEEE Transactions on Information Theory* IT-31, 348-359.

A. S. Fraenkel and S. T. Klein [1985]. "Novel Compression of Sparse Bit-Strings - Preliminary Report", *Combinatorial Algorithms on Words*, Springer-Verlag (A. Apostolico and Z. Galil, editors), 169-183.

A. S. Fraenkel and S. T. Klein [1986]. "Improved hierarchical bit-vector compression in document retrieval systems", *Proceedings Ninth ACM-SIGIR Conference*, Pisa, Italy, 88-97.

A. S. Fraenkel and M. Mor [1983]. "Combinatorial compression and partitioning of large dictionaries", *The Computer Journal* 26:4, 336-344.

A. S. Fraenkel, M. Mor, and Y. Perl [1983]. "Is Text Compression by Prefixes and Suffixes Practical?", *Acta Informatica* 20:4, 371-375.

P. A. Franasek and T. J. Wagner [1974]. "Some Distribution-Free Aspects of Paging Algorithm Performance", *JACM*, 31-39.

Frank Fallside and William A. Woods, editors [1985]. *Computer Speech Processing*, Prentice-Hall. Englewood Cliffs, NJ.

W. D. Frazer [1972]. "Compression parsing of computer file data", *Proceedings First USA-Japan Computer Conference*, October 1972, Session 19-1, 609-615.

G. H. Freeman [1990]. "Source and channel entropy coding", *Abstr. IEEE Int. Symp. Inform. Theory*, San Diego, CA, 59.

G. H. Freeman [1991]. "Asymptotic Convergence of Dual-Tree Entropy Codes", *Proceedings IEEE Data Compression Conference*, Snowbird, Utah, 208-217.

G. H. Freeman, I. F. Blake, and J. W. Mark [1988]. "Trellis source code design as an optimization problem", *IEEE Transactions on Information Theory IT-34*, 1226-1241.

G. H. Freeman, J. W. Mark, and I. F. Blake [1988]. "Tellis source codes designed by conjugate gradient optimization", *IEEE Transactions on Communications COM-36*, 1-12.

K. A. Frenkel [1989]. "HDTV and the computer industry", *Communications of the ACM* 32:11, 1301-1312.

J. H. Friedman, F. Baskett, and L. J. Shustek [1975]. "An algorithm for finding nearest neighbors", *IEEE Trans. on Comput. , C-24(10)*, 1000-1006.

R. L. Frost, C. F. Barnes, and F. Xu [1991]. "Design and Performance of Residual Quantizers", *Proceedings IEEE Data Compression Conference*, Snowbird, Utah, 129-138.

T. Fujio [1980]. "High Definition Wide Screen Television System for the Future", *IEEE Trans. on Broadcasting BC-26:4*, 113-124.

H. Fujiwara, K. Kinoshita [1978]. "On Testing Schemes for Test Data Compression", *Systems-Comput.-Controls* 9:3, 72-78.

H. Fujiwara, K. Kinoshita [1979]. "Testing Logic Circuits with Compressed Data", *J. Design Automation and Fault-Tolerant Computing*, 3:3-4, 211-225.

R. G. Gallager [1968]. *Information Theory and Reliable Communication*, Wiley, New York, NY.

R. G. Gallager [1978]. "Variations on a Theme by Huffman", *IEEE Transactions on Information Theory* 24:6, 668-674.

R. G. Gallager and D. C. Van Voorhis [1975]. "Optimal Source Codes for Geometrically Distributed Integer Alphabets", *IEEE Transactions on Information Theory*, 228-230.

J. Gallant [1982]. "String Compression Algorithms", *Ph.D. Thesis*, Dept. EECS, Princeton University.

J. Gallant, D. Maier, and J. A. Storer [1980]. "On Finding Minimal Length Superstrings", *Journal of Computer and System Sciences* 20, 50-58.

A. Gamal and A. Orlitsky [1984]. "Interactive Data Compression", *Proceedings Twenty-Fifth Annual IEEE Symposium on the Foundations of Computer Science*, Singer Island, FL, 100-108.

A. A. E. Gamal, L. A. Hemachandra, I. Shperling, and V. Wei [1987]. "Using simulated annealing to design good codes", *IEEE Trans. Inform. Theory IT-33*, 116-123.

A. Gammerman and A. Bellotti [1992]. "Experiments Using Minimal-Length Encoding to Solve Machine Learning Problems", *Proceedings IEEE Data Compression Conference*, Snowbird, Utah, 359-367.

N. Garcia, C. Munoz, and A. Sanz [1985]. "Image compression based on hierarchical coding", *SPIE Image Coding 594*, 150-157.

M. Gardner [1976]. "Mathematical Games: Monster Curves", *Scientific American*, 124-133.

M. R. Garey [1972]. "Optimal Binary Identification Procedures", *SIAM Journal on Applied Mathematics* 23:2, 173-186.

M. R. Garey [1974]. "Optimal Binary Search Trees with Restricted Maximum Depth", *SIAM Journal on Computing* 3, 101-110.

I. Gargantini [1982]. "An Effictive Way to Represent Quadtrees", *Communications of the ACM* 25:12, 905-910.

L. Gerencser [1991]. "Asymptotics of Predictive Stochastic Complexity", *Proceedings IEEE Data Compression Conference*, Snowbird, Utah, 228-238.

L. Gerencser [1992]. "On Rissanen's Predictive Stochastic Complexity for Stationary ARMA Processes", *Journal of Information Processing and Management.*

A. Gersho [1979]. "Asymptotically Optimal Block Quantization", *IEEE Transactions on Information Theory IT-25*, 373-380.

A. Gersho [1982]. "On the structure of vector quantizers", *IEEE Transactions on Information Theory* 28:2, 157-162.

A. Gersho [1990]. "Optimal nonlinear interpolative vector quantization", *IEEE Trans. Comm. , COM-38(9-10)*, 1285-1287.

A. Gersho and V. Cuperman [1983]. "Vector Quantization: A Pattern-Matching Technique for Speech Coding", IEEE Communications Magazine, 21:9, 15-21.

A. Gersho and R. M. Gray [1990]. *Vector Quantization and Signal Compression*, Kluwer Academic Publishers, Norwell, MA.

A. Gersho and N. Ramamurthi [1982]. "Image coding using vector quantization", *Proc. IEEE Int. Conf. Acoust., Speech, Signal Processing*, 428-431.

A. Gersho and M. Yano [1985]. "Adaptive vector quantization by progressive code-vector replacement", *Proc. IEEE Int. Conf. Acoust., Speech, Signal Processing*, 133-136.

I. Gertner and Y. Y. Zeevi [1991]. "Generalized Scanning and Multiresolution Image Compression", *Proceedings IEEE Data Compression Conference*, Snowbird, Utah, 434.

V. D. Gesu [1989]. "An Overview of Pyramid Machines for Image Processing", *Information Sciences 47:1*, 17-34.

M. Ghanbari [1989]. "Two-layer coding of video signals for VBR networks", *IEEE Journal on Selected Areas in Communications 7:5*, 771-781.

H. Gharavi and A. N. Netravali [1983]. "CCITT Compatible Coding of Multilevel Pictures", *BSTJ, 62:9*, 2765-2778.

H. Gharavi and A. Tabatabai [1987]. "Application of quadrature mirror filters to the coding of monochrome and color images", *Proceedings ACASSP*, 32. 8. 1-32. 8. 4.

J. D. Gibson [1980]. "Adaptive Prediction in Speech Differential Encoding Systems", *Proc. of IEEE, 68:4*, 488-525.

J. D. Gibson [1974]. "Adaptive prediction in speech differential encoding systems", *Proc. IEEE, 68*, 1789-1797.

J. D. Gibson and T. R. Fischer [1982]. "Alaphabet-constrained data compression", *IEEE Trans. Inform. Theory, IT-28*, 443-457.

J. D. Gibson and K. Sayood [1988]. "Lattice Quantization", *Advances in Electronics and Electron Physics*, 72.

E. N. Gilbert [1971]. "Codes Based on Inaccurate Source Probabilities", *IEEE Transactions on Information Theory* 17:3, 304-314.

H. Gish and J. N. Pierce [1968]. "Aysmptotically Efficient Quantizing", *IEEE Transactions on Information Theory IT-14*, 676-683.

D. D. Giusto, C. S. Regazzoni, S. B. Serpico, and G. Vernazza [1990]. "A new adaptive approach to picture coding", *Annales des Télécommunications*, 503-518.

L. A. Glasser and D. W. Dobberpuhl [1985]. *The Design and Analysis of VLSI Circuits*, Addison-Wesley, Reading, MA.

D. Glover and S. C. Kwatra [1992]. "Subband Coding for Image Data Archiving", *Proceedings IEEE Data Compression Conference*, Snowbird, Utah, 436.

T. J. Goblick, Jr. and J. L. Holsinger [1967]. "Analog source digitization: a comparison of theory and practice", *IEEE Transactions on Information Theory* IT-13, 323-326.

Y. Goldberg and M. Sipser [1985]. "Compression and ranking", *Proceedings Seventeeth ACM Symposium on Theory of Computing*, 440-448.

M. Goldberg and H. F. Sun [1986]. "Image sequence coding using vector quantization", *IEEE Transactions on Communications* COM-34, 703-710.

M. Goldberg, P. R. Boucher, and S. Shlien [1986]. "Image compression using adaptive vector quantization", *IEEE Trans. on Commun. COM-34:2.*

J. D. Golic and M. M. Obradovic [1987]. "A lower bound on the redundancy of D-ary Huffman codes", *IEEE Trans. Inform. Theory IT-33:6*, 910-911.

S. W. Golomb [1966]. "Run-Length Encodings", *IEEE Transactions on Information Theory* 12, 399-401.

S. W. Golomb [1980]. "Sources which Maximize the Choice of a Huffman Coding Tree", *Information and Control 45*, 263-272.

M. Gonzalez and J. A. Storer [1985]. "Parallel Algorithms for Data Compression", *Journal of the ACM* 32:2, 344-373.

R. C. Gonzalez and P. Wintz [1977]. *Digital Image Processing*, Addison-Wesley, Reading, MA.

I. J. Good [1969]. "Statistics of Language", *Encyclopedia of Information, Linguistics and Control* (A. R. Meetham and R. A. Hudson, editors), Pergamon, Oxford, England, 567-581.

D. Gordon [1987]. "Efficient Embeddings of Binary Trees in VLSI Arrays", *IEEE Transactions on Computers* 36:9, 1009-1018.

D. Gordon, I. Koren, and G. Silberman [1984]. "Embedding Tree Structures in VLSI Hexagonal Arrays", *IEEE Transactions on Computers* 33:1, 104-107.

D. Gotlieb, S. A. Hagerth, P. G. H. Lehot, H. S. Rabinowitz [1975]. "A Classification of Compression Methods and their Usefulness for a Large Data Processing Center", *National Computer Conference 44*, 453-458.

D. N. Graham [1967]. "Image Transmission by Two-Dimensional Contour Coding", *Proc. of the IEEE, 55:3*, 336-346.

R. M. Gray [1975]. "Sliding-block source coding", *IEEE Trans. Inform. Theory, IT-21(4)*, 357-368.

R. M. Gray [1977]. "Time-invariant trellis encoding of ergodic discrete-time sources with a fidelity criterion", *IEEE Trans. Inform. Th. IT-23*, 71-83.

R. M. Gray [1984]. "Vector quantization", *IEEE ASSP Magazine* 1, 4-29.

R. M. Gray [1988]. *Probability, Random Processes, and Ergodic Properties*, Springer-Verlag, New York, NY.

R. M. Gray [1990]. *Source Coding Theory*, Kluwer Academic Press.

R. M. Gray [1990b]. "Quantization noise spectra", *IEEE Trans. Inform. Theory, IT-36*, 1220-1244.

R. M. Gray and L. D. Davisson [1986]. *Random Processes: A Mathematical Approach for Engineers*, Prentice-Hall, Englewood Cliffs, NJ.

R. M. Gray and A. H. Gray, Jr. [1977]. "Asymptotically optimal quantizers", *IEEE Transactions on Information Theory* IT-23:1, 143-144.

R. M. Gray and E. D. Karnin [1982]. "Multiple local optima in vector quantizers", *IEEE Transactions on Information Theory* IT-28:2, 256-261.

R. M. Gray and Y. Linde [1982]. "Vector quantizers and predictive quantizers for Gauss-Markov sources", *IEEE Transactions on Communications* COM-30:2, 381-389.

R. M. Gray and D. S. Ornstein [1976]. "Sliding-block joint source/noisy-channel coding theorems", *IEEE Trans. Inform. Theory, IT-22*, 682-690.

R. M. Gray, A. Buzo, A. H. Gray, Jr., and Y. Matsuyanma [1980], "Distortion measures for speech processing", *IEEE Trans. on Asoustics, Speech, and Signal Processing* ASSP-28:4, 367-376.

R. M. Gray, W. Chou, and P. W. Wong [1989]. "Quantization noise in single-loop sigma-delta modulation with sinusoidal inputs", *IEEE Trans. Inform. Theory, IT-35*, 956-968.

R. M. Gray, P. C. Cosman, and E. A. Riskin [1992]. "Image Compression and Vector Quantization", *Image and Text Compression*, Kluwer academic Press, Norwell, MA.

R. M. Gray, M. O. Dunham and R. Gobbi [1987]. "Ergodicity of Markov Channels", *IEEE Transaction on Information Theory IT-33*, 656-664.

R. M. Gray, A. H. Gray, Jr., G. Rebolledo, and J. E. Shore [1981], "Rate-distortion speech coding with a minimum discrimination information distortion measure", *IEEE Transactions on Information Theory* IT-27:6, 708-721.

A. H. Gray, Jr. and J. D. Markel [1976]. "Distance measures for speech processing", *IEEE Trans. on Asoustics, Speech, and Signal Processing* ASSP-24:5, 380-391.

R. M. Gray, J. C. Kieffer, and Y. Linde [1980]. "Locally optimal block quantizer design", *Inform. Contr.* 45, 178-198.

R. M. Gray, D. L. Neuhoff, and J. K. Omura [1975]. "Process definitions of distortion-rate functions and source coding theorems", *IEEE Transactions on Information Theory* IT-21:5, 524-532.

R. M. Gray, and L. D. Davidson [1986]. *Random Processes: A Mathematical Approach for Engineers*, Prentice Hall, Englewood Cliffs, NJ.

K. G. Gray, and R. S. Simpson [1972]. "Upper Bound on compression Ratio for Run Length Coding", *Proc. IEEE, 60:1*, 148.

J. A. Greefkes [1970]. "A Digitally Controlled Data Code for Speech Transmission", *Proc. IEEE Int. Conf. on Comm.* , 7. 33-7. 48.

J. A. Greszczuk and M. Deshon [1992]. "A Study of Methods for Reducing Blocking Artifacts Caused by Block Transform Coding", *Proceedings IEEE Data Compression Conference*, Snowbird, Utah, 435.

M. Guazzo [1980]. "A General Minimum-Redundancy Source-Coding Algorithm", *IEEE Transactions on Information Theory* 26, 15-25.

R. K. Guha and A. I. Roy [1992]. "Cascading Modified DPCM with LZW for Lossless Image Compression", *Proceedings IEEE Data Compression Conference*, Snowbird, Utah, 383.

S. Guiasu [1977]. *Information Theory with Applications.*, McGraw-Hill.

L. Guibas [1985]. "Periodicities in Strings", *Combinatorial Algorithms on Words*, Springer-Verlag, (A. Apostolico and Z. Galil, editors), 257-270.

L. Guibas and A. M. Odlyzko [1978]. "Maximal Prefix-Synchronized Codes", *SIAM Journal on Applied Mathematics* 35, 401-418.

M. Gutman [1987]. "On uniform quantization with various distortion measures", *IEEE Transactions on Information Theory* IT-33:1, 169-171.

A. Habibi [1971]. "Comparison of nth-order DPCM Encoder with Linear Transformations and Block Quantization Techniques", *IEEE Trans. on Comm. Tech. COM-19:6*, 948-956.

A. Habibi [1974]. "Hybrid Coding of Pictorial Data", *IEEE Trans. on Comm. COM-32:5*, 614-624.

O. S. Haddadin, V. J. Mathews, and T. G. Stockham [1992]. "Subband Vector Quantization of Images Using Hexagonal Filter Banks", *Proceedings IEEE Data Compression Conference*, Snowbird, Utah, 2-11.

W. D. Hagamen, D. J. Linden, H. S. Long, and J. C. Weber [1972]. "Encoding Verbal Information as Unique Numbers", *IBM Systems Journal* 11.

B. Hahn [1974]. "A New Technique for Compression and Storage of Data", *Communications of the ACM* 17:8, 434-436.

R. F. Haines and S. L. Chuang [1992]. "The Effects of Video Compression on Acceptability of Images for Monotoring Life Sciences Experiments", *Proceedings IEEE Data Compression Conference*, Snowbird, Utah, 395.

F. Halsall [1985]. *Introdcution to Data Communications and Computer Networks*, Addison-Wesley, Reading, MA.

R. W. Hamming [1980]. *Coding and Information Theory*, Prentice-Hall, Englewood Cliffs, NJ.

H. M. Hand and J. W. Woods [1985]. "Predictive vector quantization of images", *IEEE Transactions on Communications* COM-33, 1208-1219.

H. M. Hang and B. G. Haskell [1988]. "Interpolative vector quantization of color images", *IEEE Transactions on Communications 36:4*, 465-470.

H. M. Hang and J. W. Woods [1985]. "Predictive vector quantization of images", *IEEE Trans. Comm. , COM-33*, 1208-1219.

M. A. Hankamer [1979]. "A modified huffman procedure with reduced memory requirements", *IEEE Transactions on Communications 27:6*, 930-932.

E. F. Harding [1967]. "The number of partitions of a set of N points in k dimensions induced by hyperplanes", *Proc. Edinburgh Math. Soc.* 15 (Series II), 285-289.

H. F. Harmuth [1972]. *Transmission of Information by Orthogonal Functions*, Springer Verlag.

K. Harney, M. Keith, G. Lavelle, L. D. Ryan, and D. J. Stark [1991]. "The i750 video processor: a total multimedia solution", *Communications of the ACM* 34:4, 64-68.

C. W. Harrison [1952]. "Experiments with linear prediction in television", *Bell System Tech. J. 31*, 764-783.

J. A. Hartigan [1975]. *Clustering Algorithms.* Wiley.

A. Hartman and M. Rodeh [1985]. "Optimal Parsing of Strings", *Combinatorial Algorithms on Words*, Springer-Verlag (A. Apostolico and Z. Galil, editors), 155-167.

J. Hartmanis [1973]. "Generalized Kolmogorov Complexity and the Structure of Feasible Computations", *Proceedings Twenty-Fourth IEEE Annual Symposium on the Foundations of Computer Science*, 439-445.

J. Hartmanis and T. P. Baker [1975]. "On Simple Godel Numberings and Translations", *SIAM Journal on Computing* 4:1, 1-11.

B. G. Haskell, P. L. Gordon, R. L. Schmidt, and J. V. Scattaglia [1977]. "Interframe Coding of 525-line, Monochrome Television at 1. 5 bits/s", *IEEE Trans. on Comm. COM-25:11*, 1339-1348.

B. G. Haskell, F. W. Mounts, and J. C. Candy [1972]. "Interframe Coding of Videotelephone Pictures", *Proc. of the IEEE 60:7*, 792-800.

J. P. Haton, editor [1982]. *Automatic Speech Analysis and Recognition*, D. Reidel Publishing, Dordrecht, Holland.

K. A. Hazboun and M. A. Bassiouni [1982]. "A Multi-Group Technique for Data Compression", *Proceeding ACM-SIGMOD International Conference on Management of Data vol. 1*, 284-292.

N. He, A. Buzo, and F. Kuhlmann [1986]. "A Frequency Domain Waveform Speech Comrpession System Based on Prodcut Vector Quantizers", *Proceedings of the ICASSP*, Tokyo, Japan.

H. S. Heaps [1972]. "Storage Analysis of a Compression coding for Document Data Bases", *Infor. 10:1.*

P. Heckbert [1982]. "Color image quantization for frame buffer display", *Computer Graphics (ACM/SIGGRAPH) 16:3*, 297-307.

C. E. Heil and D. F. Walnut [1989]. "Continuous and discrete wavelet transforms", *SIAM Review 31:4*, 628-666.

G. Held [1979]. "Eliminating those Blanks and Zeros in Data Transmission", *Data Communications 8:9*, 75-77.

G. Held [1983]. *Data Compression: Techniques and Applications, Hardware and Software Considerations*, John Wiley and Sons, New York, NY.

D. R. Helman and G. G. Langdon [1988]. "Data Compression", *IEEE Potentials 2*, 25-28.

D. R. Helman, G. G. Langdon, G. N. N. Martin, and S. J. P. Todd [1982]. "Statistics Collection for Compression Coding with Randomizing Feature", *IBM Technical Disclosure Bulletin 24:10*, 4917.

W. J. Hendricks [1972]. "The Stationary Distribution of an Interseting Markov Chain", *Journal of Applied Probability* 9, 231-233.

W. J. Hendricks [1973]. "An Extension of a Theorem Concerning an Interseting Markov Chain", *Journal of Applied Probability* 10, 886-890.

S. Henriques and N. Ranganathan [1990]. "A parallel architecture for data compression", *Proceedings IEEE Symposium on Parallel and Distributed Processing*, 262-266.

E. E. Hilbert [1977]. "Cluster Compression Algorithm: A Joint Clustering / Data Compression Concept", *Ph.D. Thesis*, University of Southern California.

F. S. Hill, S. Walker, and F. Gao [1983]. "Interactive query using progressive transmission", *Computer Graphics (ACM/SIGGRAPH) 17:3*, 323-330.

D. S. Hirschberg and D. A. Lelewer [1992]. "Context Modeling for Text Compression", *Image and Text Compression*, Kluwer academic Press, Norwell, MA.

L. Hodges, S. Demko, and B. Naylor [1985]. "Construction of fractal objects with iterated function systems", *ACM SIGGRAPH.*

K. Hoffman and R. Kunze [1971]. *Linear Algebra*, Prentice Hall, Englewood Cliffs, NJ.

M. J. J. Holt and C. S. Xydeas [1986]. "Recent Developments in Image Data Compression for Digital Fascimile", *ICL Technical Journal*, 123-146.

T. Hopper [1992]. "Compression of Grey-Scale Fingerprint Images", *Proceedings IEEE Data Compression Conference*, Snowbird, Utah, 309-318.

R. N. Horspool [1991]. "Improving LZW", *Proceedings IEEE Data Compression Conference*, Snowbird, Utah, 332-341.

R. N. Horspool and G. V. Cormack [1984]. "A General Purpose Data Compression Technique with Practical Applications", *Proceedings of the CIPS*, Session 84, Calgary, Canada, 138-141.

R. N. Horspool and G. V. Cormack [1986]. "Dynamic Markov Modelling - A Prediction Technique", *Proceedings of the* Nineteenth Hawaii International Conference on System Sciences, Honolulu, 700-707.

R. N. Horspool and G. V. Cormack [1987]. "A locally adaptive data compression scheme", *Commun. ACM 16:2*, 792-794.

R. N. Horspool and G. V. Cormack [1992]. "Constructing Word-Based Text Compression Algorithms", *Proceedings IEEE Data Compression Conference*, Snowbird, Utah, 62-71.

P. G. Howard and J. S. Vitter [1991]. "Analysis of Arithmetic Coding for Data Compression", *Proceedings IEEE Data Compression Conference*, Snowbird, Utah, 3-12; also in *Journal of Information Processing and Management* (1992).

P. Howard and J. S. Vitter [1991b]. "New Methods for Lossless Image Compression Using Arithmetic Coding", *Proceedings IEEE Data Compression Conference*, Snowbird, Utah, 257-266; also in *Journal of Information Processing and Management* (1992).

P. Howard and J. S. Vitter [1992a]. "Error Modeling for Hierarchical Lossless Image Compression", *Proceedings IEEE Data Compression Conference*, Snowbird, Utah, 269-278.

P. Howard and J. S. Vitter [1992b]. "Parallel Lossless Image Compression Using Huffman and Arithmetic Coding", *Proceedings IEEE Data Compression Conference*, Snowbird, Utah, 299-308.

P. G. Howard and J. S. Vitter [1992c]. "Practical Implementations of Arithmetic Coding", *Image and Text Compression*, Kluwer academic Press, Norwell, MA.

T. C. Hu [1982]. *Combinatorial Algorithms*, Addison-Wesley, Reading, MA.

T. C. Hu and K. C. Tan [1972]. "Path Length of Binary Search Trees", *SIAM Journal on Applied Mathematics* 22, 225-234.

T. C. Hu and C. Tucker [1971]. "Optimal Computer Search Trees and Variable-Length Alphabetical Codes", *SIAM Journal on Applied Mathematics* 21:4, 514-532.

T. S. Huang [1965]. "PCM picture transmission", *IEEE Spectrum* 12:12, 57-60.

B. H. Huang and A. H. Gray [1982]. "Multiple stage vector quantization for speech coding", *Proceedings IEEE International Conference on Acoustics, Speech, and Signal Processing*, 597-600.

C. M. Huang and R. W. Harris [1992]. "Large Vector Quantization Codebook Generation Problems and Solutions", *Proceedings IEEE Data Compression Conference*, Snowbird, Utah, 413.

J. J. Y. Huang and P. M. Schulthesis [1963]. "Block Quantization of Correlated Gaussian Random Variables", *IEEE Trans. on Comm. Systems CS-11:3*, 289-296.

S. S. Huang, J. S. Wang, and W. T. Chen [1992]. "FASVQ: The Filtering and Seeking Vector Quantization", *Proceedings IEEE Data Compression Conference*, Snowbird, Utah, 400.

A. K. Huber, S. E. Budge, and R. W. Harris [1992]. "Variable-Rate, Real Time Image Compression for Images Dominated by Point Sources", *Proceedings IEEE Data Compression Conference*, Snowbird, Utah, 440.

D. A. Huffman [1952]. "A Method for the Construction of Minimum-Redundancy Codes", *Proceedings of the IRE* 40, 1098-1101.

R. Hummel [1977]. "Image enhancement by histogram modification", *Computer Graphics and Image Processing 6:2*, 184-195.

R. Hunter and A. H. Robinson [1980]. "International Digital Facsimile Coding Standards", *Proc. Institute of Electrical and Electronic Engineers 68:7*, 854-867.

A. Huseyin, B. Tao, and J. L. Smith [1987]. "Vector Quantizer Architectures for Speech and Image Coding", *Proc. ICASSP*, 756-759.

S. E. Hutchins [1971]. "Data compression in context-free languages", *Proceedings IFIP Conference Ljubljana*, v.1: Foundations and Systems, North-Holland, 104-109.

D. T. Huynh [1986]. "Resource-Bounded Kolmogorov Complexity of Hard Languages", *Structure in Complexity Theory Conference, Lecture Notes in Computer Science* 223, 184-195.

IEEE Transactions on Information Theory 28:2, special two volume issue on quantization.

D. Indjic [1991]. "Reduction in Power System Load Data Training Sets Size Using Fractal Approximation Theory", *Proceedings IEEE Data Compression Conference*, Snowbird, Utah, 446.

F. M. Ingels [1971]. *Information Theory and Coding Theory*, Intext, Scranton, PA.

H. Inose and Y. Yasuda [1963]. "A unity bit coding method by negative feedback", *Proc. IEEE 51*, 1524-1535.

T. Ishiguro and K. Iinuma [1982]. "Television Bandwidth Compression Transmission by Motion-Compensated Interframe Coding", *IEEE Communications Magazine*, 24-30.

T. Ishiguro, K. Iinuma, Y. Iijima, T. Koga, S. Azami, and T. Mune [1976]. "Composite Interframe Coding of NTSC Color Television Signals", *Proceedings National Telemetering Conference*, 6:4. 1-6:4. 5.

P. Israelsen [1991]. "VLSI Implementation of a Vector Quantization Processor", *Proceedings IEEE Data Compression Conference*, Snowbird, Utah, 463.

E. L. Ivie [1991]. "Techniques for Index Compression", *Proceedings IEEE Data Compression Conference*, Snowbird, Utah, 451.

L. B. Jackson [1989]. *Digital Filters and Signal Processing*, Kluwer Academic Press, Boston, MA.

G. Jacobson [1992]. "Random Access in Huffman-coded Files", *Proceedings IEEE Data Compression Conference*, Snowbird, Utah, 368-377.

M. Jacobsson [1978]. "Huffman coding in Bit-Vector Compression", *Information Processing Letters* 7:6, 304-307.

A. Jacquin [1989]. "A fractal theory of iterated Markov operators with applications to digital image coding", *Doctoral Thesis*, Georgia Institute of Technology.

A. Jacquin [1990]. "A novel fractal block-coding technique for digital images", *IEEE ICASSP Proc. 4*, 2225.

A. K. Jain [1981]. "Advances in Mathematical Models for Image Processing", *Proceeding of the IEEE 69*, 502-528.

A. K. Jain [1981b]. "Image Data Compression: A Review", *Proceedings of the IEEE 69:3*, 349-389.

A. K. Jain [1989]. *Fundamentals of Digital Image Processing*, Prentice-Hall.

A. K. Jain and R. C. Dubes [1989]. *Algorithms for Clustering Data*, Prentice-Hall, New Jersey.

J. R. Jain and A. K. Jain [1981]. "Displacement Measurement and its Applications in Interframe Image Coding", *IEEE Transactions on Communications COM-29:12*, 1799-1808.

A. K. Jain, P. M. Farrelle, and V. R. Algazi [1984]. "Image Data Compression", *Digital Image Processing Techniques* (M. Ekstrom, editor), Prentice-Hall, Englewood Cliffs, NJ.

M. Jakobsson [1978]. "Huffman coding in bit-vector compression", *Inf. Processing Letters 7*, 304-307.

M. Jakobsson [1982]. "Evaluation of a Hierarchical Bit-Vector Compression Technique", *Information Processing Letters 14:4*, 147-149.

M. Jakobsson [1985]. "Compression of Character Strings by an Adaptive Dictionary", *BIT 25:4*, 593-603.

D. Jamison and K. Jamison [1968]. "A Note on the Entropy of Partially-Known Languages", *Information and Control 12*, 164-167.

N. Jardine and R. Sibson [1969]. "The construction of hierarchic and non-hierarchic classifications", *Computer*, 177-184.

N. S. Jayant [1970]. "Adaptive Delta Modulation with a One-bit Memory", *BSTJ 49:3*, 321-343.

N. S. Jayant [1977]. "Pitch-Adaptive Speech with Two-Bit Quantization and Fixed Spectrum Prediction", *BSTJ 56:3*, 439-454.

N. S. Jayant and P. Noll [1984]. *Digital Coding of Waveforms: Principles and Applications to Speech and Video*, Prentice-Hall, Englewood Cliffs, NJ.

N. S. Jayant, editor [1976]. *Waveform Quantization and Coding*, IEEE Press, Piscatatway, NJ.

F. Jelinek [1968]. *Probabilistic Information Theory*, McGraw-Hill, New York, NY.

F. Jelinek [1969]. "Tree encoding of memoryless time-discrete sources with a fidelity criterion", *IEEE Transactions on Information Theory* IT-15, 584-590.

F. Jelinek and J. B. Anderson [1971]. "Instrumentable Tree Encoding of Information Sources", *IEEE Trans. Info. Theory IT-17*, 118-119.

F. Jelinek and K. S. Schneider [1972]. "On variable-length-to-block coding", *IEEE Trans. Inform. Theory IT-18:6*, 765-774.

J. Jeuring [1992]. "Incremental Data Compression", *Proceedings IEEE Data Compression Conference*, Snowbird, Utah, 411.

G. C. Jewell [1976]. "Text Compaction for Information Retrieval Systems", *IEEE Systems, Man and Cybernetics Sociey Newsletter 5:1*, 47.

O. Johnsen [1980]. "On the Redundancy of Binary Huffman Codes", *IEEE Transactions on Information Theory* 26:2, 220-222.

C. B. Jones [1981]. "An Efficient Coding System for Long Source Sequences", *IEEE Transactions on Information Theory* 27:3, 280-291. D. W. Jones [1988]. "Application of Splay Trees to Data Compression", *Commun. ACM 31:8*, 996-1007.

D. W. Jones [1991]. "Practical Evaluation of a Data Compression Algorithm", *Proceedings IEEE Data Compression Conference*, Snowbird, Utah, 372-381.

S. C. Jones and R. J. Moorhead [1992]. "Image-Sequence Compression of Computational Fluid Dynamic Animations", *Proceedings IEEE Data Compression Conference*, Snowbird, Utah, 409.

B. Juang [1988]. "Design and performance of trellis vector quantizers for speech signals", *IEEE Transactions on Acoustics, Speech, and Signal Processing 36:9*, 1423-1431.

R. K. Jurgen [1991]. "Digital HDTV", *Spectrum* 28:1, 65.

R. K. Jurgen and W. F. Schreiber [1991]. "The challenges of digital HDTV", *Spectrum* 28:4, 28.

H. Jurgensen and M. Kunze [1984]. Redundance-Free Codes as Cryptocodes, *Technical Report*, Computer Science Dept., University of Western Ontario, London, Canada.

H. Jurgensen and D. E. Matthews [1983]. Some Results on the Information Theoretic Analysis of Cryptosystems, *Technical Report*, Computer Science Dept., University of Western Ontario, London, Canada.

D. Kahn [1967]. *The Code-Breakers*, MacMillan, New York, NY.

T. Kamae [1973]. "On Kolmogorov's Complexity and Information", *Osaka Journal of Mathematics* 10, 305-307.

H. Kaneko and T. Ishiguro [1980]. "Digital Television Transmission Using Bandwidth Compression Techniques", *IEEE Communications Magazine*, 14-22.

A. N. C. Kang, R. C. T. Lee, C. Chang, and S. Chang [1977]. "Storage Reduction through Minimal Spanning Trees and Spanning Forests", *IEEE Transactions on Computers* 26:5, 425-434.

E. Karnin and E. Walach [1986]. "A fractal based approach to image compession", *IEEE Transactions on ASSP*.

R. M. Karp [1961]. "Minimum-Redundancy Coding for the Discrete Noisless Channel", *IRE Transactions on Information Theory* 7, 27-39.

J. Karush [1961]. "A Simple Proof of an Inequality by MacMillian", *IRE Transactions on Information Theory* 7, 118.

J. Katajainen and E. Makinen [1990]. "Tree Compression and Optimization with Applications", *International Journal of Foundations of Computer Science 1:4*, 425-447.

J. Katajainen and T. Raita [1987]. "An Analysis of the Longest Match and the Greedy Heuristics for Text Encoding", Technical Report, Department of Computer Science, University of Turku, Turku, Finland.

J. Katajainen, M. Penttonen and J. Teuhola [1986]. "Syntax-Directed Compression of Program Files", *Software-Practice and Experience 16:3*, 269-276.

H. P. Katseff and M. Sipser [1977]. "Several Results in Program Size Complexity", *Proceedings Eighteenth Annual IEEE Symposium on Foundations of Computer Science*, Providence, RI, 82-89.

D. Kauffman, M. Johnson, and G. Knight [1976]. "The Empirical Derivation of Equations for Predicting Subjective Textual Information", *Instructional Science 5*, 253-276.

V. H. Kautz [1965]. "Fibonacci Codes for Synchronization Control", *IEEE Transactions on Information Theory* 11, 284-292.

T. Kawabata and H. Yamamoto [1991]. "A New Implementation of the Ziv-Lempel Incremental Mental Parsing Algorithm", *IEEE Trans. Inform Theory IT-37*, 1439-1440.

E. Kawaguchi and T. Endo [1980]. "On a Method of Binary-Picture Representation and its Application to Data Compression", *IEEE Trans. Machine Intel. Patt. Anal. PAMI-2:1*, 27-35.

K. Keeler and J. Westbrook [1992]. "Short Encodings of Planar Graphs and Maps", *Proceedings IEEE Data Compression Conference*, Snowbird, Utah, 432.

M. Khansari, I. Widjaja, and A. Leon-Garcia [1992]. "Convolutional Interpolative Coding Algorithms", *Proceedings IEEE Data Compression Conference*, Snowbird, Utah, 209-218.

J. C. Kieffer [1982]. "Exponential rate of convergence for Lloyd's method I", *IEEE Transactions on Information Theory* IT-28, 205-210.

J. C. Kieffer [1982b]. "Sliding-Block Coding for Weakly Continuous Channels", *IEEE Trans. Inform. Theory IT-28*, 2-10.

J. C. Kieffer, T. M. Jahns, and V. A. Obuljen [1988]. "New results on optimal entropy-constrained quantization", *IEEE Transactions on Information Theory* IT-34:5.

J. W. Kim and S. U. Lee [1989]. "Discrete cosine transform - classified VQ technique for image coding", *Proc. of ICASSP*, Glasgow, Scotland.

G. A. King [1992]. "Data Compression Using Pattern Matching, Substitution, and Morse Coding", *Proceedings IEEE Data Compression Conference*, Snowbird, Utah, 389.

R. A. King and N. M. Nasrabadi [1983. "Image coding using vector quantization in the transform domain", *Pattern Recognition Letters 1*, 323-329.

O. Kiselyov and P. Fisher [1992]. "Wavelet Compression with Feature Preservation and Derivative Definition", *Proceedings IEEE Data Compression Conference*, Snowbird, Utah, 442.

V. Klee [1980]. "On the complexity of D-dimensional Voronoi diagrams", *Archiv der Mathematik* 34, 75-80.

S. T. Klein, A. Bookstein and S. Deerwester [1989]. "Storing text retrieval systems on CD-ROM: compression and encryption considerations", *ACM Trans. Information Systems 7:3*, 230-245.

S. C. Knauer [1976]. "Real-Time Video Compression Algorithm for Hadamard Transformation Processing", *IEEE Trans. on Electromagnetic Compatibility EMC-18:1*, 28-36.

G. Knowles [1990]. "VLSI architecture for the discrete wavelet transform", *Electron. Lett. 26:15*, 1184-1185.

K. Knowlton [1980]. "Progressive Transmission of Grey-Scale and Binary Pictures by SImple, Efficent, and Lossless Encoding Schemes", *Proc. Institute of Electircal and Electronic Engineers 68:7*, 885-896.

D. E. Knuth [1982]. "Huffman's Algorithm via Algebra", *Journal Combinatorial Theory Series A* 32, 216-224.

D. E. Knuth [1985]. "Dynamic Huffman Coding", *Journal of Algorithms* 6, 163-180.

D. E. Knuth [1987]. "Digital halftones by dot diffusion", *ACM Trans. on Graphics* 6:4, 245-273.

D. E. Knuth, J. H. Morris, and V. R. Pratt [1977]. "Fast Pattern Matching in Strings", *SIAM Journal on Computing* 6:2, 323-349.

C. Ko and W. Chung [1991]. "2-D Discrete Cosine Transform Array Processor Using Non-Planar Connections", *Proceedings IEEE Data Compression Conference*, Snowbird, Utah, 456.

H. Kobayashi and L. R. Bahl [1974]. "Image Data Compression by Predictive Coding I: Prediction Algorithms", *IBM Journal of Research and Development 18:2*, 164-171.

E. Koch and M. Sommer [1991]. "Data Compression in View of Data Sealing", *Proceedings IEEE Data Compression Conference*, Snowbird, Utah, 427.

E. Koch and M. Sommer [1992]. "Design and Analysis of Self-Synchronising Codeword Sets Using Various Models", *Proceedings IEEE Data Compression Conference*, Snowbird, Utah, 391.

A. N. Kolmogorov [1965]. "Three approaches to the Quantitative Definition of Information", *Problems of Information Transmission* 1, 1-7.

A. N. Kolmogorov [1969]. "On the Logical Foundation of Information Theory", *Problems of Information Transmission* 5, 3-7.

L. G. Kraft [1949]. "A Device for Quantizing, Grouping, and Coding Amplitude Modulated Pulses", M.S. Thesis, Dept. of Electrical Engineering, Massachusetts Institute of Technology, Cambridge, MA.

H. P. Kramer and M. V. Matthews [1956]. "A Linear Coding for Transmitting Set of Correlated Signals", *IRE Trans. on Info. Theory*, 41-46.

R. M. Krause [1962]. "Channels which Transmit Letters of Unequal Duration", *Information and Control* 5:1, 13-24.

F. Kretz [1983]. "Edges in Visual Scenes and Sequences: Application to Filtering, Sampling, and Adaptive dpcm Coding", *Image Sequence Processing and Dynamic Scene Analysis* (T. S. Huang, editor), Springer-Verlag, New York, NY.

R. E. Krichevsky and V. K. Trofimov [1981]. "The performance of Universal Encoding", *IEEE Trans. Inform. Theory IT-27*, 199-207.

A. S. Krishnakumar, J. L. Karpowicz, N. Belic, D. H. Singer, and J. M. Jenkins [1981]. "Microprocessor-Based Data Compression Scheme for Enhanced Digital Transmission of Holter Recordings", *IEEE Conf. on Computers in Cardiology*, 435-437.

A. K. Krishnamurthy, S. C. Ahalt, D. Melton, and P. Chen [1990]. "Neural networks for vector quantization of speech and images", *IEEE J. on Selec. Areas in Commun. 8:8*, 1449-1457.

R. Kronland-Martinet, J. Morlet, and A. Grossmann [1987]. "Analysis of sound patterns through wavelet transforms", *International Journal of Pattern Recognition and Artificial Intelligence* 1:2, 273-302.

J. B. Kruskal [1983]. "An Overview of Sequence Comparison: Time Warps, String Edits, and Macromolecules", *SIAM Review* 25:2, 201-237.

H. Kucera and W. N. Francis [1967]. *Computational Analysis of Present-Day American English*, Brown University Press, Providence, RI.

F. Kuhlmann and J. A. Bucklew [1988]. "Piecewise uniform vector quantizers", *IEEE Transactions on Information Theory* 34, no, 5, 1259-1263.

M. Kunt [1978]. "Source Coding of X-Ray Pictures", *IEEE Trans. on Biomedical Engineering, BME-25:2*, 121-138.

M. Kunt, M. Benard, and R. Leonardi [1987]. "Recent result in high-compression image coding", *IEEE Tran. Circuits and Systems* CAS-34:11, 1306-1336.

M. Kunt, A. Ikonomopoulos, and M. Kocher [1985], "Second-generation image-coding techniques", *Proc. of the IEEE* 73:4, 549-574.

M. R. Lagana, G. Turrini, and G. Zanchi [1991]. "Experiments on the Compression of Dictionary Entries", *Proceedings IEEE Data Compression Conference*, Snowbird, Utah, 432.

K. B. Lakshmanan [1981]. "On Universal Codeword Sets", *IEEE Transactions on Information Theory* 27, 659-662.

G. M. Landau and U. Vishkin [1985]. "Efficient String Matching in the Presence of Errors", *Proceedings Twenty-Sixth Symposium on the Foundations of Computer Science*, Portland, OR, 126-136.

G. M. Landau and U. Vishkin [1986]. "Introducing Efficient Parallelism into Approximate String Matching and a New Serial Algorithm", *Proceedings Eighteenth Annual ACM Symposium on the Theory of Computing*, Berkeley, CA, 220-230.

G. M. Landau, B. Schieber, and U. Vishkin [1987]. "Parallel Construction of a Suffix Tree", *Proc. Fourteenth ICALP, Lecture Notes in Computer Science 267*, Springer-Verlag, New York, NY, 314-325.

G. G. Langdon [1981]. "A Note on the Lempel-Ziv Model for Compressing Individual Sequences", *Technical Report RJ3318*, IBM Research Lab., San Jose, CA.

G. G. Langdon [1981b]. "Tutorial on Arithmetic Coding", *Technical Report RJ3128*, IBM Research Lab., San Jose, CA.

G. G. Langdon [1984]. "On parsing versus mixed-order model structures for data compression", *IBM Research Report RJ-4163 (46091)*, IBM Research Laboratory, San Jose, CA.

G. G. Langdon [1984b]. "An introduction to arithmetic coding", *IBM Journal of Research and Development 28:2*, 135-149.

G. G. Langdon [1986]. "Compression of Multilevel Signals", *U. S. Patent No. 4,749,983.*

G. G. Langdon [1991]. "Sunset: A Hardware-Oriented Algorithm for Lossless Compression of Gray Scale Images", *SPIE Medical Imaging V: Image Capture, Formatting, and Display 1444*, 272-282.

G. G. Langdon [1991b]. "Probabilistic and Q-Coder Algorithms for Binary Source Adaption", *Proceedings IEEE Data Compression Conference*, Snowbird, Utah, 13-22.

G. G. Langdon and J. Rissanen [1981]. "Compression of black-white images with arithmetic coding", *IEEE Transactions on Commuications COM-29:6*, 858-867.

G. G. Langdon and J. Rissanen [1982]. "A simple binary source code", *IEEE Transactions on Information Theory IT-28:5*, 800-803.

G. Langdon and J. Rissanen [1983]. "A Double Adaptive File Compression Algorithm", *IEEE Trans. on Comm. , COM-31:11*, 1253-1255.

G. Langdon, A. Gulati, and E. Seiler [1992]. "On the JPEG Context Model for Lossless Image Compression", *Proceedings IEEE Data Compression Conference*, Snowbird, Utah, 172-180.

L. L. Larmore [1987]. "Height-Restricted Optimal Binary Search Trees", *SIAM Journal on Computing* (December).

L. L. Larmore and D. S. Hirschberg [1987]. "A Fast Algorithm for Optimal Length-Limited Codes", *Technical Report*, Dept. of Information and Computer Science, University of California, Irvine, CA.

R. H. Lathrop, T. A. Webster, and T. F. Smith [1987]. "Ariadne: Pattern-Directed Inference and Hierarchical Abstraction in Protein Structure Recognition", *CACM* 30:11, 909-921.

J. C. Lawrence [1977]. "A New Universal Coding Scheme for the Binary Memoryless Source", *IEEE Transactions on Information Theory 23:4*, 466-472.

D. Le Gall [1991]. "MPEG: A video compression standard for multimedia applications", *Communications of the ACM* 34:4, 46-58.

D. J. Le Gall and A. Tabatabai [1988]. "Sub-band coding of digital images using symmetric short dernel filters and arithmetic coding techniques", *Proceedings ICASSP 88*, NY.

R. M. Lea [1978]. "Text Compression with an Associative Parallel Processor", *Computer Journal* 21:1, 45-56.

E. T. Lee [1991]. "Pictorial Data Compression Using Array Grammars", *Proceedings IEEE Data Compression Conference*, Snowbird, Utah, 444.

E. T. Lee and S. Harjardi [1989]. "Volume/surface octree generation and applications", *Advances in Artificail Intelligence Research*, 251-264.

H. J. Lee and D. T. L. Lee [1986]. "A Gain-Shape Vector Quantization for Image Coding", *Proc. of ICASSP*, 141-144.

T. Lee and A. M. Peterson [1990]. "Adaptive Vector Quantization Using a Self-development Neural Network", *IEEE J. Select. Areas Commun.*, 1458-1471.

D. S. Lee and K. H. Tzou [1990]. "Heirarchical DCT coding of HDTV for ATM networks", *Proceedings ICASSP 90*, IEEE Acoustics, and Signal Processing Society, NM.

H. Lee, Y. Kim, A. H. Rowberg, and E. A. Riskin [1991]. "3-D Image Compression for X-ray CT Images Using Displacement Estimation", *Proceedings IEEE Data Compression Conference*, Snowbird, Utah, 453.

P. J. Lee, S. H. Lee, and R. Ansari [1990]. "Cell loss detection and recovery in variable rate video", *Proceedings Third International Workshop on Packet Video*, Morristown, NJ.

D. S. Lee, K. H. Tzou [1990]. "Hierarchical DCT coding of HDTV for ATM networks", *Proc. of ICASSP*, 2249-2252.

T. L. R. Lei, N. Scheinberg, and D. L. Schilling [1977]. "Adaptive Delta Modulation System for Video Encoding", *IEEE Trans. on Comm. COM-25.*

D. A. Lelewer and D. S. Hirschberg [1987]. "Data Compression", *ACM Computing Surveys 19:3*, 261-296.

D. A. Lelewer and D. S. Hirschberg [1991]. "Streamlining Context Models for Data Compression", *Proceedings IEEE Data Compression Conference*, Snowbird, Utah, 313-322.

A. Lempel and J. Ziv [1976]. "On the Complexity of Finite Sequences", *IEEE Transactions on Information Theory* 22:1, 75-81.

A. Lempel and J. Ziv [1985]. "Compression of Two-Dimensional Images", *Combinatorial Algorithms on Words*, Springer-Verlag (A. Apostolico and Z. Galil, editors), 141-154.

A. Lempel, S. Even, and M. Cohn [1973]. "An Algorithm for Optimal Prefix Parsing of a Noiseless and Memoryless Channel", *IEEE Transactions on Infromation Theory* 19:2, 208-214.

M. E. Lesk [1970]. "Compressed Text Storage", *Technical Report*, Bell Laboratories, Murray Hill, NJ.

W. H. Leung and S. Skiena [1991]. "Inducing Codes from Examples", *Proceedings IEEE Data Compression Conference*, Snowbird, Utah, 267-276.

V. E. Levenshtein [1968]. "On the Redundancy and Delay of Separable Codes for the Natural Numbers", *Problems of Cybernetics* 20, 173-179.

B. Leverett and T.G. Szymanski [1979]. "Chaining Span-Dependent Jump Instructions", *Technical Report*, Bell Laboratories, Murray Hill, NJ.

A. S. Lewis and G. Knowles [1989]. "Image Compression using the 2-D Wavelet Transform", *IEEE Trans. on Signal Processing.*

A. S. Lewis and G Knowles [1990]. "Video Compression using 3D Wavelet Transforms", *Electronics Letters 26:6*, 396-398.

A. S. Lewis and G. Knowles [1991]. "A 64 Kb Video Codec Using the 2-D Wavelet Transform", *Proceedings IEEE Data Compression Conference*, Snowbird, Utah, 196-201.

A. S. Lewis and G. Knowles [1991b]. "A VLSI architecture for the 2D Daubechies wavelet transform wsithout multipliers", *Electron. Lett. 27:2*, 171-173.

M. Li [1985]. "Lower Bounds by Kolmogorov-Complexity", Technical Report TR85-666, Computer Science Dept., Cornell University, Ithaca, NY.

W. Li [1992]. "Image Data Compression Using Vector Transformation and Vector Quantization", *Proceedings IEEE Data Compression Conference*, Snowbird, Utah, 406.

X. Li and Z. Fang [1986]. "Parallel algorithms for clustering on hypercube SIMD computers", *Proc. of the 1986 Conference on Computer Vision and Pattern Recognition*, 130-133.

S. Li and M. H. Loew [1987]. "The Quadcode and its Arithmetic", *Communications of the ACM* 30:7, 621-626.

S. Li and M. H. Loew [1987b]. "Adjacency Detection Using Quadcodes", *Communications of the ACM* 30:7, 627-631.

X. Li and B. Woo [1992]. "A High Performance Data Compression Coprocessor with Fully Programmable ISA Interface", *Proceedings IEEE Data Compression Conference*, Snowbird, Utah, 404.

K. M. Liang, S. E. Budge, and R. W. Harris [1991]. "Rate Distortion Performance of VQ and PVQ Compression Algorithms", *Proceedings IEEE Data Compression Conference*, Snowbird, Utah, 445.

H. H. J. Liao [1977]. "Upper Bound, Lower Bound, and Run-Length Substitution Coding", *National Telecommunications Conference*, 1-6.

M. Liebhold and E. M. Hoffert [1991]. "Towards an open environment for digital video", *Communications of the ACM* 34:4, 103-112.

J. J. Lim [1990]. *Two Dimensional Signal and Image Processing*, Prentice Hall.

J. O. Limb, C. B. Rubinstein and K. A. Walsh [1971]. "Digital Coding of Color Picture-Phone Signals by Element-Differential Quantization", *IEEE Trans. on Comm. Tech COM-19*, 992-1006.

J. O. Limb, C. B. Rubinstein, and J. E. Thompson [1977]. "Digital Coding of Color Video Signals- A Review", *IEEE Trans. on Comm. COM-25:11*, 1349-1385.

J. Lin [1988]. *Information-Theoretic Divergence Measures and Their Application*, Master's Thesis, University of Regina, Canada.

J. Lin [1991]. "Divergence measures based on the Shannon entropy", *IEEE Transactions on Information Theory* IT-37:1, 145-151.

J. Lin and J. A. Storer [1991]. "Improving Search for Tree-Structured Vector Quantization", *Proceedings IEEE Data Compression Conference*, Snowbird, Utah.

J. Lin and J. A. Storer [1991b]. "Resolution Constrained Tree Structured Vector Quantization for Image Compression", *Proceedings IEEE Symposium on Information Theory*, Budapest, Hungary, 1991 (coauthored with J. Lin).

J. Lin and J. A. Storer [1992]. "Improving Search for Tree-Structured Vector Quantization", *Proceedings IEEE Data Compression Conference*, Snowbird, Utah, 339-348.

J. Lin, J. A. Storer, and M. Cohn [1991]. "On the Complexity of Optimal Tree Pruning for Source Coding", *Proceedings IEEE Data Compression Conference*, Snowbird, Utah, 63-72.

J. Lin, J. A. Storer, and M. Cohn [1992]. "Optimal Pruning for Tree-Structured Vector Quantization", *Information Processing and Management*, 1992.

J. H. Lin and J. S. Vitter [1992]. "Nearly Optimal Vector Quantization via Linear Programming", *Proceedings IEEE Data Compression Conference*, Snowbird, Utah, 22-31.

J. Lin and S. K. M. Wong [1988]. "On the characterization of a new directed divergence measure", *1988 IEEE International Symposium on Information Theory*, Japan.

J. Lin and S. K. M. Wong [1990]. "A new directed divergence measure and its characterization", *International Journal of General Systems* 17:1, 73-81.

Y. Linde and R. M. Gray [1978]. "A fake process approach to data compression", *IEEE Trans. Comm. COM-26*, 702-710.

Y. Linde, A. Buzo, and R. M. Gray [1980]. "An algorithm for vector quantizer design", *IEEE Transactions on Communications* 28, 84-95.

R. Lindsay and D. E. Abercrombie [1991]. "Restricted Boundary Vector Quantization", *Proceedings IEEE Data Compression Conference*, Snowbird, Utah, 159-165.

R. A. Lindsay and D. M. Chabries [1986]. "Data Compression of Color Images Using Vector Quantization", *Technical Report*, The Unisys Corporation, Salt Lake City, Utah.

M. Liou [1991]. "Overview of the px64 kbit/s Video Coding Standard", *Communications of the ACM 34:4*, 59-63.

A. Lippman [1990]. "HDTV sparks a digital revolution", *Byte*, 297-305.

A. Lippman [1991]. "Feature sets for interactive images", *Communications of the ACM* 34:4, 92-102.

A. Lippman and W. Butera [1989]. "Coding image sequences for interactive retrieval", *Communications of the ACM* 32:7, 852-860.

R. J. Lipton and D. Lopresti [1985]. "A Systolic Array for Rapid String Comparison", *Proceedings Chapel Hill Conference on VLSI.*

T. M. Liu and Z. Hu [1986]. "Hardware Realization of Multistage Speech Waveform Vector Quantizer", *ICASSP 86*, 3095-3097.

Y. Liu and H. Ma [1991]. "W-Orbit Finite Automata for Data Compression", *Proceedings IEEE Data Compression Conference*, Snowbird, Utah, 166-175.

J. A. Llewellyn [1987]. "Data Compression for a Source with Markov Characteristics", *The Computer Journal* 30:2, 149-156.

J. A. Llewellyn [1987b]. *Information and Coding*, Chartwell-Bratt.

S. P. Lloyd [1982]. "Least squares quantization in PCM", *IEEE Trans. Inform. Theory IT-28*, 127-135.

T. L. Lookabaugh [1988]. "Variable Rate and Adaptive Frequency Domain Vector Quantization of Speech", *Ph. D. Dissertation*, Stanford University.

T. D. Lookabaugh and R. M. Gray [1989]. "High-resolution quantization theory and the vector quantizer advantage", *IEEE Transactions on Information Theory* IT-35:5, 1020-1033.

T. Lookabaugh, E. A. Riskin, P. A. Chou, and R. M. Gray [1991]. "Variable rate vector quantization for speech, image, and video compression", *IEEE Transactions on Communications.*

D. W. Loveland [1969]. "A Variant of the Kolmogorov Concept of Complexity", *Information and Control* 15, 510-526.

D. W. Loveland [1969b]. "On Minimal-Program Complexity Measures", *Proceedings First Annual ACM Symposium on Theory of Computing*, Marina Del Rey, California, 61-65.

G. Loveria and D. Kinstler [1990]. "Multimedia: DVI arrives", *Byte IBM Special Edition*, 105-108.

A. Lowry, S. Hossain, and W. Millar [1987]. "Binary search trees for vector quantization", *Proceedings International Conference on Acoustics, Speech, and Signal Processing*, Dallas, 2206-2208.

C. C. Lu and N. Choong [1991]. "Adaptive Source Modeling Using Conditioning Tree", *Proceedings IEEE Data Compression Conference*, Snowbird, Utah, 423.

Luc Longpre [1986]. "Resource Bounded Kolmogorov Complexity, a Link Between Computational Complexity and Information Theory", *Ph.D. Thesis*, Cornell University, Ithaca, NY.

T. J. Lynch [1966]. "Sequence time coding for data compression", *Proc. IEEE 54*, 1490-1491.

M. F. Lynch [1973]. "Compression of Bibliographic Files Using an Adaptation of Run-Length Coding", *Information Storage and Retrieval 9*, 207-214.

T. J. Lynch [1985]. *Data Compression: Techniques and Applications*, Lifetime Publications, Belmont, CA.

M. F. Lynch and D. Cooper [1978]. "Compression of Wiswesser Line Notations Using Variety Generation", *Journal of Chemical Information and Computer Science 19:3*, 165-169.

R .E. Machol, editor [1960]. *Information and Decision Processes*, McGraw-Hill.

B. MacMillian [1956]. "Two Inequalities Implied by Unique Decipherability", *IRE Transactions on Information Theory* 2, 115-116.

J. MacQueen [1965]. "On convergence of K-means and partitions with minimum average variance (abstract)", *Ann. Math. Statist.* 36, 1084.

F. J. MacWilliams and N. J. A. Sloane [1978]. *The Theory of Error-Correcting Codes*, North-Holland, New York, NY.

R. A. Madisett, R. Jain, and R. Baker [1992]. "The Real Time Implementation of Pruned Tree Vector Quantization", *Proceedings IEEE Data Compression Conference*, Snowbird, Utah, 152-161.

A. Maeder [1988]. "Local block pattern methods for binary image encoding", *Proc. AusGraph88*, 251-256.

D. Maier [1978]. "The Complexity of Some Problems on Subsequences and Supersequences", *Journal of the ACM* 25:2, 322-336.

D. Maier and J. A. Storer [1978]. "A Note Concerning the Superstring Problem", *Proceedings Twelfth Annual Conference on Information Sciences and Systems*,

M. E. Majster [1979]. "Efficient On-Line Construction and Correction of Position Trees", *Technical Report TR79-393*, Dept. of Computer Science, Cornell University, Ithaca, NY.

J. Makhoul [1975]. "Linear prediction: A tutorial review", *Proc. IEEE 63*, 561-580.

J. Makoul, S. Roucos, and H. Gish [1985]. "Vector Quantization in Speech Coding", *Proceedings of the IEEE* 73:11, 1551-1588.

M. Malak and J. Baker [1991]. "An Image Database for Low Bandwidth Communication Links", *Proceedings IEEE Data Compression Conference*, Snowbird, Utah, 53-62.

S. G. Mallat [1989]. "A Theory for Multiresolution Signal Decomposition - the Wavelet Representation", *IEEE Transactions on Pattern Analysis and Machine Intelligence 11:7*, 674-693.

S. G. Mallat [1991]. "Zero-Crossings of a Wavelet Transform", *IEEE Trans. on Information Theory 37:4*.

H. S. Malvar and D. H. Staelin [1989]. "The LOT: transform coding without blocking effects", *IEEE Trans. on Asoustics, Speech, and Signal Processing* 37:4, 553-559.

K. Maly [1976]. "Compressed tries", *Communications of the ACM 19:7*, 409-415.

G. Mandelbrot [1982]. *Fractal Geometry of Nature*, W. H. Freeman, Salt Lake City, Utah.

R. L. Manfrino [1970]. "Printed Portugese (Brazilian) Entropy Statistical Calculation", *IEEE Trans. Information Theory IT-16*, 122.

M. Manohar and J. C. Tilton [1992]. "Progressive Vector Quantization of Multispecial Image Data Using a Massively Parallel SIMD Machine", *Proceedings IEEE Data Compression Conference*, Snowbird, Utah, 181-190.

D. Manstetten [1992]. "Tight bounds on the redundancy of Huffman codes", *IEEE Trans. Inform. Theory IT-38:1*, 144-151.

M. W. Marcellin [1990]. "Transform coding of images using trellis coded quantization", *Proceedings International Conference on Acoustics, Speech and Signal Process.* , 2241-2244.

M. W. Marcellin and T. R. Fischer [1990]. "Trellis coded quantization of memoryless and Gauss-Markov sources", *IEEE Trans. Commun. COM-38*, 82-93.

K. V. Mardia [1972]. *Statistics of Directional Data*, Academic Press, New York, NY.

T. Markas and J. Reif [1991]. "Image Compression Methods with Distortion Controlled Capabilities", *Proceedings IEEE Data Compression Conference*, Snowbird, Utah, 93-102.

T. Markas and J. H. Reif [1992b]. "Quad Tree Structures for Image Compression Applications", *Journal of Information Processing and Management.*

T. Markas, J. Reif, and J. A. Storer [1992]. "On Parallel Implementations and Experimentations of Lossless Data Compression Algorithms", *Proceedings IEEE Data Compression Conference*, Snowbird, Utah, 425.

J. D. Markel and A. H. Gray, Jr. [1976]. *Linear Prediction of Speech*, Springer-Verlag, NY.

B. A. Marron and P.A.D. De Maine [1967]. "Automatic Data Compression", *Communications of the ACM* 10:11, 711-715.

J. Martin [1977]. *Computer Data Base Ogranization*, Prentice-Hall, Englewood Cliffs, NJ.

G. N. N. Martin [1979]. "Range Encoding: An Algorithm for Removing Redundancy from a Digitized Message", *Proceedings Video and Data Recording Conference*, Southhampton, Hampshire, England.

G. N. N. Martin, G. G. Langdon, and S. J. P. Todd [1983]. "Arithmetic Codes for Constrained Channels", *IBM Journal of Research and Development 27:2*, 94-106.

P. Martin-Lof [1966]. "The Definition of Random Sequences", *Information and Control* 9, 602-619.

T. Masui [1991]. "Keyword Dictionary Compression Using Efficient Trie Implementation", *Proceedings IEEE Data Compression Conference*, Snowbird, Utah, 438.

Y. Matsuyana and R. M. Gray [1982]. "Voice Coding and Tree Encoding Speech Compression Systems Based upon Inverse Filter Matching", *IEEE Trans. on Comm. 30:4*, 711-720.

W. D. Maurer [1969]. "File compression using Huffman coding", In *Computing Methods in Optimization Problems 2*, Academic Press, 247-256; also in *Computing Reviews* 11:10, October 1970, 944.

W. D. Maurer [1992]. "Proving the Correctness of a Data Compression Program", *Proceedings IEEE Data Compression Conference*, Snowbird, Utah, 427.

J. Max [1960]. "Quantizing for minimum distortion", *IEEE Transactions on Information Theory* IT-6:2, 7-12.

A. Mayne and E. B. James [1975]. "Information Compression by Factorizing Common Strings", *The Computer Journal* 18:2, 157-160.

J. McCabe [1965]. "On Serial Files with Relocatable Records", *Operations Research* 12, 609-618.

J. P. McCarthy [1973]. "Automatic File Compression", *Proceedings International Computing Symposium 1973*, North Holland Publishing Company, Amsterdam, 511-516.

E. M. McCreight [1976]. "A Space-Economical Suffix Tree Construction Algorithm", *Journal of the ACM* 23:2, 262-272.

T. McCullough [1977]. "Data Compression in High-Speed Digital Facsimile", *Telecommunications*, 40-43.

R. A. McDonald [1966]. "Signal-to-quantization noise ratio and idle channel performance of DPCM system with particular application to voice signal", *Bell Syst. Tech. J. 45*, 1123-1151.

R. J. McEliece [1977]. *The Theory of Information and Coding*, Addison-Wesley, Reading, MA.

D. R. McIntyre and M. A. Pechura [1985]. "Data Compression Using Static Huffman Code-Decode Tables", *Journal of the ACM* 28:6, 612-616.

B. McMillan [1953]. "The basic theorems of information theory", *Ann. Math. Stat.* 24, 196-219.

B. McMillan [1969]. "Communication systems which minimize coding noise", *Bell Syst. Tech. J. 48:9*, 3091-3112.

L. McMillan and L. Westover [1992]. "A Forward-Mapping Realization of the Inverse Discrete Cosine Transform", *Proceedings IEEE Data Compression Conference*, Snowbird, Utah, 219-228.

K. Mehlhorn [1980]. "An Efficient Algorithm for Constructing Nearly Optimal Prefix Codes", *IEEE Transactions on Information Theory* 26:5, 513-517.

N. D. Memon and K. Sayood [1992]. "Lossless Image Compressoin Using a Codebook of Prediction Trees", *Proceedings IEEE Data Compression Conference*, Snowbird, Utah, 414.

N. D. Memon, S. S. Magliveras, and K. Sayood [1991]. "Prediction Trees and Lossless Image Compression: An Extended Abstract", *Proceedings IEEE Data Compression Conference*, Snowbird, Utah, 83-92.

P. Mermelstein [1991]. "Trends in Audio and Speech Compression for Storage and Real-Time Communication", *Proceedings IEEE Data Compression Conference*, Snowbird, Utah, 455.

Michael P. Ekstrom, editor [1984]. *Digital Image Processing Techniques*, Academic Press, New York, NY.

K. Miettinen [1992]. "Compression of Spectral Meteorological Imagery", *Proceedings IEEE Data Compression Conference*, Snowbird, Utah, 388.

V. S. Miller and M. N. Wegman [1985]. "Variations on a Theme by Lempel and Ziv", *Combinatorial Algorithms on Words*, Springer-Verlag (A. Apostolico and Z. Galil, editors), 131-140.

R. P. Millett and E. L. Ivie [1991]. "Index Compression Method with Compressed Mode Bolean Operators", *Proceedings IEEE Data Compression Conference*, Snowbird, Utah, 437.

S. Minami and A. Zakhor [1991]. "An Optimization Approach for Removing Blocking Effects in Transform Coding", *Proceedings IEEE Data Compression Conference*, Snowbird, Utah, 442.

J. L. Mitchell and W. B. Pennebaker [1988]. "Optimal hardware and software arithmetic coding procedurew for the Q-coder", *IBM J. Res. Develop. 32*, 727-736.

J. W. Modestino, D. D. Harrison, and N. Farvardin [1990], "Robust adaptive buffered-instrumented entropy-coded quantization of stationary sources", *IEEE Transactions on Communications* 38:6, 856-867.

A. Moffat [1987]. "Predictive text compression based upon the future rather than the past", *Australian Computer Science Communications 9*, 254-261.

A. Moffat [1988]. "A Data Structurefor Arithmetic Encoding on Large Alphabets", *Proceedings Eleventh Australian Computer Science Conference*, Brisbane, Australia, 309-317.

A. Moffat [1989]. "Word-based text compression", *Software-practice and experience 19*, 185-198.

A. Moffat [1990]. "Linear time adaptive arithmetic coding", *IEEE Trans. Information Theory IT-36*, 401-406.

A. Moffat [1990b]. "Implementing the PPM data compression scheme", *IEEE Trans. Communications COMM-38.*

A. Moffat [1991]. "Two-Level Context Based Compression of Binary Images", *Proceedings IEEE Data Compression Conference*, Snowbird, Utah, 382-391.

A. Moffat amd J. Zobel [1992]. "Coding for Compression in Full-Text Retrieval Systems", *Proceedings IEEE Data Compression Conference*, Snowbird, Utah, 72-81.

U. Moilanen [1978]. "Information Preserving Codes Compress Binary Pictorial Data", *Computer Design* , 134-136.

J. H. Mommers, J. Raviv [1974]. "Coding for data compaction", IBM Watson Res. Report RC5150, November 1974.

B. L. Montgomery and J. Abrahams [1987]. "On the redundancy of optimal binary prefix-condition codes for finite and infinite sources", *IEEE Trans. Inform. Theory IT-33:1*, 156-160.

T. H. Morrin [1976]. "Chain-Link Compression of Arbitrary Black-White Images", *Computer Graphics and Image Processing 5*, 172-189.

R. Morris and K. Thompson [1974]. "Webster's Second on the Head of a Pin", *Technical Report*, Bell Laboratories, Murray Hill, NJ.

D. R. Morrison [1968]. "PATRICIA - A Practical Algorithm to Retrieve Information Coded in Alphanumeric", *Journal of the ACM* 15:4, 514-534.

D. J. Muder [1988]. "Minimal trellises for block codes", *IEEE Transactions on Information Theory 34:5*, 1049-1053.

A. Mukherjee [1989]. "Huffman code translator", *U. S. Patent m=no. 038039.*

A. Mukherjee and M. A. Bassiouni [1987]. "On-the-Fly Algorithms for Data Compression", *Proceedings ACM/IEEE Fall Joint Computer Conference.*

A. Mukherjee, H. Bheda, M. A. Bassiouni, and T. Acharya [1991]. "Multibit Decoding - Encoding of Binary Codes Using Memory Based Architectures", *Proceedings IEEE Data Compression Conference*, Snowbird, Utah, 352-361.

A. Mukherjee, J. Flieder, and N. Ranganathan [1992]. "MARVLE: A VLSI Chip of a Memory-Based Architecture for Variable-Length Encoding and Decoding", *Proceedings IEEE Data Compression Conference*, Snowbird, Utah, 416.

A. Mukherjee, N. Ranganathan, and M. A. Bassiouni [1990]. "Adaptive and pipelined VLSI designs for tree-based codes", *IEEE Trans. on Circuits and Systems.*

A. Mukherjee, N. Ranganathan, and M. Bassiouni [1991]. "Efficient VLSI Designs for Data Transformation of Tree-Based Codes", *IEEE Transactions on Circuits and Systems 38:3*, 306-314.

J. B. Mulford and R.K. Ridall [1971]. "Data Compression Techniques for Economic Processing of Large Commercial Files", *ACM Symposium on Information Storage and Retrieval*, 207-215.

T. Murakami, K. Asai, and E. Yamazaki [1982]. "Vector quantizer of Video signals", *Electronic Letters*, 1005-1006.

H. Murakami, S. Matsumoto, and H. Yamamoto [1984]. "Algorithm for Construction of Variable Length Code with Limited Maximum Word Length", *IEEE Trans. Commun. COM-32*, 1157-1159.

H. G. Musmann [1979]. "Predictive Image Coding", *Image Transmission Techniques* (W. K. Pratt, editor), Academic Press, New York, NY.

H. G. Musmann and D. Preuss [1973]. "A Redundancy Reducing Facsimile Coding Scheme", *NTZ*, 91-94.

H. G. Musmann and D. Preuss [1977]. "Comparison of Redundancy Reducing Codes for Facsimile Transmission of Documents", *IEEE Trans. on Comm. , COM-25:11*, 1425-1433.

H. G. Musmann, P. Pirsch, and H. J. Grallert [1985], "Advances in picture coding", *Proc. of the IEEE* 73:4, 523-548.

N. Nakatsu [1978]. "Bounds on the Redundancy of Binary Alphabetical Codes", *IEEE Trans. Inform. Theory IT-24*, 1225-1229.

A. Narayan and T. V. Ramabadran [1992]. "Predictive Vector Quantization of Grayscale Images", *Proceedings IEEE Data Compression Conference*, Snowbird, Utah, 426.

N. M. Nasrabadi and Y. Feng [1988]. "Vector quantization of images based upon the Kohonen self-organizing feature maps", *Proceedings 1988 International Joint Conference on Neural Networks 1*, San Diego, CA, 101-108.

N. M. Nasrabadi and Y. Feng [1990]. "Image compression using address-vector quantization", *IEEE Trans. Comm. 38*, 2166-2173.

N. M. Nasrabadi and Y. Feng [1990b]. "A multi-layer address-vector quantization", *IEEE Trans. Circuits and Systems CAS-37*, 912-921.

N. Nasrabadi and R. King [1988]. "Image Coding Using Vector Quantization: A Review", *IEEE Trans. on Communication COM-28:1*, 84-95.

J. Naylor and K. P. Li [1988]. "Analysis of a neural network algorithm for vector quantization for speech parameters", *Proceedings First Annual INNS Meet.* , Boston, MA, 310.

B. E. Nelson and C. J. Read [1986]. "A Bit-Serial VLSI Vector Quantizer", *Proc. ICASSP.*

A. N. Netravali [1977]. "On quantizers for DPCM coding of picture signals", *IEEE Transactions on Information Theory* IT-23, 360-370.

A. N. Netravali and J. O. Limb [1980]. "Picture coding: a review", *Proc. IEEE* 68, 366-406.

A. Netravali and F. Mounts [1980]. "Ordering Techniques for Facsimile Coding: A Review", *Proc. IEEE 68*, 796-807.

A. Netravali and B. Prasada [1977]. "Adaptive Quantization of Picture Signals Using Spatial Masking", *Proc. IEEE 65*, 536-548.

A. N. Netravali and J. D. Robbins [1979]. "Motion Compensated Television Coding: Part I", *Bell System Technical J. 58:3*, 631-670.

A. N. Netravali and R. Saigal [1976]. "Optimum quantization design using a fixed point algorithm", *Bell Syst. Tech. J.* 55:9, 1423-1435.

D. L. Neuhoff, R. M. Gray, and L. D. Davisson [1975]. "Fixed rate universal block source coding with a fidelity criterion", *IEEE Trans. Inform. Theory 21*, 511-523.

E. B. Newman and N. C. Waugh [1960]. "The Redundancy of Texts in Three Languages', *Information and Control 3*, 141-153.

H. Nguyen and J. W. Mark [1991]. "Concentric-Shell Partition Vector Quantization with Application to Image Coding", *Proceedings IEEE Data Compression Conference*, Snowbird, Utah, 119-128.

H. Nguyen and J. W. Mark [1992]. "Interframe Coding Using Quadtree Decomposition and Concentric-Shell Partition Vector Quantization", *Proceedings IEEE Data Compression Conference*, Snowbird, Utah, 412.

L. M. Ni and A. K. Jain [1985]. "A VLSI systolic architecture for pattern clustering", *IEEE Trans. on Pattern Analysis and Machine Intell.* PAMI-7, 80-89.

K. Nickels and C. Thacker [1991]. "Satellite Data Archives Algorithm", *Proceedings IEEE Data Compression Conference*, Snowbird, Utah, 447.

W. P. Niedringhaus [1979]. "Scheduling Without Queuing, the Space Factory Problem", *Technical Report 253*, Dept. of Electrical Engineering and Computer Science, Princeton University, Princeton, NJ.

R. Nix [1981]. "Experience with a Space Efficient Way to Store a Dictionary", *Communications of the Association for Computing Machinery 24:5*, 297-298.

P. Noll and R Zelinski [1978]. "Bounds on quantizer performance in the low bit-rate region", *IEEE Transactions on Communications* COM-26:2, 300-304.

N. D. Nonobashvili [1976]. "Some Questions on the Optimality of Representing Information by a Four-Letter Alphabet on a Transposition Scheme for Entering and Compressing Discrete Information", *Sakharth. SSR. Mecn. Akad. Moambe* 83:2, 317-320.

N. D. Nonobashvili [1977]. "An algorithm for compressing discrete information in a four-valued coding system", *Sakharth. SSR. Mecn. Akad. Moambe* 85:3, 565-568.

K. Nordling [1982]. "A Data Compression Modem", *Telecommunications*, 67-70.

D. A. Novik and J. C. Tilton [1992]. "Adjustable Lossless Image Compression Based on a Natural Splitting of an Image into Drawing, Shading, and Fine-Grained Components", *Proceedings IEEE Data Compression Conference*, Snowbird, Utah, 380.

J. B. O'Neal [1966]. "Predictive quantizing differential pulse code modulation for the transmission of television signals", *Bell Syst. Tech. J. 45:5*, 689-721.

A. M. Odlyzko [1985]. "Enumeration of Strings", *Combinatorial Algorithms on Words*, Springer-Verlag, (A. Apostolico and Z. Galil, editors), 205-228.

T. Ohira, M. Hayakawa, and K. Matsumoto [1977]. "Orthogonal Transform Coding System for NTSC Color Television Signal", *Proceedings of the I. C. C. 4B*, 3-86 to 3-90.

J. R. Ohm and P. Noll [1990]. "Predictive tree encoding of still images with vector quantization", *Annales des Télécommunications*, 465-470.

B. M. Oliver, J. R. Pierce, and C. E. Shannon [1948]. "The philosophy of PCM", *Proc. IRE* 36, 1324-1331.

R. A. Olshen, P. C. Cosman, C. Tseng, C. Davidson, L. Moses, R. M. Gray, and C. Bergin [1991]. "Evaluating compressed medical images", *Proceedings Third International Conference on Advances in Communication and Control Systems*, Victoria, British Columbia.

A. Omodeo, M. Pugassi, and N. Scarabottolo [1991]. "Implementing JPEG Algorithm on INMOS Transputer Equipped Machines", *Proceedings IEEE Data Compression Conference*, Snowbird, Utah, 435.

A. V. Oppenheim and R. W. Schafer [1989]. *Discrete-Time Signal Processing*, Prentice Hall.

M. Ostendorf Dunham and R. M. Gray [1985]. "An algorithm for the design of labeled-transition finite-state vector quantizers", *IEEE Trans. Comm. , COM-33*, 83-89.

G. Ott [1967]. "Compact Encoding of Stationary Markov Sources", *IEEE Transactions on Information Theory*, 82-86.

J. P. Hayes [1976]. "Check Sum Methods for Test Data Compression", *J. Design Automat. Fault-Tolerant Comput.* 1:1, 3-17.

M. D. Paez and T. H. Glisson [1972]. "Minimum Mean Squared Error Quantization in Speech", *IEEE Trans. on Comm. COM-20*, 225-230.

W. Paik [1990]. "Digicipher – All digital channel compatible HDTV broadcast system", *IEEE Trans. Broadcasting* 36:4.

A. J. Palay [1985]. *Searching with Probabilities*, Pitman, Boston, MA.

P. F. Panter and W. Dite [1951]. "Quantization distortion in pulse-count modulation with nonuniform spacing of levels", *Proc. IRE* 39, 44-48.

H. Park, V. K. Prasanna, and C. L. Wang [1992]. "VLSI Architectures for Vector Quantization Based on Clustering", *Proceedings IEEE Data Compression Conference*, Snowbird, Utah, 441.

D. S. Parker [1978]. "Combinatorial Merging and Huffman's Algorithm", *Technical Report*, Dept. of Computer Science, University of Illinois at Urbana-Champaign, Urbana, Ill.

D. S. Parker [1978b]. "On when Huffman's Algorithm is Optimal", Technical Report, Dept. of Computer Science, University of Illinois at Urbana-Champaign, Urbana, Ill.

R. Pasco [1976]. "Source coding algorithms for fast data compression", *Ph. D. Dissertation*, Stanford University.

E. A. Patrick, D. R. Anderson, and F. K. Bechtel [1968]. "Mapping Multidimensional Space to One Dimension for Computer Output Display", *IEEE Transactions on Computers* 17:10, 949-953.

D. Paul [1982]. "A 500-800 bps adaptive vector quantization vocoder using a perceputally motivated distance measure", *Conf. Record, IEEE Globecom*, 1079-1082.

T. Pavlidis [1982]. *Algorithms for Graphics and Image Processing*, Addison-Wesley, Reading, MA.

J. Pearl [1973]. "On Coding and Filtering Stationary Signals by Discrete Fourier Transforms", *IEEE Trans. on Info. Theory*, 229-232.

D. Pearson [1990]. "Texture mapping in model-based image coding", *Image Commun.* 2:4, 377-397.

D. E. Pearson and J. A. Robinson [1985]. "Visual Communication at very low data rates", *Proceedings of the IEEE 73:4*, 795-812.

P. K. Pearson [1990]. "Fast hashing of variable-length text strings", *Communications of the ACM 33:6*, 677-680.

M. Pechura [1982]. "File Archival Techniques Using Data Compression", *Communications of the ACM* 25:9, 605-609.

W. B. Pennebaker and J. L. Mitchell [1988]. "Probability estimation for the Q-coder", *IBM J. Res. Develop. 32*, 737-752.

W. B. Pennebaker and J. L. Mitchell [1988b]. "Software implementations of the Q-coder", *IBM J. Res. Develop. 32*, 753-774.

W. B. Pennebaker, J. L. Mitchell, G. G. Langdon, and R. B. Arps [1988]. "An overview of the basic principles of the Q-coder", *IBM Journal of Research and Development 32:6*, 717-726.

A. P. Pentland [1986]. "Fractal based description of natural scenes", *IEEE Trans. PAMI 6:6*, 661-674.

A. Pentland and B. Horowitz [1991]. "A Practical Approach to Fractal-Based Image Compression", *Proceedings IEEE Data Compression Conference*, Snowbird, Utah, 176-185.

Y. Perl and A. Mehta [1990]. "Cascading LZW algorithm with Huffman coding method: a variable to variable length compression algorithm", *Proceedings First Great Lakes Computer Science Conference*, Kalamazoo.

Y. Perl, M. R. Garey, and S. Even [1975]. "Efficient Generation of Optimal Prefix Code: Equiprobable Words Using Unequal Cost Letters", *Journal of the ACM* 22:2, 202-214.

Y. Perl, V. Maram, and N. Kadakuntla [1991]. "The Cascading of the LZW Compression Algorithm with Arithmetic Coding", *Proceedings IEEE Data Compression Conference*, Snowbird, Utah, 277-286.

K. Petersen [1983]. *Ergodic Theory*, Cambridge University Press, Cambridge.

J. L. Peterson [1979]. "Text Compression", *Byte 12:4*, 106-118.

G. Peterson [1980]. "Succinct Representations, Random Strings, and Complexity Classes", *Proceedings Twenty-First Annual IEEE Symposium on the Foundations of Computer Science*, Syracuse, NY, 86-95.

W. W. Peterson and E. J. Weldon [1972]. *Error-Correcting Codes*, MIT Press, Cambridge, MA.

J. L. Peterson, J.R. Bitner, and J.H. Howard [1978]. "The Selection of Optimal Tab Settings", *Communications of the ACM 21:12*, 1004-1007.

J. Pike [1981]. "Text Compression Using a 4-Bit Coding Scheme", *The Computer Journal* 24:4, 324-330.

R. S. Pinkham [1961]. "On the Distribution of First Significant Digits", *The Annals of Mathematical Statistics* 32:4, 1223-1230.

P. P. Polit and N. M. Nasrabadi [1991]. "A New Transform Domain Vector Quantization Technique for Image Data Compression in an Asynchronous Transfer Mode Network", *Proceedings IEEE Data Compression Conference*, Snowbird, Utah, 149-158.

D. Pollard [1981]. "Strong consistency of -means clustering", *Annals of Statistics* 9, 135-140.

M. Porat and Y. Y. Zeevi [1988]. "The Generalized Gabor Scheme of Image Representation in Biological and Machine Vision', *IEEE Trans. Pattern Anal. Machine Intell. PAMI-10*, 1169-1179.

E. C. Posner, E. R. Rodemich [1971]. "Epsilon Entropy and Data Compression", *Ann. Math. Statist.* 42, 2079-2125.

K. A. Prabhu and A. N. Netravali [1983]. "Motion Compensated Composite Color Coding", *IEEE Trans. on Comm. COM-31:2*, 216-223.

W. K. Pratt [1971]. "Spatial Transform Coding of Color Images", *IEEE Trans. on Comm. Tech. COM-19*, 980-992.

V. R. Pratt [1975]. "Improvements and Applications for the Weiner Repetition Finder", lecture notes (third revision).

W. K. Pratt [1978]. *Digital Image Processing*, John Wiley, New York, NY.

W. K. Pratt [1979]. *Image Transmission Techniques*, Academic Press, NY.

W. K. Pratt, W. H. Chen, and L. R. Welch [1974]. "Slant transform image coding", *IEEE Transactions on Communications* COM-22, 1075-1093.

D. Preuss [1975]. "Two Dimensional Facsimile Source Encoding Based on a Markov Model", *NTZ, 28*, 358-363.

J. G. Proakis [1983]. *Digital Communications*, McGraw-Hill, New York, NY.

G. Promhouse and M. Bennett [1991]. "Semantic Data Compression", *Proceedings IEEE Data Compression Conference*, Snowbird, Utah, 323-331.

A. Puri, H. M. Hang, and D. L. Shilling [1987]. "An Efficient Block Matching Algorithm for Motion Compensated Coding", *Proceedings of ICASSP*, 25. 4. 1-25. 4. 4.

M. B. Pursley and L. D. Davisson [1976]. "Variable rate coding for nonergordic sources and classes of ergodic sources subject to a fidelity constraint", *IEEE Transactions on Information Theory* IT-22, 324-337.

R. D. Cullum [1972]. "A Method for the Removal of Redundancy in Printed Text", *NTIS AD751*, 407.

M. Rabbaani, P. W. Jones [1991]. *Digital Imagte Compression Techniques*, SPIE Optical Engineering Press, Wellingham, WA.

L. R. Rabiner and R. W. Schafer [1978]. *Digital Processing of Speech Signals*, Prentice-Hall, Englewood Cliffs, NJ.

L. Rabiner, S. E. Levinson, and M. M. Sondhi [1983]. "On the application of vector quantization and hidden Markov models to speaker-independent isolated word recognition", *Bell Syst. Tech. J.* 62, 1075-1105.

R. A. Raimi [1969]. "On the Distribution of First Significant Figures", *American Mathematical Monthly 76*, 342-348.

R. A. Raimi [1969b]. "On the Peculiar Distribution of First Digits", *Scientific American* 221 (December), 109-120.

R. A. Raimi [1976]. "The First Digit Problem", *American Mathematical Monthly*, 521-538.

T. Raita [1987]. "An Automatic System for File Compression", *Computer Journal 30:1*, 80-86.

T. Raita and J. Teuhola [1987]. "Predictive Text Compression by Hashing", *Proceedings ACM Conference on Information Retrieval*, New Orleans.

K. R. Rajagopalan [1965]. "A Note on Entropy of Kannada Prose", *Information and Control 8*, 640-644.

T. V. Ramabadran [1991]. "Some Results on Adaptive Statistics Estimation for the Reversible Compression of Sequences", *Proceedings IEEE Data Compression Conference*, Snowbird, Utah, 449.

T. V. Ramabadran and D. L. Cohn [1989]. "An Adaptive Algorithm for the Compression of Computer Data", *IEEE Trans. Commun. 37*, 317-324.

P. A. Ramamoorthy, B. Potu, and T. Tran [1989]. "Bit-Serial VLSI Implementation of Vector Quantizer for Real-Time Image Coding", *IEEE Trans. on Circuits and System*, 1281-1290.

B. Ramamurthi and A. Gersho [1986]. "Classified vector quantization of images", *IEEE Transactions on Communications* COM-34, 1105-1115.

H. K. Ramapriyan, J. C. Tilton, and E. J. Seiler [1985]. "Impact of Data Compression on Spectral / Spatial Classification of Remotely Sensed Data", *Advances in Remote Sensing Retrieval Methods* (H. E. Fleming and M. T. Chahine, editors), Deepak Publishing.

V. Ramasubramanian and K. K. Paliwal [1988]. "An optimized k-d tree algorithm for fast vector quantization of speech", *Proceedings European Signal Processing Conference*, Grenoble, 875-878.

N. Ranganathan and S. Henriques [1991]. "A Systolic Architecture for LZ Based Decompression", *Proceedings IEEE Data Compression Conference*, Snowbird, Utah, 450.

S. Ranka and S. Sahni [1990]. *Hypercube Algorithms with Applications to Image Processing and Pattern Recognition*, Springer-Verlag.

K. R. Rao and P. Yip [1990]. *Discrete Cosine Transform - Algorithms, Advantages, Applications*, Academic Press, London.

A. Razavi [1992]. "A New VLSI Image Codec for Digital Still Video Camera Applications", *Proceedings IEEE Data Compression Conference*, Snowbird, Utah, 422.

C. J. Read, D. M. Chabries, R. W. Christiansen, and J. K. Flanagan [1989]. "A method for computing the DFT of vector quantized data", *Proceedings of the ICASSP*, 1015-1018.

T. R. Reed, V. R. Algazi, G. E. Ford, and I. Hussain [1992]. "Perceptually Based Coding of Monochrome and Color Still Images", *Proceedings IEEE Data Compression Conference*, Snowbird, Utah, 142-151.

H. Reeve and J. S. Lim [1984]. "Reduction of blocking effects in image coding", *Optical Engineering* 23:1, 34-37.

A. P. Reeves [1984]. "Parallel Computer Architectures for Image Processing", *Computer Vision, Graphics, and Image Processing 25*, 68-88.

H. K. Reghbati [1981]. "An Overview of Data Compression Techniques", *IEEE Computer 14:4*, 71-75.

J. H. Reif A. Yoshida [1992]. "Optical Techniques for Image Compression", *Proceedings IEEE Data Compression Conference*, Snowbird, Utah, 32-41; also in *Image and Text Compression* (J. A. Storer, editor), Kluwer academic Press, Norwell, MA.

D. Revuz and M. Zipstein [1992]. "DZ: A Universal Compression Algorithm Specialized in Text", *Proceedings IEEE Data Compression Conference*, Snowbird, Utah, 394.

F. M. Reza [1951]. *Introduction to Information Theory*, McGraw Hill.

R. F. Rice and J. R. Plaunt [1971]. "Adaptive variable-length coding for efficient compression of spacecraft television data", *IEEE Transactions on Comm. Tech. COM-19*, 889-897.

K. D. Rines and N. C. Gallagher, Jr. [1982]. "The design of two-dimensional quantizers using pre-quantization", *IEEE Transactions on Information Theory* IT-28, 232-238.

E. A. Riskin [1990]. "Variable Rate Vector Quantization of Images", *Ph. D. Dissertation*, Stanford University.

E. A. Riskin [1991]. "Optimal bit allocation via the generalized BFOS algorithm", *IEEE Trans. Inform. Theory 37*, 400-402.

E. A. Riskin and R. M. Gray [1991]. "A greedy tree growing algorithm for the design of variable rate vector quantizers", *IEEE Trans. Signal Process.*

E. A. Riskin, T. Lookabaugh, P. A. Chou, and R. M. Gray [1990]. "Variable vector quantization for medical image compression", *IEEE Transactions on Medical Imaging 9*, 290-298.

J. Rissanen [1976]. "Generalized Kraft Inequality and Arithmetic Coding", *IBM Journal of Research and Development* 20, 198-203.

J. Rissanen [1978]. "Modelling by shortest data description", *Automatica 14*, 465-471.

J. J. Rissanen [1979]. "Arithmetic coding as number representations", *Acta Polytech Scand. Math 31*, 44-51.

J. Rissanen [1982]. "Optimum Block Models with Fixed-Length Coding", Technical Report, IBM Research Center, San Jose, CA.

J. Rissanen [1983]. "A Universal Data Compression System", *IEEE Transactions on Information Theory* 29:5, 656-664.

J. Rissanen [1983b]. "A universal prior for integers and estimation by minimum description length", *Ann. Statist. 11*, 416-432.

J. Rissanen [1986]. "Complexity of strings in the class of Markov sources", *IEEE Trans. Infor. Theory IT-32*, 526-532.

J. Rissanen [1986b]. "Stochastic complexity and modeling", *Annals of Statistics 14*, 1080-1100.

J. Rissanen and G. Langdon [1979]. "Arithmetic coding", *IBM J. Res. Dev. 23*, 149-162.

J. Rissanen and G. G. Langdon [1981]. "Universal Modeling and Coding", *IEEE Transactions on Information Theory* 27:1, 12-23.

L. G. Roberts [1962]. "Picture Coding Using Pseudo-Random Noise", *Applications of Walsh Functions Symposium Proceedings*, Washington, DC, 229-234.

E. L. Robertson [1977]. "Code Generation for Short/Long Adress Machines", Technical Report 1779, Mathematics Research Center, University of Wisconsin, Madison, Wisconsin.

J. A. Robinson [1986]. "Low data-rate visual communication using cartoons: a comparison of data compression techniques", *IEEE Proceedings 133:3*, 236-256.

J. A. Robinson [1991]. "Compression of Natural Images Using Thread-like Visual Primitives", *Proceedings IEEE Data Compression Conference*, Snowbird, Utah, 307-312.

M. Rodeh, V. R. Pratt, and S. Even [1980]. "Linear Algorithms for Compression Via String Matching", *Journal of the ACM* 28:1, 16-24.

G. M. Roe [1964]. "Quantizing for minimum distortion", *IEEE Transactions on Information Theory* IT-10:4, 384-385.

M. Rollins and F. Carden [1992]. "Possible Harmonic-Wavelet Hybrids in Image Compression", *Proceedings IEEE Data Compression Conference*, Snowbird, Utah, 191-199.

H. C. Romesburg [1990]. *Cluster Analysis for Researchers*, Robert E. Krieger Publishers, Malabar, Florida.

P. Roos, M. A. Viergever, M. C. A. Van Dijke, and J. H. Peters [1988]. "Reversible Intraframe Compression of Medical Images", *IEEE Trans. Medical Imaging 7:4*, 328-336.

H. Rosche [1992]. "Reducing Information Loss during spatial Compression of Digital Map Image Data", *Proceedings IEEE Data Compression Conference*, Snowbird, Utah, 430.

K. M. Rose and A. Heiman [1989]. "Enhancement of One-Dimensional Variable-Length DPCM Images Corrupted by Transimission Errors", *IEEE Trans. Commun. 37*, 373-379.

A. Rosen [1987]. "Colormap: A Color Image Quantizer", *Technical Report 87-845*, Dept. of Computer Science, Cornell University, Ithaca, NY.

A. Rosenfeld [1982]. *Digital Picture Processing*, Academic Press, New York, NY.

M. C. Rost and K. Sayood [1988]. "The root lattices as low bit rate vector quantizers", *IEEE Transactions on Information Theory* 34:5, 1053-1058.

D. Rotem and Y. Y. Zeevi [1986]. "Image Reconstruction from Zero Crossings", *IEEE Trans. on Acoustics, Speech, and Signal Processing ASSP-34:5.*

S. Roucos, R. Schwartz, and J. Makhoul [1982]. "Segment quantization for very-low-rate speech coding", *Proc. IEEE Int. Conf. Acoust., Speech, Signal Processing*, 1565-1569.

I. Rubin [1976]. "Data compression for Communication Networks: The Delay-Distortion Function", *IEEE Transactions on Information Theory* 22:6, 655-665.

F. Rubin [1976]. "Experiments in Text File Compression", *Communications of the ACM* 19:11, 617-623.

F. Rubin [1979]. "Arithmetic Stream Coding Using Fixed Precision Registers", *IEEE Trans. Information Theory IT-25:6*, 672-675.

R. Rubinstein [1986]. "A Note on Sets with Small Generalized Kolmogorov Complexity", *Technical Report TR86-4*, Iowa State University.

S. S. Ruth and P. J. Kreutzer [1972]. "Data Compression for Large Business Files", *Datamation* 18:9, 62-66.

B. Y. Ryabko [1980]. "Data Compression by means of a 'Book Stack"', *Problemy Peredachi Informatsii 16:4*.

B. Y. Ryabko [1987]. "A Locally Adaptive Data Compression Scheme", *Communications of the ACM 30:9*, 792.

M. J. Sabin [1989]. "Fixed-shape adaptive-gain vector quantization for speech waveform coding", *Speech Communication 8*, 177-183.

M. J. Sabin and R. M. Gray [1984]. "Product code vector quantizers for waveform and voice coding", *IEEE Trans. on Asoustics, Speech, and Signal Processing* ASSP-32:3, 474-488.

M. J. Sabin and R. M. Gray [1984]. "Product code vector quantizers for waveform and voice coding", *IEEE Trans. Acoust. Speech Signal Process. ASSP-32*, 474-488.

M. J. Sabin and R. M. Gray [1986]. "Global convergence and empirical consistency of the generalized Lloyd algorithm", *IEEE Transactions on Information Theory.*

S. Sabri [1984]. "Movement compensated interframe prediction for NTSC color TV signals", *IEEE Transactions on Communications* COM-32:8, 954-968.

I. Sadeh [1992]. "On Digital Data Compression - The Asymptotic Large Deviations Approach", *Proceedings IEEE Data Compression Conference*, Snowbird, Utah, 392.

I. Sadeh and A. Averbuch [1992]. "Image Encoding by Polynomial Approximation", *Proceedings IEEE Data Compression Conference*, Snowbird, Utah, 408.

R. J. Safranek and J. D. Johnson [1989]. "A perceptually tuned subband image coder with image dependent quantization and post-quantization data compression", *Proc. IEEE ASSP89 3*, 1945-1948.

D. J. Sakrison [1968]. *Communication Theory: Transmission of Waveforms and Digital Information*, Wiley, NY.

E. Salari and W. A. Whyte, Jr. [1991]. "Compression of Stereoscopic Image Data", *Proceedings IEEE Data Compression Conference*, Snowbird, Utah, 425.

K. Salem [1992]. "MR-CDF: Managing Multi-Resolution Scientific Data", *Proceedings IEEE Data Compression Conference*, Snowbird, Utah, 419.

H. Samet [1983]. "A Quadtree Medial Axis Transform", *Communications of the ACM* 26:9, 680-693.

M. Samet [1984]. "The quadtree and related hierarchical data structures", *Computing Surveys 16:2*, 187-230.

H. Samet [1985]. "Data Structures for Quadtree Approximation and Compression", *Communications of the ACM* 28:9, 973-1004.

D. Sankoff and J. B. Kruskal, editors [1983]. *An Overview of Sequence Comparison:* Time Warps, String Edits, and Macromolecules, Addison-Wesley, Reading, MA.

S. A. Savari and R. G. Gallager [1992]. "Arithmetic Coding for Memoryless Cost Channels", *Proceedings IEEE Data Compression Conference*, Snowbird, Utah, 92-101.

K. Sayood, J. D. Gibson, and M. C. Rost [1984]. "An algorithm for uniform vector quantizer design", *IEEE Transactions on Information Theory* IT-30, 805-814.

J. P. M. Schalwijk [1972]. "An Algorithm for Source Coding", *IEEE Transactions on Information Theory* 18, 395-399.

G. Schay Jr. and F. W. Dauer [1957]. "A Probabilistic Model of a Self-Organizing File System", *SIAM Journal on Applied Mathematics* 15, 874-888.

W. D. Schieber and G. W. Thomas [1971]. "An Algorithm for Compaction of Alphanumeric Data", *J. Library Automation 4*, 198-206.

R. W. Scheifler [1977]. "An Analysis of In-Line Substitution for a Structured Programming Language", *Communications of the ACM* 20:9, 647-654.

J. R. Schiess and K. Stacy [1992]. "Compression of Object Boundaries in Digital Images by Using Fractional Brownian Motion", *Proceedings IEEE Data Compression Conference*, Snowbird, Utah, 387.

C. P. Schnorr [1974]. "Optimal Enumerations and Optimal Godel Numberings", *Mathematical Systems Theory* 8:2, 182-190.

O. S. Schoepke [1992]. "Executing Compressed Code: A New Approach", *Proceedings IEEE Data Compression Conference*, Snowbird, Utah, 399.

W. F. Schreiber [1956]. "The Measurement of Third Order Probability Distributions of Television Signals", *IRE Trans. on Info. Theory IT-2*, 94-105.

W. F. Schreiber, C. F. Knapp, and D. Kay [1959]. "Synthetic Highs: An experimential TV band-width reduction system", *Journal of the SMPTE 68*, 525-547.

M. R. Schroeder [1966]. "VOCODERS: Analysis and Synthesis of Speech", *Proc. IEEE 54:5*, 720-734.

L. Schuchman [1964]. "Dither Signals and their Effects on Quantization Noise", *IEEE Transactions on Communication Technology COM-12*, 162-165.

E. F. Schuegraf and H.S. Heaps [1973]. "Selection of Equifrequent Word Fragments for Information Retrieval", *Inform. Stor. Retr. 9*, 697-711.

E. F. Schuegraf and H.S. Heaps [1974]. "A Comparison of Algorithms for Data Base Compression by the Use of Fragments as Language Elements", *Inform. Stor. Retr. 10*, 309-319.

E. S. Schwartz [1963]. "A Dictionary for Minimum Redundancy Eroding", *Journal of the Association for Computing Machinery 10*, 413-439.

E. S. Schwartz [1964]. "An Optimum Encoding with Minimum Longest Code and Total Number of Digits", *Information and Control* 7, 37-44.

E. S. Schwartz and B. Kallick [1964]. "Generating a Canonical Prefix Encoding", *Communications of the ACM* 7:3, 166-169.

Schwartz, A. J. Kleiboemer [1967]. "A Language Element for Compression Coding", *Information and Control* 10, 315-333.

A. J. Scott and M. J. Symons [1971]. "Clustering methods based on likelihood ratio criteria", *Biometrics* 27, 387-397.

J. B. Seery and J. Ziv [1977]. "A Universal Data Compression Algorithm: Description and Preliminary Results", *Technical Memorandum 77-1212-6*, Bell Laboratories, Murray Hill, N.J.

J. B. Seery and J. Ziv [1978]. "Further Results on Universal Data Compression", *Technical Memorandum 78-1212-8*, Bell Laboratories, Murray Hill, N.J.

J. Seiferas [1977]. "Subword Trees", lecture notes, Pennsylvania State University, University Park, PA.

S. Z. Selim and M. A. Ismail [1984]. "K-means-type algorithms: a generalized convergence theorem and characterization of local optimality", *IEEE Trans. on Pattern Analysis and Machine Intell.* PAMI-6, 81-87.

S. C. Seth [1976]. "Data Compression Techniques in Logic Testing: An Extension of Transition Counts", *Journal of Design Automation and Fault-Tolerant Computing* 1:2, 317-320.

D. G. Severance [1983]. "A Practitioner's Guide to Database Compression", *Information Systems* 8:1, 51-62.

I. Shah, O. Akimwumi-Assani, and B. Johnson [1991]. "A Chip Set for Lossless Image Compression", *IEEE J. Solie-State Circuits SC-26:3*, 237-244.

C. E. Shannon [1948]. "A mathematical theory of communication", *Bell Syst. Tech. J.* 27, 379-423, 623-656; also in *The Mathematical Theory of Communication*, C. E. Shannon and W. Weaver, University of Illinois Press, Urbana, IL (1949).

C. E. Shannon [1951]. "Prediction of Entropy of Printed English Text", *Bell System Technical Journal* 30, 50-64; also in *Key Papers in the Development of Information Theory* (D. Slepian, editor), IEEE Press, New York, NY (1973).

C. E. Shannon [1959]. "Coding Theorems for a Discrete Source with a Fidelity Criterion", *Proceedings IRE National Conference*, 142-163; also in *Key Papers in the Development of Information Theory* (D. Slepian, editor), IEEE Press, New York, NY (1973).

C. E. Shannon and W. Weaver [1949]. *The Mathematical Theory of Communication*, University of Illinois Press, Urbana, IL.

S. D. Shapiro [1980]. "Use of the Hough Transform for Image Data Compression", *Pattern Recognition 12*, 333-337.

D. K. Sharma [1978]. "Design of absolutely optimal quantizers for a wide class of distortion measures", *IEEE Transactions on Information Theory* IT-24:6, 693-702.

D. Sheinwald [1992]. "On Alphabetical Binary Codes", *Proceedings IEEE Data Compression Conference*, Snowbird, Utah, 112-121.

D. Sheinwald, A. Lempel, and J. Ziv [1990]. "Two-dimensional encoding by finite state encoders", *IEEE T-COM 38*, 341-347.

D. Sheinwald, A. Lempel, and J. Ziv [1991]. "On Compression with Two-Way Head Machines", *Proceedings IEEE Data Compression Conference*, Snowbird, Utah, 218-227.

D. Sheinwald "Finite State Two-Dimensional Compressibility", *Image and Text Compression*, Kluwer academic Press, Norwell, MA.

L. Shepp, S. Slepian, and A. D. Wyner [1980]. "On prediction of moving average processes", *Bell Syst. Tech. J.* , 367-415.

J. S. Shiau, J. C. Tseng, and W. P. Yang [1992]. "A New compression-Based Signature Extraction Method", *Proceedings IEEE Data Compression Conference*, Snowbird, Utah, 420.

Y. Shoham and A. Gersho [1984]. "Pitch synchronous transform coding of speech at 9. 6kb/s based on vector quantization", *Proceedings Conference on Communications*, Amsterdam, 1179-1182.

Y. Shoham and A. Gersho [1985]. "Efficient codebook allocation for an arbitrary set of vector quantizers", *Proceedings International Conference on Acoustics, Speech, and Signal Processing*, 1696-1699.

J. E. Shore, D. Burton, and J. Buck [1983]. "A generatization of isolated word recognition using vector quantization", *Proc. IEEE Int. Conf. Acoust., Speech, Signal Processing*, 1021-1024.

F. Sijstermans and Jan van der Meer [1991]. "CD-I full-motion video encoding on a parallel computer", *Communications of the ACM* 34:4, 81-91.

M. Sipser [1983]. "A Complexity Theoretic Approach to Randomness", Proceedings Fifteenth Annual ACM Symposium on the Theory of Computing, 330-335.

G. Siromoney [1963]. "Entropy of Tamil Prose", *Information and Control 6*, 297-300.

V. S. Sitaram and C. M. Huang [1992]. "Efficient Codebooks for Vector Quantization Image Compression", *Proceedings IEEE Data Compression Conference*, Snowbird, Utah, 396.

D. Slepian, editor [1973]. *Key Papers in the Development of Information Theory*, IEEE Press, New York, NY.

A. D. Sloan [1992]. "Fractal Image Compression: A Resolution Independent Representation for Imagery", *Proceedings IEEE Data Compression Conference*, Snowbird, Utah, 401.

D. L. Slotnick, W. C. Bork, and R. C. McReynolds [1962]. "The Solomon Computer", *Proc. AFIPS Fall Joint Computer Conference*, Spartan Books, Washington, DC, 97-107.

B. Smith [1957]. "Instantaneous companding of quantized signals", *Bell Syst. Tech. J.* 36, 653-709.

C. P. Smith [1969]. "Perception of vocoder speech processed by pattern matching", *J. Acoust. Soc. Amer.* 46:6 (pt. 2), 1562-1571.

A. J. Smith [1976]. "A Quening Network Model for the Effect of Data Compression on System Efficiency", *AFIPS 45*, 457-465.

M. Snyderman and B. Hunt [1970]. "The Myriad Virtues of Text Compaction", *Datamation*, 36-40.

J. Soo Ma [1978]. *Data Compression.* Ph.D. Dissertation, University of Massachusetts.

G. Southworth [1977]. "Single Frame Digital Television Transmission", *Proceedings National Telemetering Conference*, 41:3. 1-41:3. 2.

H. Spath [1980]. *Cluster Analysis Algorithms for Data Reduction and Classification of Objects*, Wiley.

R. F. Sproull [1991]. "Refinements to Nearest-Neighbor Searching in k-d Trees", *Algorithmica 6*, 579-589.

R. F. Sproull and I. E. Sutherland [1992]. "A Comparison of Codebook Generation Techniques for Vector Quantization", *Proceedings IEEE Data Compression Conference*, Snowbird, Utah, 122-131.

V. S. Srinivasan and L. Chandrasekhar [1992]. "Image Data Compression with Adaptive Discrete Cosine Transform", *Proceedings IEEE Data Compression Conference*, Snowbird, Utah, 407.

R. Srinivasan and K. R. Rao [1984]. "Predictive Coding Based on Efficient Motion Estimation", *ICC Proceedings.*

A. B. Sripad and D. L. Snyder [1977]. "A Necessary and Sufficient Condition for Quantization Errors to be Uniform and White", *IEEE Trans. on ASSP ASSP-25*, 442-448.

W. Stallings [1985]. *Data and Computer Communications*, MacMillian Publishing Co., New York, NY.

L. M. Stauffer and D. S. Hirschberg [1992]. "Transpose Coding on the Systolic Array", *Proceedings IEEE Data Compression Conference*, Snowbird, Utah, 162-171.

S. D. Stearns [1991]. "Waveform Data Compression With Exact Recovery", *Proceedings IEEE Data Compression Conference*, Snowbird, Utah, 464.

S. D. Stearns and D. R. Hush [1990]. *Digital Signal Analysis*, Prentice-Hall.

R. J. Stevens, A. F. Lehar, and F. H. Preston [1983]. "Manipulation and Presentation of Multidimensional Image Data Using the Peano Scan", *IEEE Transactions on Pattern Analysis and Machine Intelligence* 5:5, 520-526.

L. C. Stewart, R. M. Gray, and Y. Linde [1982]. "The design of trellis waveform coders", *IEEE Transactions on Communications IT-20*, 702-710.

J. J. Stiffler [1971]. *Theory of Synchronous Communications*, Prentice-Hall, Englewood Cliffs, NJ.

T. G. Stockham, Jr. [1972]. "Image processing in the context of a visual model", *Proc. of the IEEE* 60:7, 828-842.

J. C. Stoffel and J. F. Moreland [1981]. "A Survey of Electronic Techniques for Pictorial Image Reproduction", *IEEE Trans. on Comm. , COM-29:12*, 1898-1925.

J. A. Storer [1977]. "NP-Completeness Results Concerning Data Compression", Technical Report 234, Dept. of Electrical Engineering and Computer Science, Princeton University, Princeton, NJ.

J. A. Storer [1979]. "Data Compression: Methods and Complexity Issues", Ph.D. Thesis, Dept. of Electrical Engineering and Computer Science, Princeton University Princeton, NJ.

J. A. Storer [1982]. "Combining Trees and Pipes in VLSI", Technical Report CS-82-107, Dept. of Computer Science, Brandeis University, Waltham, MA.

J. A. Storer [1983]. "An Abstract Theory of Data Compression", *Theoretical Computer Science* 24 221-237; see also "Toward an Abstract Theory of Data Compression", *Proceedings Twelfth Annual Conference on Information Sciences and Systems,* The Johns Hopkins University, Baltimore, MD, 391-399 (1978).

J. A. Storer [1985]. "Textual Substitution Techniques for Data Compression", *Combinatorial Algorithms on Words,* Springer-Verlag (A. Apostolico and Z. Galil editors), 111-129.

J. A. Storer [1988]. *Data Compression: Methods and Theory,* Rockville, MD: Computer Science Press.

J. A. Storer [1988b]. "Parallel Algorithms for On-Line Dynamic Data Compression", *IEEE International Conference on Communications* (ICC), Philadelphia, PA.

J. S. Storer [1989]. "System for Dynamically Compressing and Decompressing Electronic Data", United States Patent 4,876,541 (55 claims).

J. Storer [1990]. "Lossy On-Line Dynamic Data Compression", *Combinatorics, Compression, Security and Transmission,* Springer-Verlag, 348-357.

J. A. Storer [1991]. "Massively Parallel System for High Speed Data Compression", patent pending.

J. A. Storer [1991]. "The Worst-Case Performance of Adaptive Huffman Codes", *Proceedings IEEE Symposium on Information Theory,* Budapest, Hungary, 1991.

J. A. Storer [1992]. "Data Compression Bibliography", *Image and Text Compression"*, Kluwer Academic Press, Norwell, MA.

J. A. Storer [1992]. "Massively Parallel Systolic Algorithms for Real Time Dictionary Based Text Compression", *Image and Text Compression,* Kluwer Academic Press, Norwell, MA.

J. A. Storer, editor [1992]. *Image and Text Compression,* Kluwer Academic Press, Norwell, MA.

J. A. Storer and M. Cohn, editors [1992]. *Proceedings IEEE Data Compression Conference,* Sponsored by the IEEE Computer Society in Cooperation with NASA/CESDIS, Snowbird, Utah.

J. A. Storer and J. H. Reif [1988]. "Real-Time Dynamic Compression of Video on a Grid-Connected Parallel Computer", Third International Conference on Super-Computing, Boston, MA, May 15-20.

J. Storer and J. H. Reif [1991]. "Adaptive Lossless Data Compression over a Noisy Channel", *Proceedings Communication, Security, and Sequences Conference,* Positano, Italy.

J. A. Storer and J. H. Reif [1991]. "A Parallel Architecture for High Speed Data Compression", *Journal of Parallel and Distributed Computing* 13, 222-227.

J. A. Storer and J. H. Reif, editors [1991]. *Proceedings IEEE Data Compression Conference,* Sponsored by the IEEE Computer Society in Cooperation with NASA/CESDIS, Snowbird, Utah.

J. A. Storer and T. G. Szymanski [1978]. "The Macro Model for Data Compression", *Proceedings Tenth Annual ACM Symposium on the Theory of Computing,* San Diego, CA, 30-39.

J. A. Storer and T. G. Szymanski [1982]. "Data Compression Via Textual Substitution", *Journal of the ACM* 29:4 928-951.

J. A. Storer, J. H. Reif, and T. Markas "A Massively Parallel VLSI Design for Data Compression using a Compact Dynamic Dictionary", *Proceedings IEEE VLSI Signal Processing Conference,* San Diego, CA.

G. Strang [1989]. "Wavelets and dilation equations - a brief introduction", *SIAM Review 31:4,* 614-627.

C. Y. Suen [1979]. "N-Gram Statistics for Natural Language Understanding and Text Processing", *IEEE Transactions on Pattern Analysis and Machine Intelligence* 1:2, 164-172.

M. I. Svanks [1975]. "Optimizing the Storage of Alphanumeric Data", *Canadian Datasystems,* 38-40.

W. Szpankowski [1991]. "A Typical Behaviour of Some Data Compression Schemes", *Proceedings IEEE Data Compression Conference,* Snowbird, Utah, 247-256.

T. G. Szymanski [1976]. "Assembling Code for Machines with Span Dependent Instructions", *Technical Report 224,* Dept. of Electrical Engineering and Computer Science, Princeton University, Princeton, NJ.

C. P. Tan [1981]. "On the Entropy of the Malay Language", *IEEE Trans. Information Theory IT-27:3,* 383-384.

H. Tanaka and A. Leon-Garcia [1982]. "Efficient Run-Length Encodings", *IEEE Transactions on Information Theory 28:6*, 880-890.

B. P. M. Tao, R. M. Gray, and H. Abut [1984]. "Hardware realization of waveform vector quantizers", *IEEE J. Selected Areas in Commun. SAC-2*, 343-352. J. T. Tou and R. C. Gonzales [1974]. *Pattern Recognition Principles*, Addison-Wesley, Reading, MA.

J. Tarhio and E. Ukkhonen [1988]. "A greedy approximation algorithm for constructing shortest common superstrings", *Theoretical Computer Science 53*.

M. Tasto and P. A. Wintz [1971]. "Image coding by adaptive block quantization", *IEEE Trans. Commun. Technol.* COM-19:6, 956-972.

R. Techo [1980]. *Data Communications*, Plenum Press, New York, NY.

S. H. Teng [1987]. "The Construction of Huffman-Equivalent Prefix Code in NC", *ACM SIGACT News* 18:4, 54-61.

A. G. Tescher [1979]. "Transform Image Coding", *Image Transmission Techniques* (W. K. Pratt, editor), Academic Press, New York, NY.

J. Teuhola [1978]. "A Compression Method for Clustered Bit-Vectors", *Information Processing Letters 7:6*, 308-311.

J. Teuhola and T. Raita [1991]. "Piecewise Arithmetic Coding", *Proceedings IEEE Data Compression Conference*, Snowbird, Utah, 33-42.

C. W. Therrien [1989]. *Decision Estimation and Classification*, Wiley.

L. H. Thiel and H.S. Heaps [1972]. "Program Design for Retrospective Searches on Large Data Bases", *Inform. Stor. Retr. 8*, 1-20.

C. Thomborson [1991]. "V. 42bis and Other Ziv-Lempel Variants", *Proceedings IEEE Data Compression Conference*, Snowbird, Utah, 460.

C. D. Thomborson and W. Y. B. Wei [1989]. "Systolic Implementations of a Move-to-Front Text Compressor", *Proc. 1989 ACM Symp. on Parallel Algorithms and Architectures*, 283-290.

J. C. Tilton [1989]. "Image segmentation by iterative parallel region growing and splitting", *Proceedings 1989 International Geoscience and Remote Sensing Symposium*, 2420-2423.

J. C. Tilton and M. Manohar [1990]. "Hierarchical data compression: Integrated browse, moderate loss, and lossless levels of data compression", *Proceedings 1990 International Geoscience and Remote Sensing Symposium*, 1655-1658.

J. C. Tilton, D. Han, M. Manohar [1991]. "Compression Experiments with AVHRR Data", *Proceedings IEEE Data Compression Conference*, Snowbird, Utah, 411-420.

M. Tinker [1989]. "DVI parallel image compression", *Communications of the ACM* 32:7, 844-851.

P. Tischer [1987]. "A Modified Lempel-Ziv-Welch Data Compression Scheme", *Australian Computer Science Communications 9:1*, 262-272.

S. J. P. Todd, G. G. Langdon, and J. J. Rissanen [1985]. "Parameter Reduction and Context Selection for Compression of Gray-Scale Images", *IBM Journal of Research and Development 29:2*, 188-193.

K. L. Tong and M. W. Marcellin [1991]. "Transform Coding of Monochrome and Color Images Using Trellis Coded Quantization", *Proceedings IEEE Data Compression Conference*, Snowbird, Utah, 454.

H. H. Torbey and H. E. Meadows [1989]. "System for lossless digital image compression", *SPIE, Visual Communication and Image Processing.*

P. J. Tourtier [1989]. "A Compatible Approach on Digital HDTV Coding Based on Subband Techniques", *Proceedings Third International Workshop on HDTV*, Torino, Italy.

A. Tran, K. M. Liu, K. H. Tzou, and E. Vogel [1987]. "An efficient pyramid image coding system", *Proceedings ICASSP*, 18. 6. 1-18. 6. 4.

D. Tran, B. W. Wei, and M. Desai [1992]. "A Single-Chip Data Compression - Decompression Processor", *Proceedings IEEE Data Compression Conference*, Snowbird, Utah, 433.

T. Tremain [1982]. "The government standard linear predictive coding algorithm: LPC-10", *IEEE Trans. Acoust. Speech Signal Process. ASSP-36:9*, 40-49.

R. Tropper [1982]. "Binary-Coded Text: A Text-Compression Method", *Byte Magazine* April issue.

A. V. Trushkin [1982]. "Sufficient conditions for uniqueness of a locally optimal quantizer for a class of convex error weighting functions", *IEEE Trans. Inform. Theory IT-28*, 187-198.

Y. T. Tsai [1991]. "Signal Processing and Compression for Image Capturing Systems Using A Single-Chip Sensor", *Proceedings IEEE Data Compression Conference*, Snowbird, Utah, 448.

B. P. Tunstall [1968]. "Synthesis of Noisless Compression Codes", *Ph.D. Thesis*, Georgia Institute of Technology.

L. F. Turner [1975]. "The On-Ground Compression of Satellite Data", *Computer Journal 18:3*, 243-7.

J. S. Turner [1986]. "The Complexity of the Shortest Common String Matching Problem", *Proceedings Allerton Conference*, Monicello, Il.

J. S. Turner [1986b]. "Approximation Algorithms for the Shortest Common Superstring Problem", *Technical Report WUCS-86-16*, Department of Computer Science, Washington University, Saint Louis, MO.

M. J. Turner [1992]. "Entropy Reduction via Simplified Image Contourization", *Proceedings IEEE Data Compression Conference*, Snowbird, Utah, 398.

E. Ukkonen [1985]. "Finding Approximate Patterns in Strings", *Journal of Algorithms* 6, 132-137.

C. K. Un and S. J. Lee [1983]. "Effect of Channel Errors on the Performance of LPC Vocoders", *IEEE Trans. on Acoustics, Speech, and Signal Processing, ASSP-31:1*, 234-237.

G. Ungerboeck [1982]. "Channel coding with multilevel/phase signals", *IEEE Trans. Inform. Theory IT-28*, 55-67.

H. Urrows and E. Urrows [1984]. "Laser Data and other Data Disks: The Race to Store and Retrieve with Optics", *Videodisc and Optical Disk* 4:2, 130.

M. J. Usher [1984]. *Information Theory for Information Technologists*, Macmillan Publishers, London and Basingstoke.

A. V Goldberg and M. Sipser [1985]. "Compression and Ranking", *Proceedings Seventeenth Annual ACM Symposium on the Theory of Computing*, Providence, RI, 440-448.

J. Vaisey and A. Gersho [1988]. "Simulated Annealing and Codebook Design", *Proc. IEEE ICASSP*, 1176-1179.

J. Vaisey and A. Gersho [1992]. "Image Compression with variable block size segmentation", *IEEE Signal Process. ASSP-40.*

K. Van Houton and P. W. Oman [1991]. "An Algorithm for Tree Structure Compression", *Proceedings IEEE Data Compression Conference*, Snowbird, Utah, 424.

C. J. Van Rijsbergen [1970]. "A clustering algorithm (Algorithm 47)", *Computer Journal* 13, 113-115.

C. J. Van Rijsbergen [1970b]. "A fast hierarchic clustering algorithm (Algorithm 52)", *Computer Journal* 13, 324-326.

L. Vandendorpe and B. Macq [1990]. "Optimum quality and progressive resolution of video signals", *Annales des Télécommunications*, 487-502.

B. Varn [1971]. "Optimal Variable Length Codes (Arbitrary Symbol Costs and Equal Code Word Probability)", *Information and Control 19*, 289-301.

N. D. Vasyukova [1977]. "On the Compact Representation of Information", *Mathematika i Kibernetika* 4, 90-93.

S. J. Vaughan-Nichols [1990]. "Saving Space", *Byte*, March, 237-243.

J. Venbrux and N. Liu [1991]. "A Very High Speed Lossless Compression - Decompression Chip Set", *Proceedings IEEE Data Compression Conference*, Snowbird, Utah, 461.

W. Verbiest and L. Pinnoo [1989]. "A variable bit rate video codec for asynchronous transfer mode networks", *IEEE Journal on Selected Areas in Communications 7:5*, 761-770.

W. Verbiest, L. Pinnoo, and B. Voeten [1988]. "The impact of the ATM concept on video coding", *IEEE J. Selected Areas of Commun.*.

J. Verhoeff [1977]. "A New Data Compression Technique", *Ann. Sys. Res. 6*, 139-148.

M. Vetterli [1984]. "Mulit-dimensional sub-band coding: some theory and algorithms", *Signal Processing 6:2*, 97-112.

M. Vetterli and A. Ligtenberg [1986]. "A discrete fourier-cosine transform chip", *IEEE Journal on Selected Areas in Communications SAC 4:1*.

U. Vishkin [1985]. "Optimal Parallel Pattern Matching in Strings", *Proceedings Twelfth ICALP, Lecture Notes in Computer Science* 194, Springer-Verlag, 497-508; also in *Information and Control.*

M. Visvalingam [1976]. "Indexing with Coded Deltas: A Data Compaction Technique", *Software Practice and Experience* 6, 397-403.

A. J. Viterbi and J. K. Omura [1974]. "Trellis encoding of memoryless discrete-time sources with a fidelity criterion", *IEEE Transactions on Information Theory IT-20*, 325-332.

A. J. Viterbi and J. K. Omura [1979]. *Principles of Digital Communication and Coding*, McGraw-Hill Book, NY.

J. S. Vitter [1987]. "Design and analysis of dynamic Huffman codes", *Journal of the ACM 34*, 825-845.

J. S. Vitter [1989]. "Dynamic Huffman coding", *ACM Trans. Math. Software 15*, 158-167.

V. A. Vittikh [1973]. "Synthesis of Algorithms for Data Compression", *Problems of Control and Information Theory* 2:3-4, 235-241.

R. A. Wagner [1973]. "Common Phrases and Minimum-Space Text Storage", *Communications of the ACM* 16:3, 148-152.

E. Walaach and E. Karnin. "A fractal based approach to image compression", *Proc. ICASSP 86*, 529-532.

G. Wallace [1990]. "Overview of the jpeg (iso/ccitt) Still Image Compression Standard", *SPIE Image Processing Algorithms and Techniques 1244*, 220-223.

G. K. Wallace [1991]. "The JPEG: Still Picture Compression Standard", *Communications of the ACM 34:4*, 31-44.

C. S. Wallace and D. M. Boulton [1969]. "An information measure for classification", *Computer* :, 185-194.

P. Wallich [1991]. "Wavelet Theory; An Analysis Technique that's Creating Ripples", *Scientific American 264:1*, 34-35.

M. A. Wanas, A. I. Zayed, M. M. Shaker, and E. H. Taha [1976]. "First, Second, and Third Order Entropies of Arabic Text", *IEEE Trans. Information Theory 22:1*, 123.

C. C. Wang and H. S. Don [1992]. "Robust Measures for Fuzzy Entropy and Fuzzy Conditioning", *Proceedings IEEE Data Compression Conference*, Snowbird, Utah, 431.

L. Wang and M. Goldberg [1989]. "Progressive image transmission using vector quantization on images in pyramid form", *IEEE Transactions on Communications* 37:12, 1339-1349.

Y. Wang and J. M. Wu [1992]. "Vector Run-length Coding of Bi-level Images", *Proceedings IEEE Data Compression Conference*, Snowbird, Utah, 289-298.

J. H. Ward [1963]. "Hierarchical grouping to optimize an objective function", *J. of the American Statistical Association* 58, 236-244.

O. Watanable [1986]. "Generalized Kolmogorov Complexity of Computations", manuscript.

M. J. Weinberger, A. Lempel, and J. Ziv [1992]. "On the Coding Delay of a General Coder", *Proceedings IEEE Data Compression Conference*, Snowbird, Utah, 102-111; also in *Image and Text Compression*, Kluwer academic Press, Norwell, MA.

M. Weinberger, J. Ziv, and A. Lempel [1991]. "On the Optimal Asymptotic Performance of Universal Ordering and Discrimination of Individual Sequences", *Proceedings IEEE Data Compression Conference*, Snowbird, Utah, 239-246.

P. Weiner [1973]. "Linear Pattern Matching Algorithms", *Proceedings Fourteenth Annual Symposium on Switching and Automata Theory*, 1-11.

L. Weiping [1991]. "Vector Transform and Image Coding", *IEEE Transactions on Circuits and Systems for Video Technology 1:4*.

S. F. Weiss and R. L. Vernor [1978]. "A Word-Based Compression Technique for Text Files", *Journal of Library Automation 11:2*, 97-105.

T. A. Welch [1984]. "A Technique for High-Performance Data Compression", *IEEE Computer* 17:6, 8-19.

E. J. Weldon, Jr. and W. W. Peterson [1971]. *Error Correcting Codes*, MIT Press, Cambridge, MA.

M. Wells [1972]. "File Compression Using Variable Length Encodings", *The Computer Journal 15:4*, 308-313.

R. L. Wessner [1976]. "Optimal Alphabetic Search Trees with Restricted Maximal Height", *Information Processing Letters* 4, 90-94.

P. H. Westerink, J. Biemond, and F. Muller [1990]. "Subband coding of images at low bit rates", *Image Commun.* 2:4, 441-448.

P. H. Westerink, D. E. Boekee, J. Biemond, and J. W. Woods [1988], "Subband coding of images using vector quantization", *IEEE Transactions on Communications* 36:6.

S. A. Weyer and A. H. Borning [1985]. "A Prototype El,ectronic Encyclopedia", *ACM Transactions on Office Information Systems* 3:1, 63-88.

N. Weyland and E. Puckett [1986]. "Optimal Binary Models for the Gaussian Source of Fixed Precision Numbers", *Technical Report*, Mitre Corporation, Bedford, MA.

H. E. White [1967]. "Printed English Compression by Dictionary Encoding", *Proc. Institute of Electical and Electronic Engineers 55:3*, 390-396.

R. L. White [1992]. "High Performance Compression of Astronomical Images", *Proceedings IEEE Data Compression Conference*, Snowbird, Utah, 403.

B. Widrow [1956]. "A study of rough amplitude quantization by means of the Nyquist sampling theorem", *Trans. IRE* CT-3, 266-276.

R. Williams [1988]. "Dynamic History Predictive Compression", *Information Systems 13:1*, 129-140.

R. N. Williams [1990]. *Adaptive Data Compression*, Kluwer Academic Press, Norwell, MA.

R. N. Williams [1991b]. "An Extremely Fast Ziv-Lempel Data Compression Algorithm", *Proceedings IEEE Data Compression Conference*, Snowbird, Utah, 362-371.

R. Wilson [1990]. "Image Compression Chips Advance on Three Fronts", *Computer Design.*

S. G. Wilson and D. W. Lytle [1982]. "Trellis encoding of continuous-amplitude memoryless sources", *IEEE Trans. Inform. Theory IT-28*, 211-226. J. W. Woods, editor [1991]. *Subband Image Coding*, Kluwer Academic Publishers, Boston.

P. A. Wintz [1972]. "Transform picture coding", *Proc. IEEE 60*, 809-820.

D. Wishart [1969]. "A algorithm for hierarchical classifications", *Biometrics* 25, 165-170.

I. H. Witten and T. C. Bell [1990]. "Source models for natural language text", *Int. J. Man-Machine Studies 32:5*, 545-579.

I. Witten and J. Cleary [1983]. "Picture coding and transmission using adaptive modelling of quad trees", *Proceedings International Electrical, Electronics Conference*, Toronto, 222-225.

I. H. Witten and J. G. Cleary [1988]. "On the Privacy Afforded by Adaptive Text Compression", *Computers and Security 7:4*, 397-408.

I. H. Witten, T. C. Bell, M. E. Harrison, M. L. James, and A. Moffat [1992]. "Textual Image Compression", *Proceedings IEEE Data Compression Conference*, Snowbird, Utah, 42-51.

I. H. Witten, T. C. Bell, and C. G. Nevill [1991]. "Models for Compression in Full-Text Retrieval Systems", *Proceedings IEEE Data Compression Conference*, Snowbird, Utah, 23-32.

I. H. Witten, R. M. Neal, and J. G. Cleary [1987]. "Arithmetic coding for data compression", *Communications of the ACM* 30:6, 520-540.

J. G. Wolff. "Recoding of the Natural Language for Economy of Transmission or Storage", *Computer Journal 21:1*, 42-44.

J. Wolfowitz [1960]. "On Channels in Which the Distribution of Error is Known Only to the Receiver or Only to the Sender", *Information and Decison Processes*, (R. E. Machol, Editor), 178-182.

M. E. Womble, J. S. Halliday, S. K. Mitter, M. C. Lancaster, and J. H. Triebwasser [1977]. "Data Compression for Storing and Transmitting ECG's and VCG's", *Proc. of the IEEE 65:5*, 702-706.

C. W. Wong [1991]. "Motion-Compensated Video Image Compression Using Luminance and Chrominance Components for Motion Estimation", *Proceedings IEEE Data Compression Conference*, Snowbird, Utah, 440.

C. W. Wong [1992]. "Color Video Compression Using Motion-Compensated Vector Quantization", *Proceedings IEEE Data Compression Conference*, Snowbird, Utah, 385.

K. L. Wong and R. K. L. Poon [1976]. "A Comment on the Entropy of the Chinese Language", *IEEE Trans. Acoustics, Speech, and Signal Processing*, 583-585.

D. Y. Wong, B. H. Juang, and A. H. Gray [1982]. "An 800 bit's Vector Quantization LPC Vocoder", *IEEE Trans. on Acoustics, Speech, and Signal Processing, ASSP-30:5*, 770-780.

A. Wong, C. T. Chen, D. J. Le Gall, F. C. Jeng, and K. M. Uz [1990]. "MCPIC: A video coding algorithm for transmission and storage applications", *IEEE Transactions on Communications.*

R. C. Wood [1969]. "On optimum quantization", *IEEE Transactions on Information Theory* IT-15:2, 248-252.

J. W. Woods [1991]. *Subband Image Coding*, Kluwer Academic Publishers.

J. W. Woods and S. D. O'Neil [1986]. "Subband coding of images", *IEEE Trans. on Asoustics, Speech, and Signal Processing* ASSP-34:5.

X. Wu [1990]. "A tree-structured locally optimal vector quantizer", *Proceedings Tenth International Conference on Pattern Recognition*, Atlantic City, NJ, 176-181.

X. Wu [1992]. "Vector Design by Constrained Global Optimization", *Proceedings IEEE Data Compression Conference*, Snowbird, Utah, 132-141.

X. Wu and Y. Fang [1992]. "Lossless Interframe Compression of Medical Images", *Proceedings IEEE Data Compression Conference*, Snowbird, Utah, 249-258.

S. W. Wu and A. Gersho [1991]. "Rate-constrained optimal block-adaptive coding for digital tape recording of hdtv", *IEEE Trans. Video Technology in Circuits and Systems 1:1*, 100-112.

X. Wu and J. Rokne [1989]. "An algorithm for optimum K-level quantization on histogram of N points", *Proceedings ACM 1989 Conference of Computer Science*, 339-343.

X. Wu and C. Yao [1991]. "Image Coding by Adaptive Tree-Structured Segmentation", *Proceedings IEEE Data Compression Conference*, Snowbird, Utah, 73-82.

X. Wu and K. Zhang [1991]. "A Better Tree-Structured Vector Quantizer", *Proceedings IEEE Data Compression Conference*, Snowbird, Utah, 392-401.

A. D. Wyner [1974]. "On the probability of buffer overflow under an arbitrary bounded input-output distribution", *SIAM J. Appl. Math 27*, 544-570.

A. Wyner and J. Ziv [1971]. "Bounds on the Rate-Distortion Function for Stationary Sources With Memory", *IEEE Transactions on Information Theory IT-17*, 508-513.

A. D. Wyner and J. Ziv [1989]. "Some asymptotic properties of the entropy of a stationary ergodic data source with applications to data compression", *IEEE Trans. Inform. Theory IT-35*, 1250-1258.

A. D. Wyner and J. Ziv [1991]. "Fixed Data Base Version of the Lempel-Ziv Data Compression Algorithm", *Proceedings IEEE Data Compression Conference*, Snowbird, Utah, 202-207.

Z. Xie and T. G. Stockham, Jr. [1989]. "Toward the unification of three visual laws and two visual models in brightness perception", *IEEE Trans. Systems, Man, and Cybernetics* 19:2, 379-387.

Z. Xie and T. G. Stockham, Jr. [1990]. "Previsualized image vector quantization with optimized pre- and post-processors", Technical Report, Dept. of Electrical Engineering, University of Utah, UT.

K. Xue and J. M. Crissey [1991]. "An Iteratively Interpolative Vector Quantization Algorithm for Image Data Compression", *Proceedings IEEE Data Compression Conference*, Snowbird, Utah, 139-148.

T. Yamada [1979]. "Edge Difference Coding- A New, Efficient Redundancy Reduction Technique for Facsimile Signals", *IEEE Trans. on Comm.* , *COM-27:8*, 1210-1217.

Y. Yamada, S. Tazaki, and R. M. Gray [1980]. "Asymptotic Performance of Block Quantizers with a Difference Distortion Measure", *IEEE Transactions on Information Theory IT-26*, 6-14.

K. M. Yang, L. Wu, and M. Mills [1990]. "Fractal based image coding scheme using peano scan", *Proc. ISCAS 99*, 2301-2304.

E. J. Yannakoudakis, P. Goyal, and J. A. Huggil [1982]. "The Generation and Use of Text Fragments for Data Compression", *Information Processing and Management* 18:1, 15-21.

A. Yariv, S. Kwong, and K. Kyuma [1986]. "Demonstration of an all-optical associative holographic memory", *Appl. Phy. Lett. 48*, 1114-1116.

M. Yau and S. N. Srihari [1983]. "A Hierarchical Data Structure for Multidimensional Digital Images", *Communications of the ACM* 26:7, 504-515.

R. Yeung [1991]. "Local redundancy and progressive bounds on the redundancy of a Huffman code", *IEEE Trans. Inform. Theory IT-37:3*, 687-691.

Z. M. Yin, K. T. Huang, R. B. Shu, and C. K. Chui [1992]. "Efficient Data Coding and Algorithms for Volume Compression", *Proceedings IEEE Data Compression Conference*, Snowbird, Utah, 397.

J. F. Young [1971]. *Information Theory*, Wiley Interscience, New York, NY.

T. Y. Young and K. S. Fu [1986]. *Handbook of Pattern Recognition and Image Processing*, Academic Press, New York.

T. Y. Young and P. S. Liu [1980]. "Overhead Storage Considerations and a Multilinear Method for Data File Compression", *IEEE Transactions on Software Engineering* 6:4, 340-347.

T. Young and P. Liu [1980]. "Overhead Storage Considerations and a Multilinear Method of Data File Compression", *IEEE Trans. on Software Engineering, SE-6:4*, 340-347.

G. S. Yovanof and J. R. Sullivan [1991]. "Lossless Coding Techniques for Color Graphical Images", *Proceedings IEEE Data Compression Conference*, Snowbird, Utah, 439.

C. L. Yu and J. L. Wu [1992]. "Modeling Adaptive Multilevel-Dictionary Coding Scheme by Cache Memory Management Policy", *Proceedings IEEE Data Compression Conference*, Snowbird, Utah, 418.

T. L. Yu and K. W. Yu [1992]. "Further Study of the DMC Data Compression Scheme", *Proceedings IEEE Data Compression Conference*, Snowbird, Utah, 402.

P. L. Zador [1982]. "Asymptotic Quantization Error of Continuous Signals and the Quantization Dimension", *IEEE Transactions on Information Theory IT-28*, 139-148.

R. Zamir and M. Feder [1992]. "Universal Coding of Band-Limited Sources by Sampling and Dithered Quantization", *Proceedings IEEE Data Compression Conference*, Snowbird, Utah, 329-338.

A. Zandi and G. G. Langdon, Jr. [1992]. "Bayesian Approach to a Family of Fast Attack Priors for Binary Adaptive Coding", *Proceedings IEEE Data Compression Conference*, Snowbird, Utah, 382.

R. Zelinski and P. Noll [1977]. "Adaptive Transform Coding of Speech Signals", *IEEE Trans. on Acoustics, Speech, and Signal Processing ASSP-25:4*, 299-309.

Q. F. Zhu, Y. Wang, and L. Shaw [1992]. "Image Reconstruction for Hybrid Video Coding Systems", *Proceedings IEEE Data Compression Conference*, Snowbird, Utah, 229-238.

G. K. Zipf [1949]. *Human Behaviour and the Principle of Least Effort*, Addison-Wesley Publishing Company, Reading, MA.

M. Zipstein [1991]. "Data Compression with Factor Automata", *Proceedings IEEE Data Compression Conference*, Snowbird, Utah, 428.

J. Ziv [1978]. "Coding Theorems for Individual Sequences", *IEEE Transactions on Information Theory* 24:4, 405-412.

J. Ziv [1980]. "Distortion-rate theory for indivdual sequences", *IEEE Trans. Inform. Theory IT-26*, 137-143.

J. Ziv [1985]. "On Universal Quantization", *IEEE Transactions on Information Theory IT-31:3*, 344-347.

J. Ziv and A. Lempel [1977]. "A Universal Algorithm for Sequential Data Compression", *IEEE Transactions on Information Theory* 23:3, 337-343.

J. Ziv and A. Lempel [1978]. "Compression of Individual Sequences Via Variable-Rate Coding", *IEEE Transactions on Information Theory* 24:5, 530-536.

INDEX

A

B

C

D

E

F

N

O

P

Q

R

S

T

U

V

W